WORDS
FROM
THE LAND

WORDS FROM THE LAND

Encounters with Natural History Writing

Edited and with an Introduction by
Stephen Trimble

EXPANDED EDITION

University of Nevada Press
Reno   Las Vegas   London

Words From the Land: Encounters with Natural History Writing by Stephen Trimble was originally published by Gibbs-Smith Publisher, Salt Lake City, Utah, in 1989. The 1995 University of
Nevada Press edition reproduces the original except for the cover design and the front matter,
which have been modified to reflect the new publisher, and the addition of a preface and epilogue by the editor and essays by five new contributors.

The paper used in this book meets the requirements of American National Standard for Information Sciences—Permanence of Paper for Printed Library Materials, ANSI Z39.48-1984. Binding
materials were selected for strength and durability.

Library of Congress Cataloging-in-Publication Data

Words from the land : encounters with natural history writing / edited
  by Stephen Trimble.
      p.   cm.
    Originally published: Salt Lake City : Peregrine Smith Books,
  c1988.
    ISBN 0-87417-264-0 (paper ed. ; alk. paper)
    1. Natural history.   2. Natural history literature.   I. Trimble,
  Stephen, 1950–   .
  QH81.W779   1995
  508—dc20   95-9831
                                                        CIP

University of Nevada Press, Reno, Nevada 89557 USA
Preface, introduction and headnotes copyright © 1988, 1995 by Stephen Trimble
New compilation copyright © 1995 by University of Nevada Press
All rights reserved
Cover design by Heather Goulding. Cover photograph copyright © Stephen Trimble
Printed in the United States of America

9 8 7 6 5 4 3 2 1

For my father, Don Trimble, who
taught me to listen to landscape

And for Edward Abbey, whose
eloquent, honest, and ornery spirit
we will sorely miss

CONTENTS

"I only went out for a walk, and finally concluded to stay out till sundown, for going out, I found, was really going in."

JOHN MUIR, *JOURNAL,* 1913

"If there is poetry in my book about the sea, it is not because I deliberately put it there, but because no one could write truthfully about the sea and leave out the poetry."

RACHEL CARSON, ACCEPTING THE NATIONAL BOOK AWARD FOR *THE SEA AROUND US,* 1952

ACKNOWLEDGMENTS

Norman Sims's fine anthology, *The Literary Journalists* (Ballantine, 1984), gave me the idea for this book.

I am grateful to all of the writers who generously gave permission to reprint their work. Thanks especially to the following people, who took time away from their writing to talk with me: Gretel Ehrlich, Robert Finch, John Hay, Barry Lopez, John Madson, John McPhee, Gary Nabhan, David Quammen, and Ann Zwinger.

Judy Nolte Lensink, director of the Writers of the Purple Sage Project at the Tucson Public Library, kindly gave me permission to quote from her interviews with Edward Abbey and Ann Zwinger and from the papers presented by Gary Nabhan and Ann Zwinger at the "Old Southwest/New Southwest" conference in November 1985. John McPhee's "Dear Mother" story also appears in *Every Night at Five: Susan Stamberg's All Things Considered Book* (Pantheon, 1982).

Two other books deserve mention: Paul Brooks's *Speaking for Nature* (Sierra Club Books, 1983) puts these contemporary naturalists in historical perspective. And the Autumn 1986 volume of *Antaeus* (also published as *On Nature* by North Point Press) provides a fine selection of essays by twenty-five writers.

I quote from several published interviews, especially those with Barry Lopez conducted by Nick O'Connell in *Seattle Review* and by Jim Aton in *Western American Literature*. Jim Aton also provided useful suggestions on sources. I was lucky to attend the first of Parker Huber's annual celebrations of natural history writing at Glen Brook, New Hampshire, in 1986, and the stimulating discussions during that weekend helped to shape this book.

I thank Joanne Slotnik, Mona Letourneau, Mike Vause, John Davis, John Tallmadge, Gary Nabhan, Jennifer Dewey, and David Johnson for their critical comments on my introduction. Gibbs Smith's enthusiasm for the book when it was only an idea made it a reality; and Madge Baird kept all the organizational details straight while the book passed through production. Finally, I wish to acknowledge the encouragement extended by Ann Zwinger, Gary Nabhan, and Barry Lopez over the many years of my own apprenticeship as a naturalist writer; I greatly value their friendship.

STEPHEN TRIMBLE

SALT LAKE CITY, UTAH

I am grateful to the following publishers for permission to reprint these selections:

"Aurora Borealis and Bells" and "In Defense of the Common Sunflower" reprinted from *Land Circle: Writings Collected from the Land* by Linda Hasselstrom, © 1991. Fulcrum Publishing, Inc., 350 Indiana St., #350, Golden, CO 80401 (303) 277-1623.

"Basin and Range" from *Basin and Range* by John McPhee. Copyright © 1980, 1981 by John McPhee. Reprinted by permission of Farrar, Straus & Giroux, Inc.

"At Crystal Mountain" from *The Snow Leopard* by Peter Matthiessen. Copyright © 1978 by Peter Matthiessen. Used by permission of Viking Penguin, a division of Penguin Books USA Inc.

"Cabeza Prieta" from *The Mysterious Lands* by Ann Zwinger. Reprinted by permission of the author. Copyright © 1989 by Ann Haymond Zwinger.

"Chambers of Memory" and "The Miracle of the Geese" reprinted by permission of David Quammen. All rights reserved. Copyright © 1987 "Chambers of Memory" by David Quammen. Copyright © 1987 "The Miracle of the Geese" by David Quammen.

"The Country of the Mind" from *Arctic Dream: Imagination and Desire in a Northern Landscape* by Barry Lopez. Copyright 1986 by Barry Holstun Lopez. Reprinted by permission of Sterling Lord Literistic, Inc.

"Down the River with Henry Thoreau" from *Down the River* by Edward Abbey. Copyright 1981 by Edward Abbey. Used by permission of Don Congdon Associates.

"Epilogue: Remembering Edward Abbey" from *Journal of Energy, Natural Resources, & Environmental Law,* volume 11:1, used with permission.

"The Gifts of Deer" reprinted from *The Island Within*. Copyright © by Richard Nelson 1989. Published by Vintage Books, a division of Random House, Inc., New York; first published in hardcover by North Point Press in 1989.

"A Grand Surprise" from *The Thunder Tree: Lessons From an Urban Wildland*. Copyright © 1993 by Robert Michael Pyle. Reprinted by permission of Houghton Mifflin Co. All rights reserved.

"Hailing the Elusory Mountain Lion" from *Walk-*

ing the Dead Diamond River, published by Lyons and Burford. Copyright ©
1971 by Edward Hoagland. Originally appeared in *The New Yorker.*

"Homing" and "Living with Trees" reprinted from
The Undiscovered Country by John Hay, by permission of W. W. Norton &
Company, Inc. Copyright © 1981 by John Hay.

"Into the Maze" and "Scratching" reprinted from
The Primal Place by Robert Finch, with the permission of the author and
W. W. Norton & Company, Inc. Copyright © 1983 by Robert Finch.

"The Journey's End" from *The Unforeseen Wilderness* by Wendell Berry. Copyright © 1971, 1981, 1991 by Wendell Berry. Reprinted by permission of North Point Press, a division of Farrar, Straus &
Giroux, Inc.

"Life on the Backside of the Moon" by John Madson, from *Audubon,* the magazine of the National Audubon Society. Copyright 1984 by John Madson. Reprinted by permission of the author.

"Life on the Rocks: The Galapagos" from *Teaching a Stone to Talk* by Annie Dillard. Copyright © 1982 by Annie Dillard.
Reprinted by permission of HarperCollins Publishers, Inc.

"On Water" and "The Smooth Skull of Winter" from *The Solace of Open Spaces* by Gretel Ehrlich. Copyright © 1985 by Gretel Ehrlich. Used by permission of Viking Penguin, a division of Penguin Books USA Inc.

"Snowy Plovers," "Screech Owls," and "Avocets and Stilts" from *Refuge: An Unnatural History of Family and Place* by Terry Tempest Williams. Copyright © 1991 by Terry Tempest Williams. Reprinted by permission of Pantheon Books, a division of Random House, Inc.

"Spring" from *A Country Year* by Sue Hubbell. Copyright © 1986 by Sue Hubbell. Reprinted by permission of Random House, Inc.

"The Scripture of Maps, the Names of Trees," from *The Geography of Childhood* by Nabhan and Trimble. Copyright © 1994 by Gary Paul Nabhan and Stephen Trimble. Reprinted by permission of Beacon Press.

"Where Has All the Panic Gone?" from *Gathering the Desert* by Gary Paul Nabhan. Copyright © 1985 by Gary Paul Nabhan. Reprinted by permission of University of Arizona Press, Tucson.

"Where the Birds Are Our Friends: A Tale of Two Oases" from *The Desert Smells Like Rain: A Naturalist in Papago Indian Country* by Gary Paul Nabhan. Copyright © 1982 by Gary Paul Nabhan. Reprinted by permission of North Point Press, a division of Farrar, Straus & Giroux, Inc.

PREFACE TO THE NEW EDITION

A generous review in the *Earth First!* journal claimed that the 1988 publication of the first edition of *Words From the Land* signaled the arrival of "contemporary American natural history writing as a distinct and expanding literary genre." At the time, only a couple of other anthologies of nature essays existed. Today, I could add new collections to my bookshelf every publishing season. We have historical surveys of nature writing; volumes of women's nature writing; gatherings of essays about rivers, deserts, Alaska, and the Caribbean; and an annual compilation of the year's best nature essays published by Sierra Club Books. Natural history writing has been growing as fast as a tropical vine curling upward into the forest canopy in search of the sun. The genre is spilling into other disciplines, corresponding with a similar broadening of general environmental awareness.

"Ecocriticism" is now an established academic specialty, and the professors of literature who practice it publish book-length analyses of Abbey, Dillard, Berry, and Lopez, with comparisons to Thoreau, Muir, Leopold, and Carson. These scholars have created a subculture within academia, with an Association for the Study of Literature and Environment (ASLE) to connect members of their far-flung community.

What makes *Words From the Land* worth keeping in print after such a flurry of publishing in the same genre? Other anthologies include poets and novelists, restrict themselves to much shorter pieces, introduce emerging writers, or span the whole history of nature writing in America. Here, I concentrate on contemporary American writers who have generated a body of work and who have chosen the nature essay as a major outlet for their authorial voices. My introduction creates a forum for these writers to discuss their craft. No other anthology chooses quite the same focus, nor does one offer such commentary about the discipline from the writer's point of view.

I continue to hold in high esteem the fifteen writers I chose in 1987. With the exception of Edward Abbey, who died in 1989, and John Madson, who died as this expanded edition went to press in 1995, all are still writing. (And, with the 1994 release of excerpts from Abbey's journal, even Abbey is still publishing, giving support to the dream of his disciples that he cannot really be dead!) These writers roam away from nature essays in their work, but the land is never far from the core of their concerns. They continue to demonstrate their generosity of spirit, to sweep readers into their imaginations, and to vigorously articulate their views of the earth. Here are a few highlights.

Annie Dillard published a memoir, *An American Childhood* (1987), and a novel, *The Living* (1992). Ann Zwinger is finishing years of work on a book about the Colorado River in Grand Canyon. David Quammen continues to write his award-winning column for *Outside* magazine. John Hay received the first John Hay Award from The Orion Society for contributions in writing, education, and conservation; Wendell Berry received the second. At 80, Hay published *A Beginner's Faith in Things Unseen* (1995). Gretel Ehrlich was struck by lightning (twice!) and wrote *Match to the Heart* (1994) about her experience. Gary Nabhan received a MacArthur Foundation Fellowship and published books about children and nature and about his walk through Tuscany as a pilgrim/naturalist. Barry Lopez's fable *Crow and Weasel* (1990) reached the bestseller lists, *Field Notes* (1994) completed the trilogy begun with *Desert Notes* and *River Notes,* and his journeys in wild places all over the globe are gradually working their way toward culmination in a big book.

Many new voices have come to prominence. I add four of them to this edition—Richard Nelson, Terry Tempest Williams, Robert Michael Pyle, and Linda Hasselstrom—and also include an essay of my own. Many other writers are producing fine work, and any list is inevitably incomplete. But my wish list of other writers covers several sheets and scraps of yellow legal paper: Gary Snyder, Scott Russell Sanders, Diane Ackerman, Kathleen Norris, Teresa Jordan, Charles Bowden, Rick Bass, Alison Deming, Reg Saner, Peter Steinhart, Kim Stafford, John Daniel, Janet Lembke, Roger Swain, John Hanson Mitchell, Douglas Chadwick, David Rains Wallace, Marcia Bonta, John Haines, William deBuys, Susan Zwinger (daughter of Ann), and Franklin Burroughs (son-in-law of John Hay). Many younger people are seeing their first stories into print. More and more women are writing and publishing natural history essays; Lorraine Anderson's *Sisters of the Earth* (1991) is a good guide to their work. A chorus of eloquent Native American writers (Simon Ortiz, Paula Gunn Allen, Leslie Silko, Linda Hogan, and James Welch, especially) incorporate the land in every sentence, though their work cannot be fitted easily into the constraining mold of "natural history essay." It is a vital time.

In choosing writers to add to the original selection, I looked for those who have taken nature essays to fresh ground. For instance, Scott Russell Sanders is a master of the personal essay, akin to Edward Hoagland. Rick Bass, Charles Bowden, and Doug Peacock are feisty and fierce, in the tradition of Abbey. Richard Nelson, however, in *The Island Within,* makes an unparalleled journey into the spirits of deer and of old-growth forest, his senses shaped by years of living with native peoples. There is no one else like him.

Terry Tempest Williams, too, has moved the genre beyond the old comfortable foundations laid by the hearty gentlemen-naturalists. She writes of her emotional life and her experience of the land as a woman, in an intimate and passionate way that forms the model for the new section I call "Testimony." The other three new pieces also "testify" to the truth articulated by Kim Stafford in the epigraph to this section: "How can I invite the heart of mountains and streams into myself? How can I carry the forest into the city? I know from my own life this is a matter of survival, not aesthetics."

The risk in writing about such emotional survival is a descent into self-indulgence—slipping over into narcissistic new-age self-therapy. The best preventative, I believe, is to keep thinking like a naturalist. Writers with egos gentled by time in wild places are more likely to find the right balance between self and earth, to remain rooted in the land while living fully in the human world. At their best, the resulting essays will continue to be the rich harvest of the naturalist's trance: an illumination of the critical connections between the natural world and American life and literature.

Stephen Trimble
Salt Lake City, 1995

INTRODUCTION

"I walked into the forest . . . until I came to a small glade that opened onto the sandy path. I narrowed the world down to the span of a few meters. Again I tried to compose the mental set—call it the naturalist's trance, the hunter's trance—by which biologists locate more elusive organisms. . . .

In a twist my mind came free and I was aware of the hard workings of the natural world beyond the periphery of ordinary attention. . . . It seemed to me that something extraordinary in the forest was very close to where I stood, moving to the surface and discovery."

EDWARD O. WILSON, *BIOPHILIA*, 1984

The Naturalist's Trance

Contemporary natural history writers speak for the earth. They articulate our neglected connections with the rest of the living world in language both passionate and thoughtful. Landscape threads through their words and their lives, while their ideas resonate far beyond their immediate subjects. They write. They live. And they forge a voice by doing both.

These writers make journeys into the landscape; they enter "the naturalist's trance." They weigh their journals against their research, spend long hours in libraries, and talk to the experts. Then they place themselves in a second trance and translate all this into lovely prose. Within and between the lines lies a wealth of natural history, basics and esoterica, that the reader absorbs painlessly—instructed, invisibly—carried on the buoyant leading edge of the writer's curiosity, skill, and enthusiasm.

Whatever else they may be—biologists, teachers, cowboys, beekeepers, farmers, artists—these people see themselves as writers first. Many write about other things besides landscape. They are essayists, novelists, poets, journalists, and critics. But their connection with the land is emotional and pervasive. They choose places to live and then listen to those places: Douglas-fir forests in the Northwest, Utah's canyon country, the Vermont woods, Virginia's Blue Ridge, Cape Cod, the tallgrass prairie, the Ozarks, the Sonoran Desert. What they hear in the earth are the voices of what Henry Beston called the "other nations" of the planet.

In their prose, their translations of these voices, they teach us, by example, how to see more clearly and feel more truly; they put into graceful words some of our most euphoric and serious experiences. They strive, as Barry Lopez puts it, "to create an environment in which thinking and reaction and wonder and awe and speculation can take place. I have to trust that in so doing, that the metaphorical depth will reverberate there, and ideas much larger than ones I could control are going to come out."

In their attention to language, they are poets.

Henry David Thoreau was the most celebrated early craftsman of personal essays about the American landscape. His vision spread with John Muir to the shining western mountains and lived on in northeastern fields and woods with John Burroughs. Then came Henry Beston and Aldo Leopold, Loren Eiseley and Rachel Carson, and a handful of others. Their tradition has been carried on by contemporary writers, but naming that tradition remains difficult. Nature writers, natural history writers, landscape essayists, literary naturalists, naturalist writers—no single tag defines their work. Bookstores find them hard to classify and may shelve their books under anything from nature, science, conservation, or travel, to anthropology, autobiography, political science, fiction, regional writing, and belles-lettres. In Britain, look for their books under the heading "countryside."

I have been reading these writers for years, and writing natural history myself for fifteen of those years. This gathering of essays grew from two desires: I hoped that a sampling of provocative pieces would lead readers to the whole shelf of books by literary naturalists—for I have been surprised at how many of these writers are seldom read. And I wanted to talk with the writers, in an effort to illuminate their methods and to try to define the heart and spirit of this wonderful body of work. So in the fall of 1986, I drove 18,000 miles across the country, and along the way I interviewed nine of the fifteen writers whose work appears here.

I have been forced to leave out some fine writers, particularly since I have not considered fiction. No single book can include everyone working in the genre. Biologists and anthropologists, too, have important things to say about landscape, but however elegant their prose, they seldom define themselves first as *writers*, as do the people I have chosen here.

The writers in this book also do not see themselves primarily as environmentalists. David Quammen says, "I am more interested in being a writer than in being an environmentalist." Gary Nabhan agrees: "People like Ann Zwinger and I back off from the term *environmentalism*. We prefer to explain what we do as going out into the boonies and interviewing plants."

"I Am Not a Naturalist"

"I would not call myself a naturalist," says John McPhee. "My interest is in doing pieces of writing. Compositions with a beginning, a middle, and an end—and people in them. If I can find something to do that with, I don't really care what it comes from." But he adds, "You're more likely to be able to do that by following developed interests than you are by simply leaping into something and saying, 'I am now going to do this about . . .' " His voice trails off and he smiles, a little sheepishly. "I tried to pick a subject in which I have no interest, but it's hard to do that."

In the breadth of his interests, John McPhee is a bit anomalous in any list of naturalist writers. He says, "The common thread in all of my work is people—and what they're up to, as it's expressed through their work. Why is there so much nature involved in all this? Because when I was a kid, my father, who was a university physician, didn't work in the summer. We went up to a camp in Vermont where he was the camp doctor, and this place specialized in sending kids out on backpacking and canoe trips. It taught what much later became known as ecology and the environmental movement. These words became current when I was in my thirties; I had been taught these same things since the time I was six.

"I know you could call me a nature writer, but it's not the nature that got me into it. It's the writing that got me into it."

This self-assessment turns up over and over again. So-called nature writers acknowledge their love of landscape, but they define themselves as writers. *Naturalist* is an outsider's term, a critic's term, like *artist*. Barry Lopez says, "The writer works on the inside and the critic works on the outside. I don't know what it looks like on the outside, sometimes. It's not that I'm not interested—it's not where I live. I live inside the story."

These writers often reserve the term *naturalist* for more highly trained scientists, or for *other* natural history writers. Robert Finch lays part of the blame for his designation as a naturalist on the subtitle of one of his books of essays, *Common Ground: A Naturalist's Cape Cod.* Finch says, "That was at the insistence of the business manager at David R. Godine, the publisher of the book, who said the words *nature* and *Cape Cod* had to appear somewhere on the front cover.

"If you look at the introduction to the book, I spend most of it disclaiming the fact that I am a naturalist. I'm almost tired of trying. I basically say I'm not a trained naturalist; I hang around naturalists."

The introduction to an essay collection is a favorite place for a writer to make his or her disclaimer. Edward Abbey introduces *The Journey Home* in this way. "I am not a naturalist. Hardly even a sportsman. True, I bagged my first robin at the age of seven, with a BB gun back on the farm in Home, Pennsylvania, but the only birds I can recognize without hesitation are the turkey vulture, the fried chicken, and the rosy-bottomed skinny dipper. My favorite animal is the crocodile. I'll never make it as a naturalist."

John Hay spent his autumns for many years teaching classes in Nature Writers and Nature and Human Values at Dartmouth College. He claims he never heard the term *nature writing* until he came to Dartmouth to teach it in 1971. Hay says, "I'm not sure I like the term. It sounds lesser, somehow." And yet, "I was quite flattered when I was first called a naturalist."

Ann Zwinger, for one, admits to the title. She is fond of quoting Jacob Bronowski's description of a naturalist's work, "that quaint Victorian profession." At the same time, she says, "I would like to take a course in natural history writing. I would like to see what I'm doing."

Zwinger says that being a "writer of natural history is a visual discipline. It requires a great deal of observation, a lot of research. Every question you ask—why is that flower blue? why is a sunset red? why is the grass green?—every time you get an answer you get ten more questions. And somehow you have to put it in a form that somebody can read, and in sentences that have some life of their own.

"Natural history writers, more than anyone else, write for themselves. They write for the sheer glory of being outdoors, in the fresh air, in the fresh breeze, and just the great pleasure and serenity of being there. Maybe writing is simply an excuse to get there."

Maybe. In the beginning. But these writers glory equally in the writing itself. As Ann Zwinger says, "Writing strikes me as addictive as salted peanuts and sex. Once you start, it's very hard to stop."

For Gary Nabhan, "writing is the major vehicle through which I sort out the universe." He also points out that "a lot of natural history writers are really students of relationships. They are more interested in the dynamics between two living beings than in a single living organism as a *thing*." When Robert Finch defines the naturalist writer, he uses words like *subjective, personal,* and *holistic.* Perhaps that is why some natural history writers prefer to call themselves simply "writers." They see the natural world in a different way than scientists do, and "naturalist" sounds a little too scientific for many of them.

David Quammen, best known for his natural history column "Natural Acts" in *Outside* magazine, says, "My first loyalty is to turning out interesting prose. A lot of the time it's got critters in it, or principles of natural science in it, but a lot of times those are just excuses for me to satisfy the editors of *Outside* that yes, this is a 'Natural Acts' column. What I'm really writing about is Faulkner or friendship or some crazy person that I want to write about." Quammen calls himself a "science journalist," a little off to the side of the "humanistic non-professional-scientist literary naturalist." He and his wife affectionately sum up his profession as "comedy entomology essayist."

John Madson believes that "there is a lot of animal in us. Maybe a naturalist is simply a little bit more animal than other people." Even Edward Abbey might go for that.

"Nature can't be summarized," says Barry Lopez. "You can't say by an enumeration of animals and minerals and forms that this is a forest. Patterns don't work that way, and natural history writers are interested in making apparent a pattern in the natural world. What we are talking about is using the material of natural history, geography, and anthropology to open up the dilemmas of the late twentieth century.''

Lopez thinks that the genre now ''includes anthropology much more than it did before.'' Gary Nabhan, the youngest writer in this anthology, points out that members of his generation are ''post Earth-Day kids. We grew up from the teens on reading environmentalists and ecologically-influenced natural history writing. Whatever pioneering any of us do will be somehow a different tangent from that. There is not just one scientific tradition regarding *life*.'' Nabhan emphasizes the contradictions in the group of people called nature writers: ''We don't all wear L. L. Bean apparel. Some of us live in cities, are addicted to drugs—coffee, computers—can't name all the creatures within reach, are allergic to what we write about, are absent-minded campers, don't grow our own food, and like electronic music.''

Annie Dillard goes even further afield from the stereotype of a nature writer. She says, ''I distrust the forest, or any wilderness, as a place to live. Living in the wilderness, you may well fall asleep on your feet, or go mad. Without the stimulus of other thinkers, you handle your own thoughts on their worn paths in your own skull till you've worn them smooth.''

Being a naturalist is a *feeling,* a conscious sense of connection to the land, to the other animals and plants. Being a writer is a creative practice that permeates one's life. Being both is a fulfilling, lively, and intensely involving challenge. And a lifework.

Tutored By the Land

Writers of natural history respond personally to the landscape. Their subject matter is not just the natural world itself; books full of such information fill the shelves of science libraries. Nor is it as exclusive as the invented world of fiction. Their subject is what writer Mary Austin called, in 1920, ''a third thing . . . the sum of what passed between me and the Land which has not, perhaps never could, come into being with anyone else.''

Experience ''in the field'' (the wilderness, the woods, the beach, the wilds, the *out-there*) is where naturalist writers begin. They walk in the desert, look out their windows into the forest, travel to the ends of the earth, ponder a black widow spider in a corner of their study. Each experience begins as raw sensation. But as soon as writers attend to it, sensation becomes perception and starts to move out of the present and into the past. The naturalists begin to ponder, analyze, and make choices.

The writers can dissect any of these experiences objectively, or react aesthetically and emotionally—a private reaction, the meat of journals, the grist of nature writing. Their choice of idiosyncratic metaphor and telling fact shapes the voice, the mood, the content of their writing. John McPhee says, "That's what writing is—selection."

In turn, their choice shapes the landscape for every reader who comes after, innocent and voiceless and looking for guidance. Thus the canyon country of Utah and Arizona becomes Abbey's country, and the Green becomes Ann Zwinger's river. The hills and farms of Kentucky belong in a literary sense to Wendell Berry, the west slope of the Big Horn Mountains in Wyoming to Gretel Ehrlich. Similarly anyone interested in understanding our relations with wolves will read Barry Lopez, or in migrations and alewives, John Hay. The writers and their perceptions have become permanently intertwined with our concepts of these animals.

Ann Zwinger writes "about home, not place." The love and intimate knowledge inherent in her attitude make whatever else such a writer may say worth listening to. Her personal vision makes it art.

To stake out this kind of literary geography is an exhilarating and creative act, but it does not capture the only truth about a place. Such truths remain beyond books, out there somewhere in the real world. Barry Lopez says, "Your responsibility is not to know the truth, because no one knows the truth, but to set the story up in such a way that truth can be revealed."

The process of "selection" starts with simple and sedentary pastimes, what John Madson calls "loafing," sitting under a juniper on a hot day and just watching. "I'll get my notebook out and begin reflecting and spend maybe an hour or two just jotting down impressions—letters and notes to myself."

Barry Lopez calls this "being tutored by the land." He says, "You try to make yourself available to the place. To be tutored—it's like making an application. Insofar as you behave well, the place will open to you. Then you have the problem of translating what you got into language."

Lopez does most of his translating in a journal, one he has kept since he was nineteen. Robert Finch and Gretel Ehrlich speak of the importance of their journals, as well. Finch knows that nature writers are perceived "as people who are out there writing down their essays on scraps of birch bark. I almost never take notes out in the field. I go into the experience with no preconceptions or no particular idea of what slant I am going to take on it, and then let the experience settle out."

Each writer works this metamorphosis from the land to the words in his or her own way. Much of Edward Abbey's *Desert*

Solitaire "came right out of the journals" he kept while he was a park
ranger at Arches National Monument (now Arches National Park) in Utah
in 1956 and 1957, when he was in his late twenties. He says, "I had nothing much to do out there but sit around, take notes, think about things."

John McPhee "scribbles all day long," too. He
says, "I put down in the notebook anything I think even faintly might
be useful. But the selection process is going on right there, eliminating
ninety-five percent or more. When David Love [the geologist who
dominates McPhee's book *Rising From the Plains*] and I go up on some
outcrop, we sit on that outcrop for hours—talking about the landscape,
looking at a 270-degree view. It takes one minute to read that page in the
book." He pointed to my tape recorder on the desk between us. "This
thing does not select."

Gary Nabhan takes notes in the field, "but I lose
most of my notebooks on the way home (or Coyote steals them). The most
pleasurable thing is just watching a landscape, trying to figure it out, watching it in different light. I learn in another way when I sit down to write
at five in the morning in the back room."

I stayed with John Hay for a night at his apart-
ment in Hanover, New Hampshire, where he was teaching his annual fall
term at Dartmouth. Just before he went upstairs to bed he brought out
a box for me to look at, bundles and stacks of pocket-sized spiral notebooks and tiny bound black books, in a carton left over from a University
of Chicago Press book order. These were the field notes for his book about
terns, a book he had published in 1974 as *Spirit of Survival* and which
he was thinking of rewriting and expanding.

I looked through his scrawled notes and found
what I would expect to find in any nature writer's raw archives. A stream
of ideas, questions, journal citations; book titles (*The Deer and the Tiger,*
by Schaller; *Notes From the Century Before,* by Hoagland; *The Birds of
Western Siberia,* by a Professor Hermann Johanssen); anecdotes about
nudists Hay surprised on the beach; bird observations; addresses of people met on walks (people from Nairobi; officials of the Massachusetts Game
Department); Ojibwa and Micmac names for terns; drafts of poems;
sketches of waves, yellowlegs posturing, circling and landing eider ducks,
directions, sensory canals in the inner ear; comments on weather; fragments of writing:

Nature—only beside us not with us or within us?

love of place is love of familiarity

'keery, keery' when carrying fish—an announcement of *having*

sky courting, loud passionate 'chutta chea'

Fish size related to the season?

ANTS

In the book, to play things back and forth against each other—part of the practice of being—practice in the terns.

Kik kik [crossed out] kikik

We are desperately concerned, consciously or otherwise, about whether the universe will accept us—the aberrant arrival. It might be better to wonder whether we have the guts to accept the universe.

April 30—terns very far out from island

answer questions of adult or child

Roseates modest about copulating in the open

Dreams are an outline that have to be filled in

Curiosity, hope, anger, love, irritation, gluttony, speed, slow time, confusion

Journeys. "I love field trips," says Hay, "especially with somebody else that knows a little more than I do. It's being out where things are working, where the planet is working."

Ann Zwinger says, "*Run River Run* is the book that means the most to me because that was the first time I went out alone. That was a real passage—to what I am now. I'm not sure what I am, but I do know what I'm not. It would be impossible for me to write about nature without being out in it up to my ears. There is no substitute for blisters and sunburn and seeing it where it is."

Seeing and listening. It helps to approach the natural world with humility. Barry Lopez says, "People who talk to whales assume that whales are interested in talking to them, which is an enormously arrogant frame of mind. You're far better off to approach it from the point of view of wanting to listen, in case whales have anything to say."

You could also choose to listen to terns.

Naturalist writers glean ideas from people, as well—from field companions, from neighbors, from natives. Gary Nabhan says of his work with Sonoran Desert Indians, "I was a vehicle through which some Papago elders and middle-aged men were able to speak about a much older, more persistent kind of wisdom, of knowing a place. I don't have the consistent day-to-day dependence that makes their impressions

9

much richer." Gretel Ehrlich joked with her friends whenever she helped them with some cowboying during the time she was writing *The Solace of Open Spaces* essay by essay: "Sure, I'll help you. I'm about running out of material!"

Ehrlich believes that "it isn't the landscape that matters so much as the way you live in it. You have to take some kind of step out of your own known world. You are not ever really going to know anything or anyone or any place unless you go there in some naked state emotionally and physically. You have to make it your life. You just have to go and live in a place and let it tell you how to live. That's what counts, that sort of Zen monk attitude."

Writers have less access to elemental connections to landscape than do Indian farmers, though they may have a more self-examined one. They are simply too busy taking notes, doing research, and writing—turning landscape and experience into carefully chosen words.

Igniting Recollection

The writing itself is magic, of course. You come home, your head full of feelings and memories, and sit in front of a blank computer screen or yellow pad of paper, surrounded by your notes and journals, your files of research, and go to work. Much of that work consists of sitting and staring out the window, or browsing in the bookshelves that line the office, waiting for that unpredictable something that moves you to begin. "Where it comes from, I don't know," says Edward Abbey.

Barry Lopez describes the waiting, the mulling over an experience until it comes clear: "You sit and think about it, or split wood and think about it, or drive to town and think about it. Just turn it over, keep turning it over until a light bursts through. Parts of it will fall away and I'll be left with what I was after all along and didn't realize."

Once back in his basement office in Illinois, back from loafing, John Madson's scrawled impressions will "immediately ignite recollection." They get him started, but what comes after isn't all easy and smooth: "Most people have more sense than to subject themselves to the uncertainty that a writer has to go through, this masochistic self-isolation that writers love to whimper about."

John McPhee says, "A writer, this writer anyway, is blocked to a certain extent every day. You sit in here and break through that, and that's a very real barrier. That's the smallest one; it's a daily barrier. Then it can get more intense, and stop you for a week or so. And then, what has never happened to me, the psychological problem of writer's block will stop writers cold for years if not forever. I think that the major problem is not unrelated to the thing that a writer goes through each day."

It is getting started that in some ways separates the writer from the would-be writer. It is the trick of "igniting recollections" and turning them into words and then crafting them into structured prose that is the writer's gift. Ann Zwinger says, "I just close my eyes and walk back. With practice I think you can have a marvelous amount of recall. Smells come back, the way the gravel felt comes back, your sense of ease or unease. . . . After I get it down out of my notes on the screen, then I can start really thinking about it. That's when I think you come up with the connections."

Getting started is sometimes the key. John McPhee describes this delicate beginning as "a very nervous-making thing." He tells the following story about participating in a writers' panel at the Princeton YMCA. Someone asked, "How do you get started?"

"And I suddenly heard myself saying, 'I'll tell you how to get started. You put a piece of paper in the typewriter and type: *Dear Mother, I am having an awful time trying to begin a piece of writing. It just won't work. It's wretched. It's dreadful. You see, it has to do with a certain grizzly bear in northwest Alaska. I can't do it. It's awful. This bear was standing on the hillside eating blueberries . . .*

" 'Keep on writing about the bear. Then keep going. Then go back and cut off the Dear Mother and all that garbage at the beginning. And there you are. You've gotten somewhere.' "

After a long period of thinking, preparation, and outlining, Barry Lopez writes first drafts of shorter pieces in one sitting. On books, "My habit is to write 2,000 to 3,000 words a day, then the same amount the next day and the next day, until I'm finished with a section or chapter."

John Hay speaks of the advantages of living in a landscape that nourishes. "On Cape Cod, if nothing happens at a dirty desk—just a hell of a lot of bad writing and lousy notes—you can always go on a walk. That's made all the difference, to be able to walk out to the tides, see the ocean, as I can walk out here in New Hampshire and look at the mountains. It frees you." After the release, though, it's back to the desk. Gary Nabhan says, "It's trying to allow something that you've distilled in yourself to emerge again, the genie-in-the-bottle kind of thing. I don't think it's an analytical problem."

Nabhan's book *Gathering the Desert* discusses a dozen edible desert plants, chapter by chapter. "I did this Method acting where I tried to immerse myself in the feel, taste, smell of a plant while I was writing about it. I would eat wild chiles or smear their resins on my hands—being careful not to wipe my eyes during the process—and try to live within their aura or influence during the time I was doing the writing. I don't think that this was just contrived exercise. I think that plants do have personae that we can feel, that are as strong as the ones most naturalists commonly associate with animals—with coyotes, with ravens.

"I immerse myself in a topic, the literature, the place, the plants, for a while, take a lot of notes and then brew over it, try to crack it.

"And then I start imagining pieces of a mosaic, not an outline. You have one keystone piece and the others are interlocking pieces that may be diagonal/tangential/curvilinear to each other rather than in a direct linear chain. What I concentrate on is linking those pieces up by images rather than with principles or historic sequences of events or anything else that looks like a story line. I rely on sensual phenomena as the glue rather than logic. When I do the writing, I have to do it in a certain length of time or it will unravel and get stale."

John McPhee is known for carefully planning the structure of his pieces, particularly in terms of restructuring chronological time. One piece he diagrammed as a lower case *e* before writing a word—beginning halfway through the story, then circling back to the chronological beginning, moving past the point where he began the piece, to the end of both.

Barry Lopez traveled in the Arctic for three years before outlining *Arctic Dreams*. "I spent 1978, 1979, and 1980 going in and out of the Arctic, taking extensive notes, doing lots of things that I didn't see clearly as being part of the structure of the book, but still not wanting to make an outline until I really had a good feel for where I was and what was going on." Once he began to write, Lopez looked back at his structure and saw that, "I tend to move in huge arcs. I make a kind of loop, and then I make the same loop again. Having done it a second time, it makes it almost like a third time."

John Hay also describes his approach to writing and to the world as circular: "I'm not a linear type. I circle like alewives and terns and herring gulls. I think it's much more interesting to be circular. You go farther. You take in more. In a sense, I never wrote consciously. I just tried to do my best to follow through what I started."

Writers can tell us about technique, but their art is ineffable and indivisible from their lives. They often become writers without knowing that is what they have chosen. An inner drive somehow motivates them to start—and keep going. Their lives and their interests and their passions simply lead them to that point, and, if all goes well, beyond.

12

The Digressive Voice

In *Arctic Dreams,* Barry Lopez talks about icebergs: "The icebergs were like pieces of Montana floating past." He relates them to luminist landscape painting of the nineteenth century, and to the "passion for light" that connects them to cathedral architecture. When he writes several pages about ice and cathedrals, he does not simply toss off a passing metaphor, but he spends time with the idea, quoting twelfth-century scholars and placing the "architecture of light" in historic context.

Lopez told me, "For me, it's not so much a matter of one thing leading to another, as much as a matter of something going the other way. I knew a lot about cathedral architecture for a long time, and what made it come alive for me was the contemplation of icebergs. It wasn't the other way around. There are things I've encountered, and it will be ten or fifteen years before I'll run into them in a way that I can make them come alive for myself."

Lopez and his wife live along the McKenzie River in Oregon. Dark-green forest surrounds their house, trees a hundred, a hundred and fifty feet tall. Other people live along the river; Lopez knows that when his neighbors drive past at three in the morning and see the light burning in his writing room, they understand a little better that this strange thing he does for a living involves hard work.

Today's settlers live year-round along the river, but Lopez believes that Indian people could not have used the McKenzie canyon much more than seasonally; it was just too rainy. They *did* use the canyon at times, however. In laying a brick walk thirty-one inches from his house, Lopez cut through a layer of charcoal. He paused and thought skeptically, "Ahh . . . a campsite." A few moments later, his shovel clinked against a chipped obsidian blade.

He says, "When you pick up something in the woods, it is not only connected to everything else by virtue of its being a set piece in an ecosystem, but it's connected to everything else by virtue of the fact that you have an imagination."

An imagination—and a broad education, for without awareness of a variety of things, the imagination has limited material to act on. Barry Lopez and David Quammen went to Jesuit high schools, and both speak with respect about what this classical education did for them. I asked Ann Zwinger how she could write with productivity and with such a rich vocabulary in her very first book, *Beyond the Aspen Grove.* Her answer: "It was the scholarly background and a Wellesley education. I know that sounds kind of silly, but being an art historian, you learn how to look. And being over forty helped—to have the discipline to get that

massive amount of material together. You probably shouldn't be allowed
to write natural history until you are forty."

"I'm a three-time college dropout," admits
Gretel Ehrlich, "but I took Latin for seven years and I still read Wallace
Stevens every morning. Maybe not every morning, but often. I love study-
ing; you can't be a nonfiction writer without studying all your life.

"I think that's why those of us who write about
landscape are so interested in science, too. It's not in order to categorize
the wildflowers, to make lists of things, but to make our vision, if we have
one at all, go deeper and deeper. The more I learn, the more I see. And
the more I see, the more I learn."

Naturalist writers tend to be people of the
humanities, products of liberal arts educations and self-educations, rather
than scientific training. Edward Abbey has a degree in philosophy. Peter
Matthiessen is a student of Zen Buddhism.

Scratch even the apparent exceptions, and under
the surface you find generalists. Gary Nabhan does technical research in
both botany and anthropology but feels a part of several traditions. In his
list of mentors, he includes Ann Zwinger, Wendell Berry, and Gary Snyder
along with pioneering scientists. David Quammen studied aquatic entomology for a time, but he wrote his first novel when he was just nineteen. His mentor is Robert Penn Warren (who, Quammen says, also started
his life wanting to be a scientist and naturalist). Quammen studied Faulkner
through undergraduate and graduate school before his brief immersion
in biology. He says, "I've been on the cusp in terms of my interests for
a long time. *The Golden Nature Guide to Insects* was the single most
influential thing I owned as a child."

John Madson was trained in wildlife biology. But
he sharply distinguishes his "training" from his "education," the latter
of which started on "a little prairie river" called the south fork of the
Skunk, and included exposure to the great ecologists Paul Errington and
Aldo Leopold. Madson spent an afternoon in 1940 cutting up muskrat carcasses in the back of Errington's lab and listening to the two professors
talk about marsh ecology. Then, after World War II, he spent two years
reading—"reading everything in the world I could put my hands on"—
while lying in the southern Arizona sun recovering from tuberculosis. He
says, "I am a dilettante and this is what I prefer to be."

But how does an educated person—a well-read
"dilettante"—decide to write? Robert Finch "loved to read more than anything. I grew up wanting to be able to do what the best writers were able
to do—to evoke that kind of response. I thought there couldn't be anything more fun or worthwhile."

John Hay says, "I write because I think I'm good
at it. There are a lot of things I'm not good at. Like playing tennis, riding

a motorbike. Passing math exams." Hay read Thoreau early, "with ela-tion." He read Walt Whitman and Robinson Jeffers, James Fenimore Cooper and Tolstoy. When he read D. H. Lawrence's *Mornings in Mexico,* he said to himself, "My god, you can practically hear that parrot call-ing after you down the street. Marvelous evocation. I thought, 'I wish I could write like that.' "

After World War II, after publishing a book of poetry, after working as a newspaper correspondent, Hay was living on Cape Cod, "trying to be a free-lance writer. I wasn't quite sure where I was going—just writing articles, reviews, and whatnot. And I went down the road and there was this run of alewives which absolutely choked the brook every year. I never knew where the fish came from. Nobody locally could give me facts on the subject. So I looked it up and wrote a book on it. I was just tremendously excited to see this phenomenon from out of the sea. And that led to all those others."

Ann Zwinger moved to Colorado in 1960, and her family bought a piece of land in the aspens below Pikes Peak. A writer friend, Nancy Wood, came up to visit one day with her agent, Marie Rodell. "I'd been keeping a notebook of the plants, and Marie kept asking me about them. Finally, she said, 'Why don't you write a book on Colorado ecology?' And I thought she was nuts. And then I realized that, of course, I would adore doing it.

"She said, 'write a full outline and a chapter.' I wrote a two-page chapter and a one-page outline. She wrote back and said, 'You have lots of work to do.' But the next thing I knew there was the phone call, 'Random House took your book.' I was abysmally lucky."

The first book grows from a passion for reading and words, mixed with discipline, general curiosity, and enough resource-fulness and luck to nail down a contract. With ideas for pieces of writing "just streaming by, by the thousands," (as John McPhee says) the next book and the book after that tend to follow right along.

Just Keep Reading

John McPhee has calculated that it takes four times as long to write about geology as it has taken him to write about anything else—first because of the research and then the translation into understanding, and finally the necessity to keep consulting with geologists for technical review. "But in the end of all that, to have a really good geologist look at something and say, 'I don't see anything wrong with that,' that's a nice moment."

In 1978, McPhee had an idea for a "Talk of the Town" piece in the *New Yorker.* Why not go to a road cut with a geolo-gist and tell the story of that road cut? Then he thought, why not several road cuts? He called his friend Ken Deffeyes, a geology professor at

Princeton, and asked him, "Why don't we go up the Adirondack Thruway? It's beautiful country up there." But Deffeyes said, "No, in this continent if you want to make a journey, go east/west, because you are cutting across structure. That's the way to see the geology vary." And so McPhee thought, why not make a cross-section of the whole continent along Interstate 80?

He began to travel with geologists and has so far published three books based on these experiences, with more to come. "The organizational principle has to do with the reactions of the scientists to plate tectonics, not to geography," he says. "I had to learn this subject to a much greater extent than I had learned any other subject in which someone I was writing about was involved. You simply can't root up one area of geology without having to know the one next door to that and the one next door to that. I was ensnared, and I didn't mind."

Each one of McPhee's geology books stands distinctly on the bookshelf, but he is well aware that they form a progression. "To go back to square one every time would be just to kill it. And so I hope that some of those passages take off. There is a way to read those things. You've just got to keep going. I can't stand there putting up flags every paragraph. If I lose readers, I just have to. I don't see an alternative.

"When I get depressed about it, I'll say, 'Rocks don't have heartbeats.' But it seems to me that if people aren't interested in the earth and how it works, they are disenfranchised."

Natural history writers told me repeatedly that their material dictates the way it must be written, as novelists speak of their characters taking over a story. The reader must go along with the material and just keep reading to see what develops. McPhee says, "In theory, a piece of nonfiction is as long as it needs to be. The ideal situation is that the thing will occupy anything from the length of a haiku to the Anglo-Saxon Chronicle, depending on the material you collect and what kind of structure and composition result."

"The way you do a book is your way," says Ann Zwinger. "That's why I think writing is such a naked occupation. You reveal so much more about yourself than you ever think you do." Gary Nabhan: "I feel like I force my readers into giving me the benefit of the doubt at the starts of chapters—it's trying to disorient both myself and the readers enough to look at something new. The reader is not necessarily going to be with me on this first page-and-a-half. Only if he makes it that far do I have the chance of really getting across to him what I had hoped to."

Nabhan adds, "I have to work to build a story, and sometimes I feel it's okay to make the reader work. Novelists do that all the time, intentionally, to pull the reader into a larger message."

Barry Lopez says, "I don't think about my work all the time, I just do it. I want, when I tell a story that becomes *Arctic Dreams,* to tell a wonderful story about that part of the world. If the ideas were foremost, or if some baroque structure was foremost, it wouldn't be a very interesting book. After a while, you would get the feeling you were being lectured to. The primary job is to just do that good job of research and to tell the story, tell the story well."

Lopez looks back at the structure of *Arctic Dreams,* and he can see why he could trust his instincts, even if he couldn't articulate this so clearly while writing. He can talk at length about the pace of his introduction of ideas, about his choice of animals that lead the reader from the land "off the edge, into the ice and sea, inverting the hierarchy of senses," and then to "light and ice, these two big animals that I saw in abstractions."

Lopez emphasizes that "the narrator is a part of me. It's a refinement of something that's inside of me that I found useful technically. I wanted the narrator to be a man you would be comfortable traveling with. He's the number two guy—he's not out there leading the dogs in the first sled. He's back a few sleds. So he's never intruding on the reader's ability to enter the environment."

In his original draft of *The Deltoid Pumpkin Seed,* John McPhee knew precisely what he wanted. "There was one place in 60,000 words where it was absolutely necessary for me to be there. And when I turned that manuscript in, my editor said that was really odd—this one first-person singular pronoun sitting there in an entire piece." At the editor's insistence, McPhee included himself in one other scene in the book, near the beginning. "But," he says, "I still think that was a mistake. I'm not going to pause and talk about myself at some point unless it makes sense. If it does make sense, I'll stop and tell a story."

Each of these writers has developed a certainty about voice. They interact with editors on a conceptual level, but the details of rewriting are the writer's responsibility. As McPhee says, "The real writing takes place between the first miserable, crude draft and the finished thing." "We learn by trial and error," says Gary Nabhan. Gretel Ehrlich values an editor who reads "well and neutrally and fairly." She believes, "Just having one reader frames your work in a way that you can see it more objectively."

Mostly, the writers self-edit; John Madson describes himself as "merciless." John Hay says, "You have to be very aware yourself of order—you know if the retaining walls are going to fall in or not." Barry Lopez revises and revises, until he finds "a point at which it's released and it is what it is. It's whole—more would be too much, less would not be enough."

Knowing you are on the right track becomes instinctive—but how much information can you mix in? Ann Zwinger compares this to "stuffing raisins into a cake. I think there is a limited capacity for a reader who isn't as enamored of a subject as you are, to read. I'm aware that there's an awful lot of information in my books. Many people say they have to read them in bits and pieces, by the spoonful."

Zwinger loves words. She values "playing with words, juggling and prodding words. Chucking them under the chin. Sautéing and scrambling and skewering words. Running and jumping and playing tag with words . . . I am being flippant because what I am talking about is too important to me to be serious about."

She can be serious indeed about those words. She tells of a dream she had, a dream of nuclear holocaust "somewhere, off the edge, off the horizon." While she waited for the disaster to reach her, she sat doing what mattered most, writing "words that would remain hanging in the dry desert air, suspended with a life of their own, witnesses to grace and coherence. That was all that mattered.

"The words."

To write well, you must have a keen ear. Gretel Ehrlich says, "I'm very particular about language, as are most people who started out writing poetry. I think about every word. I care a lot about how it sounds . . . I always read everything to my husband, Press, usually pretty soon after I've written it. If you are embarrassed to read it to your husband, you sure as hell don't want the rest of the world reading it."

"Words and writing—it's a verbal, oral thing," says John McPhee. "I've never published a syllable I didn't read out loud—ever." Gary Nabhan works with "spoken starts" when he begins a project. "If it sounds right to me, I think I can proceed. It has to sustain itself as spoken word, because literature emanates from there."

Humanistic Studies 406:
The Literature of Fact

John McPhee believes that "writing is a private thing," but he spends spring semester, two years out of three, teaching Humanistic Studies 406: The Literature of Fact to sixteen students at Princeton University.

McPhee says it is "just a nuts-and-bolts course in factual writing," but he knows what it does for him: "It takes me out of my own shell. It gets me involved in something related, but different. It's very hard work, but it is not the same kind of work. It's a dimension, a kind of flywheel in my world, that I greatly value. And I hope the students get something out of it."

John Hay becomes particularly exasperated with his Dartmouth students "who don't know what a landscape is, who don't know the difference between a white pine and a sugar maple.'' The morning before I talked with him, he had been trying to teach one student the "connection with his natural speaking voice and what he has written down on paper." The sound of the words.

Not many beginning writers have the benefit of such illustrious teachers. Writers are forged through random combinations of family, education, and experience. Hay's own identity as a writer started at his grandfather's summer home on Lake Sunapee. He took me there on a fall day soft with cool air and smoky New Hampshire light. In the 1880s, his grandfather, also John Hay, came to New Hampshire to "start some kind of a summer colony for intellectuals. But nobody else moved in. My grandfather was left with this large place."

Hay and I scuffled through the maple leaves, looking for a good-sized American chestnut tree that the caretaker had found—in spite of the chestnut blight. Hay talked about what childhood summers at the lake had taught him.

"Lake Sunapee is eleven miles long. My introduction to space was up here when I was a kid. I had a houseboat which I copied out of Dan Beard's *American Boy's Handybook,* built like a scow with flared ends and a cabin in the middle. I put an outboard in the end of it and drove it slowly down the lake. I wandered around in the houseboat, fished, spent the night on it, and climbed the local mountains.

"My grandfather wrote poetry, romantic Victorian poetry. I suppose all my family was interested in writing. But in the society I was brought up in, you weren't necessarily in favor of the imagination. Nobody was reckless in my house in talking about ideas. They were a little scared of ideas, like a lot of American families are. In this kind of setting, you were never going to be allowed to go as far as you wanted to. So I had to break out in some way.

"I did it by being an enlisted man in World War II, and by living with an eccentric poet on Cape Cod for a month or two just before being drafted into the war. Between twenty and thirty, I needed a little extravagance in my life."

The "eccentric" poet with whom Hay served an apprenticeship was Conrad Aiken. "I don't know how much writing I did with Conrad; it was more like drinking. There was a long, long cocktail hour, during which he would emote various matters to do with consciousness and so forth. I was set to writing sonnets. The next day, to recover from the evening, I would go out and clear the woods for him.

"But I found a piece of land I could buy for twenty-five dollars an acre on the Cape. I got my idea of independence. I guess I absorbed something through osmosis of one kind or another.

And to find this guy who was a little more extravagant than what I knew helped me quite a lot."

Learning to be extravagant, learning to take risks, learning confidence. "Not to be afraid to take certain kinds of leaps," as Gretel Ehrlich puts it. "Learning to soar," John Hay says.

And it continues throughout one's life. Gary Nabhan believes that "we can't assume that our topic, our feeling of the magic of the landscape, is simply good enough to sustain people being interested in everything we do. Unless we are really working with the writing in the same way that any other writer should, we are eventually going to run into dead ends."

Nabhan singles out Peter Matthiessen as someone who shows "incredible growth as a human being and as a writer, taking risks all over," someone who moves natural history writing beyond its "one little clustered set of styles and ideas."

Learning to be a naturalist writer begins with learning to be comfortably alone in the landscape. Edward Hoagland has written about this unusual, even odd, predilection in his essay "Writing Wild."

How long will these readers continue to miss walking in the woods enough to employ odd-balls like me and Edward Abbey and Peter Matthiessen and John McPhee to do it for them? Not long, I suspect. We're a peculiar lot: McPhee long bent to the traces of the *New Yorker,* Matthiessen an explorer in remote regions that would hound most people into a nervous breakdown, Abbey angry, molded by what is nowadays euphemistically called "Appalachia." As a boy, I myself was mute for years, forced either to become acutely intuitive or to take to the woods. By default, we are the ones the phone rings for, old enough to have known real cowboys and real woods. McPhee and I were classmates at prep school. I used to watch him star at basketball; attendance at the games was required, and if we in the bleachers didn't cheer, the headmaster's assistant wrote our names down. But cheered though he was, he too somewhere must have picked up a taste for solitude.

. . . At home in Vermont, lounging on my lawn, I look up at a higher tier of land where only black bears live. I'm on the brink of embarking upon still another trip to some bleak national swamp or public forest, and I think, Good God, who needs it? Like anybody else, I'm lonely enough right in the bosom of family and friends. But excitement, the hope of visions and some further understanding—that old, old boondoggle perpetrated by the wilderness—draws me on.

The "taste for solitude" so much a part of this work may be one reason why so few women have become nature writers while many have become novelists. I have been surprised at how consistently women tell me that they really do not feel safe traveling alone in the wilderness. Obviously, there are exceptions, and some of them are in this book. Still, there is a dearth of female natural history writers.

David Quammen suggests that it simply has to do with the "historical bias and socialization that is the answer to so many of the questions about the state of women now. If it was uncool for a young boy to be turning over rocks and looking at salamanders instead of playing Little League, certainly it was even more uncool for a young girl." Young girls were not encouraged to be interested in science or to wander in the woods by themselves.

Has this changed? If new generations of women have begun to get around this "historical bias," the ones who did so are only now just coming into maturity as writers. Gretel Ehrlich is forty, and at the beginning of her career. Sue Hubbell is ten years older, but also newly published. Others are out there—Vicki Hearne, who writes mostly about our relationship with domestic animals; Terry Tempest Williams, who is making the transition from children's literature to books for adults. But there are only a few. And I wish I knew more about why.

Gretel Ehrlich has realized that she is more than one person: "I lead several lives. I'm one person to myself and my writing, and I'm a different person when I'm out cowboying and ranching, and I'm another person in my marriage. But I don't see that as a problem."

Ann Zwinger says, "I've never gone into nature as a woman; I've gone as a person. I'm not sure but what in other guises men and women perceive the landscape very differently. But I think as nature writers you have a focus, and you have a discipline, and that is a very precise one—to find out what's going on—and you put aside all feelings of femininity or masculinity. I don't see it as a sex difference. I just see it as a bent of personality."

Writers who share that "bent of personality" are filled with generosity for each other, both in practice and in print, and without regard to gender. There is a sense of community among them.

21

Edward Hoagland and Annie Dillard were very encouraging to Gretel Ehrlich. Wendell Berry gave Jack Shoemaker, publisher at North Point Press, a present for his anniversary—editing Gary Nabhan's *The Desert Smells Like Rain,* gratis.

When John Hay sold Robert Finch a piece of his land on Cape Cod, Finch was a little concerned about being so close to a writer with similar interests. "It was sort of like moving into Walden and saying, 'What am I going to write about?' " But Finch hasn't found it a problem, and he calls Hay "the nature writer's writer. *The Run* was really a groundbreaking book. It gave me a sense of the possibilities of language, of writing using natural history as a focus."

David Quammen says, "Barry Lopez reminds me of the need for seriousness and care and scrupulosity, and I remind him of the need for a certain amount of frivolity and smart-aleckyness. We complement each other." Gary Nabhan speaks with amazement of Ann Zwinger's "technical proficiency as a naturalist. Her taxonomic knowledge is so well developed for an extraordinarily wide range of organisms." Zwinger herself told me she was saving *Arctic Dreams* to read at a special time: "I'm intimidated by Barry Lopez, I think he is so good."

She also speaks of the "tremendous impression" made on her by Rachel Carson, particularly Carson's *The Edge of the Sea,* which Zwinger thinks is "the perfect book." "The integrity and fineness of her work are a guide, always."

Reintegrating the Disciplines

In seeing nature as a teacher, and yourself not as an expert but as a listener, naturalists part company from many scientists.

What makes Rachel Carson's *The Edge of the Sea* "perfect" is its combination of careful observation, careful writing, and, to use Carson's own phrase, "a sense of wonder." In integrating science with art, and still retaining her capacity for innocence, Carson managed to do what Barry Lopez also aims for as a goal. He says that a writer must try "to cultivate a sense of yourself as the one who doesn't know. As long as you do this, you don't run the risk of writing the story as though you were the expert. The authority will always be with the subject."

Gary Nabhan suggests that the "philosophical bias, that nature counts as a model, a guide to how we should live in the world," is the primary defining trait of the naturalist. Nabhan and I talked as we climbed a hot, dry trail up Camelback Mountain, overlooking Phoenix, where he works as assistant director of the Desert Botanical Garden. We looked out from the sharp rock of the mountain over the miles and miles of sprawling city, and Nabhan articulated the contrasts inherent in this mix of desert and metropolis. "We've got little more than thrashers, mockingbirds, and grackles in our yards—a humanized landscape, something different than what you see in a wild desert ecosystem. Wilderness areas aren't islands. Their value is that of contrast, of making ourselves confront those differences rather than trying to ignore them."

Nabhan traces the roots for his attitude to growing up in Gary, Indiana. "My most euphoric experiences as a child were in the dunes and bogs where Henry Cowles worked out some of the principles of ecology—but where you never left the sound and the sight of the steel mills. And so, I grew up thinking about nature as a number of juxtapositions with pervasive civilization. You always carried one with you as you dealt with the other."

Naturalist writers thrive on such juxtapositions, on a mix of experience and knowledge, adventure and contemplation, humanism and science. Robert Finch calls it "reintegrating the disciplines." Writers strive to remain deliberately uncommitted to preconceptions; in this way, they differ from biologists intent on supporting or rejecting hypotheses. Finch points out that in this "risking of the self," a writer stays ready for sudden changes "right in the middle of things—from what you expected to find to what you do find." This "element of surprise" is a key to the discovery of stories, a key to what it is that naturalists are after.

Delight in surprise leads to acceptance of unresolved questions. Biologists hope for clear answers. Natural history writers may be able to tell better stories with no clear answers at all. Barry Lopez puts it this way: 'I think it is one of the strengths of human wisdom to protect mystery. The answers don't always serve as well as the questions, and unresolved mystery serves us very well."

Looking for surprises may have to do with why nature writers choose to live in out-of-the-way places. Such homes make for diversity. David Quammen lives in a small frame house on a shaded street not far from Montana State University in Bozeman. Old friends from graduate school ask him, "Why are you hiding out in Montana, isolating yourself from what's going on?" Quammen refers to island biogeography—the remarkable adaptations made by creatures living in isolation—when he answers. For writers as well as Galápagos finches, "a lot that is unique happens when you have isolation and evolution. And I think that's good. Otherwise you have a flattening out and homogenization."

Living where not many other writers live can be a small problem when you have the urge to work in other places. When assignment editors call, they ask you to do a piece about your known territory. Somehow you must break out of the role of "regional writer" and shatter your typecast image. As John Madson says, "There are all kinds of writers that want to go to Papeete or Nairobi, but who in the hell is willing to write about a vacant lot outside Des Moines?'' The vacant lot, of course, can prepare a good writer for Nairobi. Robert Finch puts it this way: "I've created the equipment here on Cape Cod that I could use elsewhere now. I carry that perspective with me."

Gretel Ehrlich speaks eloquently about how her home informs her work. "Instead of becoming jaded by the place, it's quite the opposite. I feel more astonished and inspired by it, the more I live there. Every time you look out the door, it's different. Somehow it seems that's how your whole life has to be to a certain extent, or it's going to be very claustrophobic. Dull and dark and dim and depressing."

Barry Lopez compares his work to the monastic traditions, where "your work is your prayer," with no distinction between workdays and weekends. "Writing gets done by not having anything else to do," says John McPhee, "being protective about your time. What you are trying to do is mark off a terrain, put a fence around it, and say 'I'm going to be in here for a while, and I've got to be alone.' "

Working for the *New Yorker* gives McPhee a consistent vehicle for his work. But "once you're off and working on your idea you might as well be in Tibet. There is no dialogue unless you want it. And nobody is giving you a nickel, either. You are paid like a farmer, not by salary."

Gretel Ehrlich says, "I've been able to develop the ability to just pop into my study and sit down and start writing without any elaborate preparation—even it it's just a half an hour." We were talking in her ranch house, just down the hill from the little cabin she uses for writing. "I could run up right now and write a couple of more pages in the chapter of my novel that I'm in the middle of, without having felt interrupted." She suggests that "maybe women are better at this, because we are sort of geared up to have children. I don't have children but I am sure that's what it's like—you have to be able to do several things at once.

"And also it's because the rest of the life here is in such a different vein. My books just go on and on; I can stop and feed the cows and it doesn't really get interrupted."

Robert Finch says, "Writing is what I do, so it's not 'finding time to write,' it's finding time for other stuff. I can afford to be distracted at any point by anything that's happening that's promising, whether it's walking out to the mailbox or lunch. There are times when you have deadlines, when you've got to shut out things. It's very frustrating. You know all sorts of things are going on and you have to say, 'No, no, I won't look.' "

Writers hate to write and they love to write. And they can be at loose ends when they finish a project. Ann Zwinger believes, "There is a real post-partum blues when a book is finished. I just crash and burn. Work is what holds your life together. It's the one thing that's the ballast. Unless you have another project coming on, there is this abyss out there.''

The work holds your life together until the very end. John Madson takes heart in that: "As long as you can put your palsied hands on a damn keyboard, you can grind out something—whether it's for publication or for your own satisfaction. I feel so sorry for other professionals who go into retirement in their seventies or even eighties, and wind up playing shuffleboard in Sun City." Gretel Ehrlich says, "You can start when you are eighty and be just as good or better, and you can do it forever—it's just a really democratic thing. The work speaks for itself.''

"You're no better than your current work," says John Madson. "There is only one person I feel in competition with—myself. Can I maintain ideas and style as fresh and in many ways as naive as that young writer I was a quarter of a century ago?" John Hay says, "Inevitably, you repeat yourself. The only thing you can do is avoid looking at your previous books so you don't become sad about it. That way, you never fall into, 'I should be writing this well now,' or, 'This isn't very good.' "

Part of what holds life together for Ann Zwinger is her drawing, for she illustrates her own books. "I have drawn ever since I could remember. Drawing is a piece of paper . . . and rubbing your hand over it the drawing comes up—it's there. Writing is pain, agony, and backache. I certainly could write without drawing, but drawing makes me see so much more. And by the same token, writing and learning makes me see so much more when I draw.''

At one time, Barry Lopez worked both as a photographer and a writer, but he stopped taking pictures to devote all his energy to the words. Robert Finch notes that books about the West "tend to be more easy about the combination of illustration and writing." I think this may have to do with the size and raw power of the western landscape, which demand visual as well as verbal expression. Finch points out, however, that "the more a book is grounded in the language, the more intrusive photographs become, because the language itself becomes suggestive and tactile and visual. In other words, the best-written books I prefer without photographs.''

Finch's comments intrigue me, because I am both a photographer and a writer. I see the two acts as different ways to interact with the landscape; photography feels more immediate and emotional, much as Ann Zwinger describes her drawing. Photography is a release for me, after the sedentary challenge of writing. But after being out photographing for a while, I crave the extended control and satisfying precision of creating a story with words. I love them both, and would never choose between them. I believe that their combination offers yet one more way to "reintegrate" the various approaches to the land.

But there are indeed books that beg for no photographs. They have passed a crucial threshold in the richness of their language and story; they are nonfiction books with the impact of fiction, and as Finch says, they demand a clear space on which imagination can work.

These issues have to do more with readers than writers. The audience is out there—whoever they are—and each writer has some sense of his or her readers and what obligations are owed them. This is the final aspect of writing, the huge one that the writer has worked toward but cannot really control: the readers.

Preaching to the Unconverted

John McPhee speaks highly of his readers. "It follows, absolutely, that if something is published by the *New Yorker* magazine and subsequently by Farrar, Straus, that there has to be a certain, and not inconsiderable, percentage of people who are reading the pieces who are smarter and more sensitive than I am, who in all sorts of ways are better educated, more alert, wiser, deeper, whatever—it just follows. I'm much aware of them; I don't want to offend them by pointing out the obvious. I'm hoping that the pieces will be acceptable to those people, whoever they are. That's about the only idea of a reader that I have."

David Quammen knows as much as any nature writer about his audience, at least for his column in *Outside* magazine. He appreciates the opportunity of "having a regular soapbox with a fairly large readership," even though he periodically feels that he is repeating himself, becoming "a slave to this thing," tiring "of chopping everything down into little Chiclets to fit into the magazine." But Quammen likes the fact that he "disagrees with a lot of the things that are implicitly valued" by his readers.

"I don't see a lot of value in preaching to the converted," he says. "I don't like being a cheerleader for a homogeneous audience. But writing for *Esquire* or *Outside,* to sneak up on some people who have very different value systems, really seems to me more valuable. And more fun."

Quammen finds "making connections" the most interesting part of constructing a piece. "Something that's just a piece about the manatee—here are the manatee, aren't they beautiful, we should save them—fine, but boring as a piece of writing. But something that's entangled with human history, the complexity of the human character, (without being too grandiose about it) some moral issues, and is connected with the natural world—can really have to me a power and an interest and a value."

Natural history writers take pains to avoid sounding didactic, but they have no problems with sounding moral. Listen to Ann Zwinger: "I accept that I write for a limited audience. But the chain of principles on which I operate is simple. If you can entice people to look at a facet of nature, they may get curious. If they get curious they may make an attempt to learn something about what they are seeing. If they learn something, that becomes an irrevocable part of their experience.

"If it becomes part of your experience, it becomes *yours* in a way that nothing else in the world can. We don't own anything; all we own is our knowledge. If you appreciate it and enjoy it, you're a lot less likely to destroy it."

John Hay says, "You sometimes feel a little hysterical about the lack of awareness, lack of response, on the other side. You are trying to convey something about the human spirit. You are trying to convey a sense of immortality. I sure don't have a reader in mind when I write—unless it's all those who speak the English language, the American language, all those who are receptive to feelings outside themselves. If I can write well enough so that people can pick up their ears and their senses . . . if I can do it in some sort of ethical, moral fashion so that people will listen—I realize they won't listen very much—it's about all you can do."

You write for yourself, and you simply have to trust your readers to come with you. David Quammen, for instance, trusts his readers to understand when he talks about crows being bored, about black widows "knowing their place." He says, "Anthropomorphism is a type of analogy, and large, large portions of human communication depend on analogy, whether visible or invisible. I think it's okay if it's clear to the reader that you're taking these kinds of liberties. You've planed out and taken flight into facetiousness or fancy or whimsy, so that the reader should know not to take everything you say literally.

"In terms of preservation of wildlands and ecological issues, at this point in this century I think that we are on the gallows. Part of what I see as a useful service I can provide is some of the gallows humor." At the same time, Quammen admits that "the great thing about having a natural science column in a yuppie magazine is that you can use it as a soapbox and secretly be [Quammen whispered this, with a mischievous twinkle in his eye]—a moral philosopher."

Edward Abbey shares Quammen's urge to provoke. Abbey says, "I write in a deliberately outrageous or provocative manner because I like to startle people. I hope to wake up people. I have no desire to simply soothe or please. I would rather risk making people angry than putting them to sleep. And I try to write in a style that's entertaining as well as provocative. It's hard for me to stay serious for more than half a page at a time.

"When I started writing *The Monkey Wrench Gang,* my original motive was, I guess, anger. Trying to get some sort of revenge for the destruction I had witnessed in the American Southwest. But I soon discovered that the anger in itself would lead to a sort of dead end, get me nowhere as an artist, as a writer. I found that I could communicate my contempt and my disgust much better by trying to make people laugh."

Abbey arouses strong reactions from reviewers, readers, and critics. Wendell Berry has written of Abbey's outrageous inconsistencies, of why Abbey is such "a problem" in the book *Resist Much, Obey Little: Some Notes on Edward Abbey:* "The trouble . . . with Mr. Abbey—a trouble, I confess, that I am disposed to like—is that he speaks insistently as himself."

Some try to make Abbey into a high priest, or at least an archdruid, of wilderness values. As Berry and Abbey, himself, point out, Abbey doesn't make much of a priest. But he stands for something, and he is proud of it: "What is both necessary and sufficient—for honest work—is to have faith in the evidence of your senses and in your common sense. To be true to your innate sense of justice. To be loyal to your family, your clan, your friends, and your community. (Let the nation-state go hang itself.) Stand up for the stupid and crazy. Love the earth and the sun and the animals."

I repeatedly heard this refrain of moral concern when talking with these writers. I would ask a question about the mechanics of writing, and the answer would quickly lead from writing to the world at large to the things in that world that each writer cares about. When writing natural history, you step out of yourself to address specific issues you have decided are more important than yourself. This balances egos, and that is a rare thing. It distinguishes these writers from journalists who write about anything, and from those novelists who are purely inward-directed.

Barry Lopez speaks of this as the primary North American contribution to literature: "an illumination of the relationship between human beings and landscape, particularly modern man trying to come to grips with a re-orientation toward landscape."

You don't have to live out there permanently. Barry Lopez does not live in the Arctic, and Peter Matthiessen does not live in Nepal. But they go there "unsheltered," to use Gretel Ehrlich's words, "as immigrants, sort of hobos, with a great deal of curiosity and openness and humility."

They bring back new understanding, and with that knowledge comes responsibility. As Edward Abbey puts it, "I think that poets and writers, essayists and novelists, have a moral obligation to be the conscience of their society. I think it's the duty of a writer, as Samuel

Johnson said, to try to make the world better, however futile that effort might be.''

Barry Lopez subscribes to Abbey's dictum that ''writers should remind society of its ideals.'' Lopez defines this responsibility: ''to find a moral, dignified, decent way of living in the world with regard to other peoples, and with regard to the landscape.

''The moral dimension of language must be brought out. All good writers are troublemakers, all good writers are sworn enemies of complacency and dogma. The storyteller's responsibility is not to be wise, a storyteller is the person who creates an atmosphere in which wisdom can reveal itself.''

Lopez says that what you do when writing a piece of nonfiction is ''to make a bow of respect toward the material and to make a bow of respect toward the reader.''

He sums up his respect toward the material in the following way: ''Listen. Pay attention. Do your research. Try to learn. Don't presume. And always imagine that there's more there than you could possibly understand or sense.''

In paying respects to the reader, Lopez says, with a bow: ''I have assembled this material. I have tried to bring order to these disparate elements. I have tried to use the language elegantly. I have sought, everywhere I could, illumination, clarity. I have tried to organize things with a proper sense of the drama of human life. I have tried to think hard about these things. I have tried to get rid of all that is unnecessary for you to understand the story.''

As readers, let us return his courtesy. On behalf of other readers, let me make a bow of respect to the writers of natural history. We thank you for your work and your passion. You make us take a stand, and you make us feel. Keep writing. We will keep reading.

I

JOURNEYS

"Never did we plan the morrow, for we had learned that in the wilderness some new and irresistible distraction is sure to turn up each day before breakfast. Like the river, we were free to wander. . . .

I am glad I shall never be young without wild country to be young in. Of what avail are forty freedoms without a blank spot on the map?"

ALDO LEOPOLD, *A SAND COUNTY ALMANAC AND SKETCHES HERE AND THERE*, 1949

ANNIE
DILLARD

In 1974, Annie Dillard received the Pulitzer Prize in nonfiction for *Pilgrim At Tinker Creek,* her first published book of prose. An intense narrative of one year's personal exploration in Virginia's Blue Ridge, the book is now a classic.

Dillard anchors her books in "information and landscape," as the following essay, taken from *Teaching A Stone to Talk: Expeditions and Encounters* (1982), shows.

Born in Pittsburgh in 1945, she attended college in Virginia and has spent her time since then writing. She says, "Your life is literature. It's all hard, conscious, terribly frustrating work. But this never occurs to people. They think it happens in a dream, that you just sit on a tree stump and take dictation from a chipmunk." Dillard wrote the last few chapters of *Tinker Creek* in fifteen- to sixteen-hour rounds, seven days a week for two months—in a college library.

In the late 1970s, Dillard lived on an island in Puget Sound, Washington;

she now lives in Middletown, Connecticut, where she is writer-in-residence at Wesleyan University. Her most recent book is *An American Childhood* (1987), a passionate memoir of her youth in Pittsburgh. Like all her books, she says, "it is about what it feels like to be alive." Since *Pilgrim at Tinker Creek,* she has published volumes of poetry, essays, criticism, and *Encounters With Chinese Writers* (1986).

Dillard wrote, in 1984, "The trick of writing, which drives previously sane people around the bend, is to locate some weird interior spot our brains don't seem well programmed for: the spot that enables you to be wholly alive while wholly alone. A few hours a day of this is quite enough. When it's over, I'm ready for lunch. Lunch with familiar people I've come to care for. . . .

"In a wilderness solitude, you can get any books you want by writing the state library—but you must know precisely which books you want. How can you do real work if you already know everything? Where is the chaos, the clash and grate of disparate materials, that makes sparks? Where's the parade?

". . . So I live, and have almost always lived, in association with a college campus. I need lots of time, a big house with room for lots of desks, and a batch of people to befriend—people who are also working among bits of paper all day.

". . . The mental landscape is as rich and various as the human landscape on Main Street, and as the forest landscape on the hillsides. It is all a parade. It is all interesting. Dig anywhere."

Life on the Rocks:
The Galápagos

I

First there was nothing, and although you know with your reason that nothing is nothing, it is easier to visualize it as a limitless slosh of sea—say, the Pacific. Then energy contracted into matter, and although you know that even an invisible gas is matter, it is easier to visualize it as a massive squeeze of volcanic lava spattered inchoate from the secret pit of the ocean and hardening mute and intractable on nothing's lapping shore—like a series of islands, an archipelago. Like: the Galápagos. Then a softer strain of matter began to twitch. It was a kind of shaped water; it flowed, hardening here and there at its tips. There were blue-green algae; there were tortoises.

The ice rolled up, the ice rolled back, and I knelt on a plain of lava boulders in the islands called Galápagos, stroking a giant tortoise's neck. The tortoise closed its eyes and stretched its neck to its greatest height and vulnerability. I rubbed that neck and when I pulled away my hand, my palm was green with a slick of single-celled algae. I stared at the algae, and at the tortoise, the way you stare at any life on a lava flow, and thought: Well—here we all are.

Being here is being here on the rocks. These Galapagonian rocks, one of them seventy-five miles long, have dried under the equatorial sun between five and six hundred miles west of the South American continent; they lie at the latitude of the Republic of Ecuador, to which they belong.

There is a way a small island rises from the ocean affronting all reason. It is a chunk of chaos pounded into visibility *ex nihilo:* here rough, here smooth, shaped just so by a matrix of physical necessities too weird to contemplate, here instead of there, here instead of not at all. It is a fantastic utterance, as though I were to open my mouth and emit a French horn, or a vase, or a knob of tellurium. It smacks of folly, of first causes.

I think of the island called Daphnecita, little Daphne, on which I never set foot. It's in half of my few photographs, though, because it obsessed me; a dome of gray lava like a pitted loaf, the size of the Plaza Hotel, glazed with guano and crawling with red-orange crabs. Sometimes I attributed to this island's cliff face a surly, infantile consciousness, as though it were sulking in the silent moment after it had just shouted, to the sea and the sky, "I didn't ask to be born." Or sometimes it aged to a raging adolescent, a kid who's just learned that the game is fixed, demanding, "What did you have me for, if you're just going to push me around?" Daphnecita: again, a wise old island, mute, leading the life of pure creaturehood open to any antelope or saint. After you've blown the ocean sky-high, what's there to say? What if we the people had the sense or grace to live as cooled islands in an archipelago live, with dignity, passion, and no comment?

It is worth flying to Guayaquil, Ecuador, and then to Baltra in the Galápagos just to see the rocks. But these rocks are animal gardens. They are home to a Hieronymus Bosch assortment of windblown, stowaway, castaway, flotsam, and shipwrecked creatures. Most exist nowhere else on earth. These reptiles and insects, small mammals and birds, evolved unmolested on the various islands on which they were cast into unique species adapted to the boulder-wrecked shores, the cactus deserts of the lowlands, or the elevated jungles of the large islands' interiors. You come for the animals. You come to see the curious shapes soft proteins can take, to impress yourself with their reality, and to greet them.

You walk among clattering four-foot marine iguanas heaped on the shore lava, and on each other, like slag. You swim with penguins; you watch flightless cormorants dance beside you, ignoring you, waving the black nubs of their useless wings. Here are nesting blue-footed boobies, real birds with real feathers, whose legs and feet are nevertheless patently fake, manufactured by Mattel. The tortoises are big as stoves. The enormous land iguanas at your feet change color in the sunlight, from gold to blotchy red as you watch.

There is always some creature going about its beautiful business. I missed the boat back to my ship, and was left behind momentarily on uninhabited South Plaza Island, because I was watching the Audubon's shearwaters. These dark pelagic birds flick along pleated seas in stitching flocks, flailing their wings rapidly—because if they don't, they'll stall. A shearwater must fly fast, or not at all. Consequently it has evolved two nice behaviors which serve to bring it into its nest alive. The nest is a shearwater-sized hole in the lava cliff. The shearwater circles over the water, ranging out from the nest a quarter of a mile, and veers gradually toward the cliff, making passes at its nest. If the flight angle is precisely right, the bird will fold its wings at the hole's entrance and stall directly onto its floor. The angle is perhaps seldom right, however; one shearwater I watched made a dozen suicidal-looking passes before it vanished into a chink. The other behavior is spectacular. It involves choosing the nest hole in a site below a prominent rock with a downward-angled face. The shearwater comes careering in at full tilt, claps its wings, stalls itself into the rock, and the rock, acting as a backboard, banks it home.

The animals are tame. They have not been persecuted, and show no fear of man. You pass among them as though you were wind, spindrift, sunlight, leaves. The songbirds are tame. On Hood Island I sat beside a nesting waved albatross while a mockingbird scratched in my hair, another mockingbird jabbed at my fingernail, and a third mockingbird made an exquisite progression of pokes at my bare feet up the long series of eyelets in my basketball shoes. The marine iguanas are tame. One settler, Carl Angermeyer, built his house on the site of a marine iguana colony. The gray iguanas, instead of moving out, moved up on the roof, which is corrugated steel. Twice daily on the patio, Angermeyer feeds them a mixture of boiled rice and tuna fish from a plastic basin. Their names are all, unaccountably, Annie. Angermeyer beats on the basin with a long-handled spoon, calling, "Here AnnieAnnieAnnieAnnie"—and the spiny reptiles, fifty or sixty strong, click along the steel roof, finger their way down the lava boulder and mortar walls, and swarm round his bare legs to elbow into the basin and be elbowed out again smeared with a mash of boiled rice on their bellies and on their protuberant, black, plated lips.

The wild hawk is tame. The Galápagos hawk is related to North America's Swainson's hawk; I have read that if you take pains, you can walk up and pat it. I never tried. We people don't walk up and pat each other; enough is enough. The animals' critical distance and mine tended to coincide, so we could enjoy an easy sociability without threat of violence or unwonted intimacy. The hawk, which is not notably sociable, nevertheless endures even a blundering approach, and is apparently as content to perch on a scrub tree at your shoulder as anyplace else.

In the Galápagos, even the flies are tame. Although most of the land is Ecuadorian national park, and as such rigidly protected, I confess I gave the evolutionary ball an offsides shove by dispatching every fly that bit me, marveling the while at its pristine ignorance, its blithe failure to register a flight trigger at the sweep of my descending hand—an insouciance that was almost, but not quite, disarming. After you kill a fly, you pick it up and feed it to a lava lizard, a bright-throated four-inch lizard that scavenges everywhere in the arid lowlands. And you walk on, passing among the innocent mobs on every rock hillside; or you sit, and they come to you.

We are strangers and sojourners, soft dots on the rocks. You have walked along the strand and seen where birds have landed, walked, and flown; their tracks begin in sand, and go, and suddenly end. Our tracks do that: but we go down. And stay down. While we're here, during the seasons our tents are pitched in the light, we pass among each other crying "greetings" in a thousand tongues, and "welcome," and "good-bye." Inhabitants of uncrowded colonies tend to offer the stranger famously warm hospitality—and such are the Galápagos sea lions. Theirs is the greeting the first creatures must have given Adam—a hero's welcome, a universal and undeserved huzzah. Go, and be greeted by sea lions.

I was sitting with ship's naturalist Soames Summerhays on a sand beach under cliffs on uninhabited Hood Island. The white beach was a havoc of lava boulders black as clinkers, sleek with spray, and lambent as brass in the sinking sun. To our left a dozen sea lions were bodysurfing in the long green combers that rose, translucent, half a mile offshore. When the combers broke, the shoreline boulders rolled. I could feel the roar in the rough rock on which I sat; I could hear the grate inside each long backsweeping sea, the rumble of a rolled million rocks muffled in splashes and the seethe before the next wave's heave.

To our right, a sea lion slipped from the ocean. It was a young bull; in another few years he would be dangerous, bellowing at intruders and biting off great dirty chunks of the ones he caught. Now this young bull, which weighed maybe 120 pounds, sprawled silhouetted in the late light, slick as a drop of quicksilver, his glistening whiskers radii of gold like any crown. He hauled his packed hulk toward us up the long beach; he flung himself with an enormous surge of fur-clad muscle onto the boulder where I sat. "Soames," I said—very quietly, "he's here because *we're* here, isn't he?" The naturalist nodded. I felt water drip on my elbow behind me, then the fragile scrape of whiskers, and finally the wet warmth and weight of a muzzle, as the creature settled to sleep on my arm. I was catching on to sea lions.

Walk into the water. Instantly sea lions surround you, even if none has been in sight. To say that they come to play with you is not especially anthropomorphic. Animals play. The bull sea lions are off patrolling their territorial shores; these are the cows and young, which range freely. A five-foot sea lion peers intently into your face, then urges her muzzle gently against your underwater mask and searches your eyes without blinking. Next she rolls upside down and slides along the length of your floating body, rolls again, and casts a long glance back at your eyes. You are, I believe, supposed to follow, and think up something clever in return. You can play games with sea lions in the water using shells or bits of leaf, if you are willing. You can spin on your vertical axis and a sea lion will swim circles around you, keeping her face always six inches from yours, as though she were tethered. You can make a game of touching their back flippers, say, and the sea lions will understand at once; somersaulting conveniently before your clumsy hands, they will give you an excellent field of back flippers.

And when you leave the water, they follow. They don't want you to go. They porpoise to the shore, popping their heads up when they lose you and casting about, then speeding to your side and emitting a choked series of vocal notes. If you won't relent, they disappear, barking; but if you sit on the beach with so much as a foot in the water, two or three will station with you, floating on their backs and saying, Urr.

Few people come to the Galápagos. Buccaneers used to anchor in the bays to avoid pursuit, to rest, and to lighter on fresh water. The world's whaling ships stopped here as well, to glut their holds with fresh meat in the form of giant tortoises. The whalers used to let the tortoises bang around on deck for a few days to empty their guts; then they stacked them below on their backs to live—if you call that living—without food or water for a year. When they wanted fresh meat, they killed one.

Early inhabitants of the islands were a desiccated assortment of grouches, cranks, and ships' deserters. These hardies shot, poisoned, and enslaved each other off, leaving behind a fecund gang of feral goats, cats, dogs, and pigs whose descendants skulk in the sloping jungles and take their tortoise hatchlings neat. Now scientists at the Charles Darwin Research Station, on the island of Santa Cruz, rear the tortoise hatchlings for several years until their shells are tough enough to resist the crunch; then they release them in the wilds of their respective islands. Today, some few thousand people live on three of the islands; settlers from Ecuador, Norway, Germany, and France make a livestock or pineapple living from the rich volcanic soils. The settlers themselves seem to embody a high degree of courteous and conscious humanity, perhaps because of their relative isolation.

On the island of Santa Cruz, eleven fellow passengers and I climb in an open truck up the Galápagos' longest road; we shift to horses, burros, and mules, and visit the lonely farm of Alf Kastdalen. He came to the islands as a child with his immigrant parents from Norway. Now a broad, blond man in his late forties with children of his own, he lives in an isolated house of finished timbers imported from the mainland, on four hundred acres he claimed from the jungle by hand. He raises cattle. He walks us round part of his farm, smiling expansively and meeting our chatter with a willing, open gaze and kind words. The pasture looks like any pasture—but the rocks under the grass are round lava ankle-breakers, the copses are a tangle of thorny bamboo and bromeliads, and the bordering trees dripping in epiphytes are breadfruit, papaya, avocado, and orange.

Kastdalen's isolated house is heaped with books in three languages. He knows animal husbandry; he also knows botany and zoology. He feeds us soup, chicken worth chewing for, green *naranjilla* juice, noodles, pork in big chunks, marinated mixed vegetables, rice, and bowl after bowl of bright mixed fruits.

And his isolated Norwegian mother sees us off; our beasts are ready. We will ride down the mud forest track to the truck at the Ecuadorian settlement, down the long road to the boat, and across the bay to the ship. I lean down to catch her words. She is gazing at me with enormous warmth. "Your hair," she says softly. I am blond. *Adiós.*

II

Charles Darwin came to the Galápagos in 1835, on the *Beagle;* he was twenty-six. He threw the marine iguanas as far as he could into the water; he rode the tortoises and sampled their meat. He noticed that the tortoises' carapaces varied wildly from island to island; so also did the forms of various mockingbirds. He made collections. Nine years later he wrote in a letter, "I am almost convinced (quite contrary to the opinion I started with) that species are not (it is like confessing a murder) immutable." In 1859 he published *On the Origin of Species,* and in 1871 *The Descent of Man.* It is fashionable now to disparage Darwin's originality; not even the surliest of his detractors, however, faults his painstaking methods or denies his impact.

Darwinism today is more properly called neo-Darwinism. It is organic evolutionary theory informed by the spate of new data from modern genetics, molecular biology, paleobiology—from the new wave of the biologic revolution which spread after Darwin's announcement like a tsunami. The data are not all in. Crucial first appearances of major invertebrate groups are missing from the fossil record—but these early forms, sometimes modified larvae, tended to be fragile either by virtue of their actual malleability or by virtue of their scarcity and rapid variation into "hardened," successful forms. Lack of proof in this direction doesn't worry scientists. What neo-Darwinism seriously lacks, however, is a description of the actual mechanism of mutation in the chromosomal nucleotides.

In the larger sense, neo-Darwinism also lacks, for many, sheer plausibility. The triplet splendors of random mutation, natural selection, and Mendelian inheritance are neither energies nor gods; the words merely describe a gibbering tumult of materials. Many things are unexplained, many discrepancies unaccounted for. Appending a very

39

modified neo-Lamarckism to Darwinism would solve many problems—and create new ones. Neo-Lamarckism holds, without any proof, that certain useful acquired characteristics may be inherited. Read C. H. Waddington, *The Strategy of the Genes,* and Arthur Koestler, *The Ghost in the Machine.* The Lamarckism/Darwinism issue is not only complex, hinging perhaps on whether DNA can be copied from RNA, but also politically hot. The upshot of it all is that while a form of Lamarckism holds sway in Russia, neo-Darwinism is supreme in the West, and its basic assumptions, though variously modified, are not overthrown.

So much for scientists. The rest of us didn't hear Darwin as a signal to dive down into the wet nucleus of a cell and surface with handfuls of strange new objects. We were still worried about the book with the unfortunate word in the title: *The Descent of Man.* It was dismaying to imagine great-grandma and great-grandpa effecting a literal, nimble descent from some liana-covered tree to terra firma, scratching themselves, and demanding bananas.

Fundamentalist Christians, of course, still reject Darwinism because it conflicts with the creation account in Genesis. Fundamentalist Christians have a very bad press. Ill feeling surfaces when, from time to time in small towns, they object again to the public schools' teaching evolutionary theory. Tragically, these people feel they have to make a choice between the Bible and modern science. They live and work in the same world as we, and know the derision they face from people whose areas of ignorance are perhaps different, who dismantled their mangers when they moved to town and threw out the baby with the straw.

Even less appealing in their response to the new evolutionary picture were, and are, the social Darwinists. Social Darwinists seized Herbert Spencer's phrase, "the survival of the fittest," applied it to capitalism, and used it to sanction ruthless and corrupt business practices. A social Darwinist is unlikely to identify himself with the term; social Darwinism is, as the saying goes, not a religion but a way of life. A modern social Darwinist wrote the slogan "If you're so smart, why ain't you rich?" The notion still obtains, I believe, wherever people seek power: that the race is to the swift, that everybody is *in* the race, with varying and merited degrees of success or failure, and that reward is its own virtue.

Philosophy reacted to Darwin with unaccustomed good cheer. William Paley's fixed and harmonious universe was gone, and with it its meticulous watchmaker god. Nobody mourned. Instead philosophy shrugged and turned its attention from first and final causes to analysis of certain values here in time. "Faith in progress," the man-in-the-street philosophy, collapsed in two world wars. Philosophers were more guarded; pragmatically, they held a very refined "faith in process''—which, it would seem, could hardly lose. Christian thinkers, too, outside of Fundamentalism, examined with fresh eyes the world's burgeoning change. Some Protestants, taking their cue from Whitehead, posited a dynamic god who lives alongside the universe, himself charged and changed by the process of becoming. Catholic Pierre Teilhard de Chardin, a paleontologist, examined the evolution of species itself, and discovered in that flow a surge toward complexity and consciousness, a free ascent capped with man and propelled from within and attracted from without by god, the holy freedom and awareness that is creation's beginning and end. And so forth. Like flatworms, like languages, ideas evolve. And they evolve, as Arthur Koestler suggests, not from hardened final forms, but from the softest plasmic germs in a cell's heart, in the nub of a word's root, in the supple flux of an open mind.

Darwin gave us time. Before Darwin (and Huxley, Wallace, et al.) there was in the nineteenth century what must have been a fairly nauseating period: people knew about fossils of extinct species, but did not yet know about organic evolution. They thought the fossils were litter from a series of past creations. At any rate, for many, this creation, the world as we know it, had begun in 4004 B.C., a date set by Irish Archbishop James Ussher in the seventeenth century. We were all crouched in a small room against the comforting back wall, awaiting the millennium which had been gathering impetus since Adam and Eve. Up there was a universe, and down here would be a small strip of man come and gone, created, taught, redeemed, and gathered up in a bright twinkling, like a sprinkling of confetti torn from colored papers, tossed from windows, and swept from the streets by morning.

The Darwinian revolution knocked out the back wall, revealing eerie lighted landscapes as far back as we can see. Almost at once, Albert Einstein and astronomers with reflector telescopes and radio telescopes knocked out the other walls and the ceiling, leaving us sunlit, exposed, and drifting—leaving us puckers, albeit evolving puckers, on the inbound curve of space-time.

III

It all began in the Galápagos, with these finches. The finches in the Galápagos are called Darwin's finches; they are everywhere in the islands, sparrowlike, and almost identical but for their differing beaks. At first Darwin scarcely noticed their importance. But by 1839, when he revised his *Journal* of the *Beagle* voyage, he added a key sentence about the finches' beaks: "Seeing this gradation and diversity of structure in one small, intimately related group of birds, one might really fancy that from an original paucity of birds in this archipelago, one species had been taken and modified for different ends." And so it was.

The finches come when called. I don't know why it works, but it does. Scientists in the Galápagos have passed down the call: you say pssssh pssssh pssssh pssssh pssssh until you run out of breath; then you say it again until the island runs out of birds. You stand on a flat of sand by a shallow lagoon rimmed in mangrove thickets and call the birds right out of the sky. It works anywhere, from island to island.

Once, on the island of James, I was standing propped against a leafless *palo santo* tree on a semiarid inland slope, when the naturalist called the birds.

From other leafless *palo santo* trees flew the yellow warblers, speckling the air with bright bounced sun. Gray mockingbirds came running. And from the green prickly pear cactus, from the thorny acacias, sere grasses, bracken and manzanilla, from the loose black lava, the bare dust, the fern-hung mouths of caverns or the tops of sunlit logs—came the finches. They fell in from every direction like colored bits in a turning kaleidoscope. They circled and homed to a vortex, like a whirlwind of chips, like draining water. The tree on which I leaned was the vortex. A dry series of puffs hit my cheeks. Then a rough pulse from the tree's thin trunk met my palm and rang up my arm—and another, and another. The tree trunk agitated against my hand like a captured cricket: I looked up. The lighting birds were rocking the tree. It was an appearing act: before there were barren branches; now there were birds like leaves.

Darwin's finches are not brightly colored; they are black, gray, brown, or faintly olive. Their names are even duller: the large ground finch, the medium ground finch, the small ground finch; the large insectivorous tree finch; the vegetarian tree finch; the cactus ground finch, and so forth. But the beaks are interesting, and the beaks' origins even more so.

Some finches wield chunky parrot beaks modified for cracking seeds. Some have slender warbler beaks, short for nabbing insects, long for probing plants. One sports the long chisel beak of a woodpecker; it bores wood for insect grubs and often uses a twig or cactus spine as a pickle fork when the grub won't dislodge. They have all evolved, fanwise, from one bird.

The finches evolved in isolation. So did everything else on earth. With the finches, you can see how it happened. The Galápagos islands are near enough to the mainland that some strays could hazard there; they are far enough away that those strays could evolve in isolation from parent species. And the separate islands are near enough to each other for further dispersal, further isolation, and the eventual reassembling of distinct species. (In other words, finches blew to the Galápagos, blew to various islands, evolved into differing species, and blew back together again.) The tree finches and the ground finches, the woodpecker finch and the warbler finch, veered into being on isolated rocks.

The witless green sea shaped those beaks as surely as it shaped the beaches. Now on the finches in the *palo santo* tree you see adaptive radiation's results, a fluorescent splay of horn. It is as though an archipelago were an arpeggio, a rapid series of distinct but related notes. If the Galápagos had been one unified island, there would be one dull note, one super-dull finch.

IV

Now let me carry matters to an imaginary, and impossible, extreme. If the earth were one unified island, a smooth ball, we would all be one species, a tremulous muck. The fact is that when you get down to this business of species formation, you eventually hit some form of reproductive isolation. Cells tend to fuse. Cells tend to engulf each other; primitive creatures tend to move in on each other and on us, to colonize, aggregate, blur. (Within species, individuals have evolved immune reactions, which help preserve individual integrity; you might reject my liver—or someday my brain.) As much of the world's energy seems to be devoted to keeping us apart as was directed to bringing us here in the first place. All sorts of different creatures can mate and pro-duce fertile offspring: two species of snapdragon, for instance, or mallard and pintail ducks. But they don't. They live apart, so they don't mate. When you scratch the varying behaviors and conditions behind reproduc-tive isolation, you find, ultimately, geographical isolation. Once the isola-tion has occurred, of course, forms harden out, enforcing reproductive isolation, so that snapdragons will never mate with pintail ducks.

Geography is the key, the crucial accident of birth. A piece of protein could be a snail, a sea lion, or a systems analyst, but it had to start somewhere. This is not science; it is merely a metaphor. And the landscape in which the protein "starts" shapes its end as surely as bowls shape water.

We have all, as it were, blown back together like the finches, and it's hard to imagine the isolation from parent species in which we evolved. The frail beginnings of great phyla are lost in the crushed histories of cells. Now we see the embellishments of random chro-mosomal mutations selected by natural selection and preserved in geographically isolate gene pools as *faits accomplis,* as the differentiated fringe of brittle knobs that is life as we know it. The process is still going on, but there is no turning back; it happened, in the cells. Geographical determination is not the cow-caught-in-a-crevice business I make it seem. I'm dealing in imagery, working toward a picture.

Geography is life's limiting factor. Speciation— life itself—is ultimately a matter of warm and cool currents, rich and bare soils, deserts and forests, fresh and salt waters, deltas and jungles and plains. Species arise in isolation. A plaster cast is as intricate as its mold; life is a gloss on geography. And if you dig your fists into the earth and crumble geography, you strike geology. Climate is the wind of the mineral earth's rondure, tilt, and orbit modified by local geological conditions. The Pacific Ocean, the Negev Desert, and the rain forest in Brazil are local geological conditions. So are the slow carp pools and splashing trout riffles of any backyard creek. It is all, God help us, a matter of rocks.

The rocks shape life like hands around swelling dough. In Virginia, the salamanders vary from mountain ridge to mountain ridge; so do the fiddle tunes the old men play. All this is because it is hard to move from mountain to mountain. These are not merely anomalous details. This is what life is all about: salamanders, fiddle tunes, you and me and things, the split and burr of it all, the fizz into particulars. No mountains and one salamander, one fiddle tune, would be a lesser world. No continents, no fiddlers. No possum, no sop, no taters. The earth, without form, is void.

The mountains are time's machines; in effect, they roll out protoplasm like printers' rollers pressing out news. But life is already part of the landscape, a limiting factor in space; life too shapes life. Geology's rocks and climate have already become Brazil's rain forest, yielding shocking bright birds. To say that all life is an interconnected membrane, a weft of linkages like chain mail, is truism. But in this case, too, the Galápagos islands afford a clear picture.

On Santa Cruz island, for instance, the saddle-back carapaces of tortoises enable them to stretch high and reach the succulent pads of prickly pear cactus. But the prickly pear cactus on that island, and on other tortoise islands, has evolved a treelike habit; those lower pads get harder to come by. Without limiting factors, the two populations could stretch right into the stratosphere.

Ça va. It goes on everywhere, tit for tat, action and reaction, triggers and inhibitors ascending in a spiral like spatting butterflies. Within life, we are pushing each other around. How many animal forms have evolved just so because there are, for instance, trees? We pass the nitrogen around, and vital gases; we feed and nest, plucking this and that and planting seeds. The protoplasm responds, nudged and nudging, bearing the news.

And the rocks themselves shall be moved. The rocks themselves are not pure necessity, given, like vast, complex molds around which the rest of us swirl. They heave to their own necessities, to stirrings and prickings from within and without.

The mountains are no more fixed than the stars. Granite, for example, contains much oxygen and is relatively light. It "floats." When granite forms under the earth's crust, great chunks of it bob up, I read somewhere, like dumplings. The continents themselves are beautiful pea-green boats. The Galápagos archipelago as a whole is surfing toward Ecuador; South America is sliding toward the Galápagos; North America, too, is sailing westward. We're on floating islands, shaky ground.

So the rocks shape life, and then life shapes life, and the rocks are moving. The completed picture needs one more element: life shapes the rocks.

Life is more than a live green scum on a dead pool, a shimmering scurf like slime mold on rock. Look at the planet. Everywhere freedom twines its way around necessity, inventing new strings of occasions, lassoing time and putting it through its varied and spirited paces. Everywhere live things lash at the rocks. Softness is vulnerable, but it has a will; tube worms bore and coral atolls rise. Lichens in delicate lobes are chewing the granite mountains; forests in serried ranks trammel the hills. Man has more freedom than other live things; anti-entropically, he batters a bigger dent in the given, damming the rivers, planting the plains, drawing in his mind's eye dotted lines between the stars.

The old ark's a moverin'. Each live thing wags its home waters, rumples the turf, rearranges the air. The rocks press out protoplasm; the protoplasm pummels the rocks. It could be that this is the one world, and that world a bright snarl.

Like boys on dolphins, the continents ride their crustal plates. New lands shoulder up from the waves, and old lands buckle under. The very landscapes heave; change burgeons into change. Gray granite bobs up, red clay compresses; yellow sandstone tilts, surging in forests, incised by streams. The mountains tremble, the ice rasps back and forth, and the protoplasm furls in shock waves, up the rock valleys and down, ramifying possibilities, riddling the mountains. Life and the rocks, like spirit and matter, are a fringed matrix, lapped and lapping, clasping and held. It is like hand washing hand. It is like hand washing hand and the whole tumult hurled. The planet spins, rapt inside its intricate mists. The galaxy is a flung thing, loose in the night, and our solar system is one of many dotted campfires ringed with tossed rocks. What shall we sing?

What shall we sing, while the fire burns down? We can sing only specifics, time's rambling tune, the places we have seen, the faces we have known. I will sing you the Galápagos islands, the sea lions soft on the rocks. It's all still happening there, in real light, the cool currents upwelling, the finches falling on the wind, the shearwaters looping the waves. I could go back, or I could go on; or I could sit down, like Kubla Khan:

Weave a circle round him thrice,
  And close your eyes with holy dread,
For he on honey-dew hath fed,
  And drunk the milk of Paradise.

EDWARD
ABBEY

Photograph © Jay Dusard

In the preface to his 1984 collection of essays, *Beyond the Wall,* Edward Abbey said, "This book is my last to be 'writ in sand.' Never again will I vandalize the slipface of a dune with my impertinent signature. I have nothing new to say about vultures, stone, scorpions, kissing bugs, alkali, silence. . . . Let other younger, more hopeful voices carry on."

Thus Abbey proclaimed the end of his years as a natural history writer, though he never admitted to them in the first place, preferring to describe himself as a novelist. Sadly, his proclamation was prophetic; Abbey died on March 14, 1989. His first three books were novels, the most widely known being *The Brave Cowboy* (1958), on which the film "Lonely Are The Brave" was based. His best-seller is *The Monkey Wrench Gang* (1975), a novel of eco-sabotage full of anger and humor. But it was *Desert Solitaire* (1968), a volume of what Abbey called "personal history" that grew from his time as a park ranger at Arches National Monument in the late 1950s, that established him as a voice for the Southwest. Abbey later claimed he didn't much like the book, but it has introduced a generation to the canyon country of Utah and Arizona, and promises to do so for the next generation, as well. His other collections of personal history include *The Journey Home* (1977), *Abbey's Road* (1979), and *Down the River* (1982), from which this piece comes.

Abbey was born in 1927 and reared on a western Pennsylvania farm. After a stint in Italy with the U.S. Army, he moved to the Southwest in 1947, and except for a year in New York and a year at Edinburgh University in Scotland, he lived there ever after. Abbey worked as a seasonal park ranger and fire lookout for sixteen years, and taught part-time in the University of Arizona English Department. He said, "I don't have a career, only a life."

Abbey called his essays "antidotes to despair." In *One Life at a Time, Please* (1988), he explained why:

"I write to entertain my friends and exasperate our enemies. I write to record the truth of our time, as best as I can see it. To investigate the comedy and tragedy of human relationships. To resist and sabotage the contemporary drift toward a technocratic, militaristic totalitarianism, whatever its ideological coloration. To oppose injustice, defy the powerful, and speak for the voiceless.

"I write to make a difference. . . . To give pleasure and promote esthetic bliss. To honor life and praise the divine beauty of the world. For the joy and exultation of writing itself. To tell my story."

Down the River with Henry Thoreau

November 4, 1980

Our river is the Green River in southeast Utah. We load our boats at a place called Mineral Bottom, where prospectors once searched for gold, later for copper, still later for uranium. With little luck. With me are five friends plus the ghost of a sixth: in my ammo can—the river runner's handbag—I carry a worn and greasy paperback copy of a book called *Walden, or Life in the Woods*. Not for thirty years have I looked inside this book; now for the first time since my school days I shall. Thoreau's mind has been haunting mine for most of my life. It seems proper now to reread him. What better place than on this golden river called the Green? In the clear tranquillity of November? Through the red rock canyons known as Labyrinth, Stillwater, and Cataract in one of the sweetest, brightest, grandest, loneliest of primitive regions still remaining in our America?

Questions. Every statement raises more and newer questions. We shall never be done with questioning, so long as men and women remain human. QUESTION AUTHORITY reads a bumper sticker I saw the other day in Moab, Utah. Thoreau would doubtless have amended that to read "Always Question Authority." I would add only the word "All" before the word "Authority." Including, of course, the authority of Henry David himself.

Here we are, slipping away in the early morning of another Election Day. A couple of us did vote this morning but we are not, really, good citizens. Voting for the lesser evil on the grounds that otherwise we'd be stuck with the greater evil. Poor grounds for choice, certainly. Losing grounds.

We will not see other humans or learn of the election results for ten days to come. And so we prefer it. We like it that way. What could be older than the news? We shall treasure the bliss of our ignorance for as long as we can. "The man who goes each day to the village to hear the latest news has not heard from himself in a long time." Who said that? Henry, naturally. The arrogant, insolent village crank.

I think of another bumper sticker, one I've seen several times in several places this year: NOBODY FOR PRESIDENT. Amen.

50

The word is getting around. Henry would have approved. Heartily. For he also said, "That government is best which governs not at all."

Year by year the institutions that dominate our lives grow ever bigger, more complicated, massive, impersonal, and powerful. Whether governmental, corporate, military, or technological—and how can any one of these be disentangled from the others?—they weigh on society as the pyramids of Egypt weighed on the backs of those who were conscripted to build them. The pyramids of power. Five thousand years later the people of Egypt still have not recovered. They remain a passive and powerless mass of subjects. Mere fellahin, expendable and interchangeable units in a social megamachine. As if the pride and spirit had been crushed from them forever.

In many a clear conclusion we find ourselves anticipated by the hoer of beans on the shores of Walden Pond. "As for the Pyramids," wrote Henry, "there is nothing to wonder at in them so much as the fact that so many men could be found degraded enough to spend their lives constructing a tomb for some ambitious booby, whom it would have been wiser and manlier to have drowned in the Nile. . . ."

Some critic has endeavored to answer this observation by claiming that the pyramid projects provided winter employment for swarms of peasants who might otherwise have been forced to endure long seasons of idleness and hunger. But where did the funds come from, the surplus grain, to support and feed these hundreds of thousands of two-legged pismires: Why, from the taxes levied on the produce of their *useful* work in the rice fields of the Nile Delta. The slaves were twice exploited. Every year. Just as the moon rides, concrete monuments, and industrial war machines of contemporary empire-states, whether capitalist or Communist, are funded by compulsory taxation, erected and maintained by what is in effect compulsory labor.

The river flows. The river will not wait. Let's get these boats on the current. Loaded with food, bedrolls, cooking gear—four gourmet cooks in a party of six (plus ghost)—they ride on the water, tethered to shore. Two boats, one an eighteen-foot rubber raft, the other an aluminum dory. Oar-powered. We scramble on board, the swampers untie the lines, and oarsmen heave at their oars. Rennie Russell (author of *On the Loose*) operates the raft; a long-connected, lean fellow named Dusty Teale rows the dory.

We glide down the golden waters of Labyrinth Canyon. The water here is smooth as oil, the current slow. The sandstone walls rise fifteen hundred feet above us, radiant with sunlight, manganese and iron oxides, stained with old tapestries of organic residues left on the rock faces by occasional waterfalls. On shore, wheeling away from us, the stands of willow glow in autumn copper; beyond the willow are the green-gold cottonwoods. Two ravens fly along the rim, talking about us. Henry would like it here.

November 5, 1980

We did not go far yesterday. We rowed and drifted two miles down the river and then made camp for the night on a silt bank at the water's edge. There had been nobody but ourselves at Mineral Bottom but the purpose, nonetheless, was to "get away from the crowd," as Rennie Russell explained. We understood. We cooked our supper by firelight and flashlight, ate beneath the stars. Somebody uncorked a bottle of wine. Rennie played his guitar, his friend Ted Seeley played the fiddle, and Dusty Teale played the mandolin. We all sang. Our music ascended to the sky, echoing softly from the cliffs. The river poured quietly seaward, making no sound but here and there, now and then, a gurgle of bubbles, a thrilling of ripples off the hulls of our half-beached boats.

Sometime during the night a deer stalks nervously past our camp. I hear the noise and, when I get up before daybreak, I see the dainty heart-shaped tracks. I kindle the fire and build the morning's first pot of black, rich, cowboy coffee, and drink in solitude the first cupful, warming my hands around the hot cup. The last stars fade, the sky becomes brighter, passing through the green glow of dawn into the fiery splendor of sunrise.

The others straggle up, one by one, and join me around the fire. We stare at the shining sky, the shining river, the high canyon walls, mostly in silence, until one among us volunteers to begin breakfast. Yes, indeed, we are a lucky little group. Privileged, no doubt. At ease out here on the edge of nowhere, loafing into the day, enjoying the very best of the luckiest of nations, while around the world billions of other humans are sweating, fighting, striving, procreating, starving. As always, I try hard to feel guilty. Once again I fail.

"If I knew for a certainty that some man was coming to my house with the conscious intention of doing me good," writes our Henry, "I would run for my life."

We Americans cannot save the world. Even Christ failed at that. We Americans have our hands full in trying to save ourselves. And we've barely tried. The Peace Corps was a lovely idea— for idle and idealistic young Americans. Gave them a chance to see a bit of the world, learn something. But as an effort to "improve" the lives of other peoples, the inhabitants of the so-called underdeveloped nations (our nation is overdeveloped), it was an act of cultural arrogance. A piece of insolence. The one thing we could do for a country like Mexico, for example, is to stop every illegal immigrant at the border, give him a good rifle and a case of ammunition, and send him home. Let the Mexicans solve their customary problems in their customary manner.

If this seems a cruel and sneering suggestion, consider the current working alternative: leaving our borders open to unlimited immigration until—and it won't take long—the social, political, economic life of the United States is reduced to the level of life in Juarez. Guadalajara. Mexico City. San Salvador. Haiti. India. To a common peneplain of overcrowding, squalor, misery, oppression, torture, and hate.

What could Henry have said to this supposition? He lived in a relatively spacious America of only 24 million people, of whom one-sixth were slaves. A mere 140 years later we have grown to a population ten times larger, and we are nearly all slaves. We are slaves in the sense that we depend for our daily survival upon an expand-or-expire agro-industrial empire—a crackpot machine—that the specialists cannot comprehend and the managers cannot manage. Which is, furthermore, devouring world resources at an exponential rate. We are, most of us, dependent employees.

What would Henry have said? He said, "In wildness is the preservation of the world." He said, somewhere deep in his thirty-nine-volume *Journal,* "I go to to my solitary woodland walks as the homesick return to their homes." He said, "It would be better if there were but one inhabitant to a square mile, as where I live." Perhaps he did sense what was coming. His last words, whispered from the deathbed, are reported to us as being "moose . . . Indians . . ."

Looking upriver toward Tidwell Bottom, a half mile away, I see a lone horse grazing on the bunch grass, the Indian rice grass, the saltbush, and sand sage of the river's old floodplain. One horse, unhobbled and untended, thirty miles from the nearest ranch or human habitation, it forages on its own. That horse, I'm thinking, may be the one that got away from me years ago, in another desert place, far from here. Leave it alone. That particular horse has found at least a temporary solution to the question of survival. Survival with honor, I mean, for what other form of survival is worth the trouble? That horse has chosen, or stumbled into, solitude and independence. Let it be. Thoreau defined happiness as "simplicity, independence, magnanimity and trust."

But solitude? Horses are gregarious beasts, like us. This lone horse on Tidwell Bottom may be paying a high price for its freedom, perhaps in some form of equine madness. A desolation of the soul corresponding to the grand desolation of the landscape that lies beyond these canyon walls.

"I never found the companion that was so companionable as solitude," writes Henry. "To be in company, even with the best, is soon wearisome and dissipating."

Perhaps his ghost will forgive us if we suspect an element of *extra-vagance* in the above statement. Thoreau had a merry time in the writing of *Walden;* it is an exuberant book, crackling with humor, good humor, gaiety, with joy in the power of words and phrases, in ideas and emotions so powerful they tend constantly toward the outermost limit of communicable thought.

"The sun is but a morning star." Ah yes, but what exactly does that mean? Maybe the sun is also an evening star. Maybe the phrase had no exact meaning even in Thoreau's mind. He was, at times, what we today might call a put-on artist. He loved to shock and exasperate; Emerson complains of Henry's "contrariness." The power of Thoreau's assertion lies not in its meaning but in its exhilarating suggestiveness. Like poetry and music, the words imply more than words can make explicit.

Henry was no hermit. Hardly even a recluse. His celebrated cabin at Walden Pond—some of his neighbors called it a "shanty"—was two miles from Concord Common. A half-hour walk from pond to post office. Henry lived in it for only two years and two months. He had frequent human visitors, sometimes too many, he complained, and admitted that his daily rambles took him almost every day into Concord. When he tired of his own cooking and his own companionship he was always welcome at the Emersons' for a free dinner. Although it seems that he earned his keep there. He worked on and off for years as Emersons' household handyman, repairing and maintaining things that the great Ralph Waldo was too busy or too incompetent to attend to himself. "Emerson," noted Thoreau in a letter, "is too much the gentleman to push a wheelbarrow." When Mrs. Emerson complained that the chickens were scratching up her flower beds, Henry attached little cloth booties to the chickens' feet. A witty fellow. Better and easier than keeping them fenced in. When Emerson was off on his European lecture tours, Thoreau would look after not only Emerson's house but also Emerson's children and wife.

We shall now discuss the sexual life of Henry David Thoreau.

November 6, 1980

Awaking as usual sometime before the dawn, frost on my beard and sleeping bag, I see four powerful lights standing in a vertical row on the eastern sky. They are Saturn, Jupiter, Mars, and, pale crescent on a darkened disc, the old moon. The three great planets seem to be rising from the cusps of the moon. I stare for a long time at this strange, startling apparition, a spectacle I have never before seen in all my years on planet Earth. What does it mean? If ever I've seen a portent in the sky this must be it. Spirit both forms and informs the universe, thought the New England transcendentalists, of whom Thoreau was one; all Nature, they believed, is but symbolic of a greater spiritual reality beyond. And within.

Watching the planets, I stumble about last night's campfire, breaking twigs, filling the coffeepot. I dip waterbuckets in the river; the water chills my hands. I stare long at the beautiful, dimming lights in the sky but can find there no meaning other than the lights' intrinsic beauty. As far as I can perceive, the planets signify nothing but themselves. "Such suchness," as my Zen friends say. And that is all. And that is enough. And that is more than we can make head or tail of.

"Reality is fabulous," said Henry; "be it life or death, we crave nothing but reality." And goes on to describe in precise, accurate, glittering detail the most subtle and minute aspects of life in and about his Walden Pond; the "pulse" of water skaters, for instance, advancing from shore across the surface of the lake. Appearance *is* reality, Thoreau implies or so it appears to me. I begin to think he outgrew transcendentalism rather early in his career, at about the same time that he was overcoming the influence of his onetime mentor Emerson; Thoreau and the transcendentalists had little in common—in the long run—but their long noses, as a friend of mine has pointed out.

Scrambled eggs, bacon, green chiles for breakfast, with hot *salsa,* toasted tortillas, and leftover baked potatoes sliced and fried. A gallon or two of coffee, tea and—for me—the usual breakfast beer. Henry would not have approved of this gourmandising. To hell with him. I do not approve of his fastidious puritanism. For one who claims to crave nothing but reality, he frets too much about *purity.* Purity, purity, he preaches, in the most unctuous of his many sermons, a chapter of *Walden* called "Higher Laws."

"The wonder is how they, how you and I," he writes, "can live this slimy, beastly life, eating and drinking. . . ." Like Dick Gregory, Thoreau recommends a diet of raw fruits and vegetables; like a Pythagorean, he finds even beans impure, since the flatulence that beans induce disturbs his more ethereal meditations. (He would not agree with most men that "farting is such sweet sorrow.") But confesses at one point to a sudden violent lust for wild woodchuck, devoured raw. No wonder; Henry was probably anemic.

He raised beans not to eat but to sell—his only cash crop. During his lifetime his beans sold better than his books. When a publisher shipped back to Thoreau 706 unsellable copies of his *A Week on the Concord and Merrimack Rivers* (the author had himself paid for the printing of the book), Henry noted in his *Journal,* "I now have a library of 900 volumes, over 700 of which I wrote myself."

Although professing disdain for do-gooders, Thoreau once lectured a poor Irish immigrant, a neighbor, on the advisability of changing his ways. "I tried to help him with my experience . . ." but the Irishman, John Field, was only bewildered by Thoreau's earnest preaching. "Poor John Field!" Thoreau concludes; "I trust he does not read this, unless he will improve by it. . . ."

Nathaniel Hawthorne, who lived in Concord for a time and knew Thoreau, called him "an intolerable bore."

On the subject of sex, as we would expect, Henry betrays a considerable nervous agitation. "The generative energy, which, when we are loose, dissipates and makes us unclean, when we are continent invigorates and inspires us. Chastity is the flowering of man. . . ." (But not of flowers?) "We are conscious of an animal in us, which awakens in proportion as our higher nature slumbers. It is reptile and sensual. . . ." "He is blessed who is assured that the animal is dying out in him day by day. . . ." In a letter to his friend Harrison Blake, Henry writes: "What the essential difference between man and woman is, that they should be thus attracted to one another, no one has satisfactorily answered."

Poor Henry. We are reminded of that line in Whitman (another great American oddball), in which our good gray poet said of women, "They attract with a fierce, undeniable attraction," while the context of the poem makes it clear that Whitman himself found young men and boys much more undeniable.

Poor Thoreau. But he could also write, in the late essay "Walking," "The wildness of the savage is but a faint symbol of the awful ferity with which good men and lovers meet." Ferity—now there's a word. What could it have meant to Thoreau? Our greatest nature lover did not have a loving nature. A woman acquaintance of Henry's said she'd sooner take the arm of an elm tree than that of Thoreau.

Poor Henry David Thoreau. His short (forty-five years), quiet, passionate life apparently held little passion for the opposite sex. His relationship with Emerson's wife, Lidian, was no more than a long brotherly-sisterly friendship. Thoreau never married. There is no evidence that he ever enjoyed a mutual love affair with any human, female or otherwise. He once fell in love with and proposed marriage to a young woman by the name of Ellen Sewall; she rejected him, bluntly and coldly. He tried once more with a girl named Mary Russell; she turned him down. For a young man of Thoreau's hypersensitive character, these must have been cruel, perhaps disabling blows to what little male ego and confidence he possessed to begin with. It left him shattered, we may assume, on that side of life; he never again approached a woman with romantic intentions on his mind. He became a professional bachelor, scornful of wives and marriage. He lived and probably died a virgin, pure as shriven snow. Except for those sensual reptiles coiling and uncoiling down in the root cellar of his being. Ah, purity!

But we make too much of this kind of thing nowadays. Modern men and women are obsessed with the sexual; it is the only realm of primordial adventure still left to most of us. Like apes in a zoo, we spend our energies on the one field of play remaining; human lives otherwise are pretty well caged in by the walls, bars, chains, and locked gates of our industrial culture. In the relatively wild, free America of Henry's time there was plenty of opportunity for every kind of adventure, although Henry himself did not, it seems to me, take advantage of those opportunities. (He could have toured the Western plains with George Catlin!) He led an unnecessarily constrained existence, and not only in the "generative" region.

Thoreau the spinster-poet. In the year 1850, when Henry reached the age of thirty-three, Emily Dickinson in nearby Amherst became twenty. Somebody should have brought the two together. They might have hit it off. I imagine this scene, however, immediately following the honeymoon:

EMILY (raising her pen)
Henry, you haven't taken out the garbage.
HENRY (raising his flute)
Take it out yourself.

What tunes did Thoreau play on that flute of his? He never tells us; we would like to know. And what difference would a marriage—with a woman—have made in Henry's life? In his work? In that message to the world by which he challenges us, as do all the greatest writers, to change our lives? He taunts, he sermonizes, he condemns, he propounds conundrums, he orates and exhorts us:

"Wherever a man goes, men will pursue and paw him with their dirty institutions. . . ."

"I found that by working six weeks a year I could meet all the expenses of living."

"Tell those who worry about their health that they may be already dead."

"When thousands are thrown out of employment, it suggests they were not well-employed."

"If you stand right fronting and face to face with a fact, you will see the sun glimmer on both its surfaces, as if it were a scimitar, and feel its sweet edge dividing you through the heart and marrow, and so you will happily conclude your mortal career."

". . . Little is to be expected of a nation when the vegetable mould is exhausted, and it is compelled to make manure of the bones of its fathers."

"Genius is a light which makes the darkness visible, like the lightning's flash, which perchance shatters the temple of knowledge itself. . . ."

"When, in the course of ages, American liberty has become a fiction of the past—as it is to some extent a fiction of the present—the poets of the world will be inspired by American mythology."

"We should go forth on the shortest walk . . . in the spirit of undying adventure, never to return."

". . . If I repent of anything, it is very likely to be my good behavior. What demon possessed me that I behaved so well?"

"No man is so poor that he need sit on a pumpkin; that is shiftlessness."

"I would rather sit on a pumpkin and have it all to myself than be crowded on a velvet cushion."

"A man is rich in proportion to the number of things which he can afford to let alone."

"We live meanly, like ants, though the fable tells us that we were long ago changed into men. . . ."

"A living dog is better than a dead lion. Shall a man go and hang himself because he belongs to the race of pygmies, and not be the biggest pygmy that he can?"

"I will endeavor to speak a good word for the truth."

"Rather than love, than money, than fame, give me truth."

"Any truth is better than make-believe."
And so forth.

November 7, 1980

On down this here Greenish river. We cast off, row south past Woodruff, Point, and Saddlehorse bottoms, past Upheaval Bottom and Hardscrabble Bottom. Wherever the river makes a bend—and this river comes near, in places, to bowknots—there is another flat area, a bottom, covered with silt, sand, gravel, grown up with grass and brush and cactus and, near shore, trees: willow, cottonwood, box elder, and jungles of tamarisk.

The tamarisk does not belong here, has become a pest, a water-loving exotic engaged in the process of driving out the cottonwoods and willows. A native of arid North Africa, the tamarisk was imported to the American Southwest fifty years ago by conservation *experts*—dirt management specialists—in hopes that it would help prevent streambank erosion. The cause of the erosion was flooding, and the primary cause of the flooding, then as now, was livestock grazing.

Oars at rest, we drift for a while. The Riverine String Band take up their instruments and play. The antique, rowdy, vibrant music from England and Ireland by way of Appalachia and the Rocky

Mountains floats on the air, rises like smoke toward the high rimrock of the canyon walls, fades by infinitesimal gradations into the stillness of eternity. Where else could it go?

Ted Seeley prolongs the pause, then fills the silence with a solo on the fiddle, a Canadian invention called "Screechin' Old Woman and Growlin' Old Man." This dialogue continues for some time, concluding with a triumphant outburst from the Old Woman.

We miss the landing off the inside channel at Wild Horse Bench and have to fight our way through thickets of tamarisk and cane to the open ground of Fort Bottom. We make lunch on crackers, canned tuna, and chopped black olives in the shade of a cottonwood by the side of a long-abandoned log cabin. A trapper, prospector, or cow thief might have lived here—or all three of them—a century ago. Names and initials adorn the lintel of the doorway. The roof is open to the sky.

We climb a hill of clay and shale and limestone ledges to inspect at close hand an ancient ruin of stone on the summit. An Anasazi structure, probably seven or eight hundred years old, it commands a broad view of river and canyon for many miles both up and downstream, and offers a glimpse of the higher lands beyond. We can see the great Buttes of the Cross, Candlestick Tower, Junction Butte (where the Green River meets the Colorado River), Ekker Butte, Grandview Point, North Point, and parts of the White Rim. Nobody human lives at those places, or in the leagues of monolithic stone between them. We find pleasure in that knowledge. From this vantage point everything looks about the same as it did when Major John Wesley Powell and his mates first saw it in 1869. Photographs made by members of his party demonstrate that nothing much has changed except the vegetation types along the river, as in the case of tamarisk replacing willow.

We return to our river. A magisterial magpie sails before us across the barren fields. Two ravens and a hawk watch our lazy procession downstream past the long straightaway of Potato Bottom. We make camp before sundown on an island of white sand in the middle of the river. A driftwood fire under an iron pot cooks our vegetable stew. Russell mixes a batch of heavy-duty cornbread in the Dutch oven, sets the oven on the hot coals, and piles more coals on the rimmed lid. The cornbread bakes. We drink our beer, sip our rum, and listen to a pack of coyotes yammering like idiots away off in the twilight.

"I wonder who won the election," says one member of our party—our boatwoman Lorna Corson.

"The coyotes can explain everything," says Rennie Russell.

It's going to be a cold and frosty night. We add wood to the fire and put on sweaters and coats. The nights are long in November; darkness by six. The challenge is to keep the fire going and conversation and music alive until a decent bedtime arrives. Ten hours is too long to spend curled in a sleeping bag. The body knows this if the brain does not. That must be why I wake up every morning long before the sun appears. And why I remain sitting here, alone on my log, after the others have crept away, one by one, to their scattered beds.

Henry gazes at me through the flames of the campfire. From beyond the veil. Edward, he says, what are you doing here? Henry, I reply, what are you doing out there?

How easy for Thoreau to preach simplicity, asceticism and voluntary poverty when, as some think, he had none but himself to care for during his forty-five years. How easy to work part-time for a living when you have neither wife nor children to support. (When you have no payments to meet on house, car, pickup truck, cabin cruiser, life insurance, medical insurance, summer place, college educations, dinette set, color TVs, athletic club, real estate investments, holidays in Europe and the Caribbean. . . .)

Why Henry never took a wife has probably more to do with his own eccentric personality than with his doctrine of independence-through-simplicity. But if he had *wanted* a partner, and had been able to find one willing to share his doctrine, then it seems reasonable to suppose that the two of them—with their little Thoreaus—could have managed to live a family life on Thoreauvian principles. Henry might have been compelled to make pencils, survey woodlots, and give public lectures for twenty-four weeks, rather than only six, each year—but his integrity as a free man would still have been preserved. There is no reason—other than the comic incongruity of imagining Henry Thoreau as husband and father—to suppose that his bachelorhood invalidates his arguments. If there was tragedy in the life of Thoreau, that tragedy lies not in any theoretical contradiction between what Henry advocated and how he lived but in his basic loneliness. He was a psychic loner all his life.

But a family man nevertheless. Except for his two years and two months at Walden Pond, his student years at Harvard, and occasional excursions to Canada, Cape Cod, and Maine, Thoreau lived most of his life in and upon the bosom of family—Emerson's family, part of the time, and the Thoreau family—mother, sister, uncles, and aunts—during the remainder.

When his father died Henry took over the management of the family's pencil-making business, a cottage industry carried on in the family home. Always a clever fellow with his hands, Henry developed a better way of manufacturing pencils, and a better product. Some think that the onset of his tuberculosis, which eventually killed him, was hastened by the atmosphere of fine powdered graphite in which he earned a part of his keep.

A part of it: Thoreau had no wish to become a businessman—"Trade curses everything it handles"—and never gave to pencils more than a small part of his time.

He was considered an excellent surveyor by his townsmen and his services were much in demand. His work still serves as the basis of many property lines in and around the city of Concord. There is a document in the Morgan Library in New York, a map of Walden Pond, signed "H. D. Thoreau, Civil Engineer."

But as with pencil-making, so with surveying—Thoreau would not allow it to become a full-time career. Whatever he did, he did well; he was an expert craftsman in everything to which he put his hand. But to no wage-earning occupation would he give his life. He had, he said, "other business." And this other business awaited him out in the woods, where, as he wrote, "I was better known."

What was this other business? It is the subject of *Walden,* of his further books and essays, and of the thirty-nine volumes of his *Journal,* from which, to a considerable extent, the books were quarried. Thoreau's subject is the greatest available to any writer, thinker and human being, one which I cannot summarize in any but the most banal of phrases: "meaning," or "the meaning of life" (meaning *all* life, of course, not human life only), or in the technical usage preferred by professional philosophers, "the significance of existence."

It is this attempt to encircle with words the essence of being itself—with or without a capital *B*—which gives to Henry's prose-poetry the disturbing, haunting, heart-opening quality that some call mysticism. Like the most ambitious poets and artists, he was trying to get it all into his work, whatever "it" may signify, whatever "all" may include. Living a life full of wonder—wonderful—Henry tries to impart that wonder to his readers.

"There is nothing inorganic. . . . The earth is not a mere fragment of dead history, stratum upon stratum, like the leaves of a book, to be studied by geologists and antiquaries chiefly, but living poetry like the leaves of a tree, which precede flowers and fruit; not a fossil earth but a living earth. . . ."

That the earth, considered whole, is a kind of living being, might well seem like nonsense to the hardheaded among us. Worse than nonsense—mystical nonsense. But let us remember that a hard head, like any dense-hulled and thick-shelled nut, can enclose, out of necessity, only a tiny kernel of meat. Thinking meat, in this case. The hard head reveals, therefore, while attempting to conceal and shelter, its tiny, soft, delicate, and suspicious mind.

The statement about earth is clear enough. And probably true. To some, self-evident, though not empirically verifiable within the present limitations of scientific method. Such verification requires a more sophisticated science than we possess at present. It requires a science with room for more than data and information, a science that includes sympathy for the object under study, and more than sympathy, love. A love based on prolonged contact and interaction. Intercourse, if possible. Observation informed by sympathy, love, intuition. Numbers, charts, diagrams, and formulas are not in themselves sufficient. The face of science as currently construed is a face that only a mathematician could love. The root meaning of "science" is "knowledge"; to see and to see truly, a qualitative, not merely quantitative, understanding.

For an example of science in the whole and wholesome sense read Thoreau's description of an owl's behavior in "Winter Visitors." Thoreau observes the living animal in its native habitat, and watches it for weeks. For an example of science in its debased sense take this: According to the L. A. *Times,* a psychologist in Los Angeles defends laboratory experimentation on captive dogs with the assertion that "little is known about the psychology of dogs." Anyone who has ever kept a dog knows more about dogs than that psychologist—who doubtless considers himself a legitimate scientist—will learn in a year of Sundays.

Or this: Researchers in San Francisco have confined chimpanzees in airtight glass cubicles (gas chambers) in order to study the effect of various dosages of chemically polluted air on these "manlike organisms." As if there were not already available five million human inhabitants of the Los Angeles basin, and a hundred other places, ready, willing, and eager to supply personally informed testimony on the subject under scrutiny. Leaving aside any consideration of ethics, morality, and justice, there are more intelligent ways to study living creatures. Or nonliving creations: rocks have rights too.

That which today calls itself science gives us more and more information, an indigestible glut of information, and less and less understanding. Thoreau was well aware of this tendency and foresaw its fatal consequences. He could see the tendency in himself, even as he partially succumbed to it. Many of the later *Journals* are filled with little but the enumeration of statistical data concerning such local Concord phenomena as the rise and fall of lake levels, or the thickness of the ice on Flint's Pond on a January morning. Tedious reading—pages and pages of "factoids," as Norman Mailer would call them—attached to no coherent theory, illuminated by neither insight nor outlook nor speculation.

Henry may have had a long-range purpose in mind but he did not live long enough to fulfill it. Kneeling in the snow on a winter's day to count the tree rings in a stump, he caught the cold that led to his death on May 6, 1862. He succumbed not partially but finally to facticity.

Why'd you do it, Henry? I ask him through the flames.

The bearded face with the large, soft, dark eyes, mournful and thoughtful as the face of Lincoln, smiles back at me but offers no answer. He evades the question by suggesting other questions in his better-known, "mystical" vein:

"There was a dead horse in the hollow by the path to my house, which compelled me sometimes to go out of my way, especially in the night when the air was heavy, but the assurance it gave me of the strong appetite and inviolable health of Nature was my compensation for this. I love to see that Nature is so rife with life that myriads can be afforded to be sacrificed and suffered to prey on one another. . . . The impression made on a wise man is that of universal innocence. Compassion is a very untenable ground. It must be expeditious. Its pleadings will not bear to be stereotyped."

Henry, I say, what the devil do you mean?

He smiles again and says, "I observed a very small and graceful hawk, like a nighthawk, alternately soaring like a ripple and tumbling a rod or two over and over, showing the underside of its wings, which gleamed like a satin ribbon in the sun. . . . The merlin it seemed to me it might be called; but I care not for its name. It was the most ethereal flight I have ever witnessed. It did not simply flutter like a butterfly, nor soar like the larger hawks, but it sported with proud reliance in the fields of the air. . . . It appeared to have no companion in the universe . . . and to need none but the morning and the ether with which it played. It was not lonely, but made all the earth lonely beneath it."

Very pretty, Henry. Are you speaking for yourself? I watch his lined, gentle face, the face of his middle age (though he had no later) as recorded in photographs, and cannot help but read there the expression, engraved, of a patient, melancholy resignation. All babies look identical; boys and adolescents resemble one another, in their bewildered hopefulness, more than they differ. But eventually the inner nature of the man appears on his outer surface. Character begins to shine through. Year by year a man reveals himself, while those with nothing to show, show it. Differentiation becomes individuation. By the age of forty, if not before, a man is responsible for his face. The same is true of women too, certainly, although women, obeying the biological imperative, strive harder than men to preserve an appearance of youthfulness—the reproductive look—and lose it sooner. Appearance *is* reality.

Henry replies not to my question but, as befits a ghostly seer, to my thought: "Nothing can rightly compel a simple and brave man to a vulgar sadness."

We'll go along with that, Henry; you've been accused of many things but no one, to my knowledge, has yet accused you of vulgarity. Though Emerson, reacting to your night in jail for refusing to pay the poll tax, called the gesture "mean and skulking and in bad taste." In bad taste! How typically Emersonian. Robert Louis Stevenson too called you a "skulker" on the grounds that you preached more strongly than you practiced, later recanting when he learned of your activity in the antislavery movement. The contemporary author Alan Harrington, in his book *The Immortalist,* accuses you of writing, at times, like "an accountant of the spirit." That charge he bases on your vague remarks concerning immorality, and on such lines as "Goodness is the only investment that never fails."

Still other current critics, taking their cue from those whom Nabokov specified as "the Viennese quacks," would deflect the force of your attacks on custom, organized religion, and the state by suggesting that you suffered from a complex of complexes, naturally including the castration complex and the Oedipus complex. Your defiance of authority, they maintain, was in reality no more than the rebelliousness of an adolescent rejecting his father—in this case the meek and mousy John Thoreau.

Whatever grain of truth may be in this diagnosis, such criticism betrays the paternalistic condescension of these critics toward human beings in general. The good citizen, they seem to be saying, is like the obedient child; the rebellious man is a bad boy. "The people are like children," said our own beloved, gone but not forgotten, Richard Nixon. The psychiatric approach to dissidence has been most logically applied in the Soviet Union, where opposition to the state is regarded and treated as a form of mental illness.

In any case, Henry cannot be compelled to confess to a vulgar sadness. The vulgarity resides in the tactics of literary Freudianism. Of the opposition. Psychoanalysis is the neurosis of the psychoanalyst—and of the psychoanalytic critic. Why should we bother any more with this garbage? I thought we stopped talking about Freud back in 1952. Sometime near the end of the Studebaker era.

Fading beyond the last flames of the fire, Henry lulls me to sleep with one of his more soporific homilies:

"The light which puts out our eyes is darkness to us. Only that day dawns to which we are awake. There is more day to dawn. The sun . . . ."

Yes, yes, Henry, we know. How true. Whatever it means. How late it is. Whatever the hour.

I rise from my log, heap the coals of the fire together, and by their glimmering light and the cold light of the stars fumble my way back and into the luxury of my goosedown nest. Staring up at mighty Orion, trying to count six of the seven Pleiades, a solemn thought comes to me: We Are Not Alone.

I nuzzle my companion's cold nose, the only part of her not burrowed deep in her sleeping bag. She stirs but does not wake. We're not alone, I whisper in her ear. I know, she says; shut up and go to sleep. Smiling, I face the black sky and the sapphire stars. Mark Twain was right. Better the savage wasteland with Eve than Paradise without her. Where she is, there is Paradise.

Poor Henry.

And then I hear that voice again, far off but clear: "All Nature is my bride."

November 8, 1980

Who won the election? What election? Mere vapors on the gelid air, like the breath from my lungs. I rebuild the fire on the embers of last night's fire. I construct the coffee, adding fresh grounds to yesterday's. One by one, five human forms reassemble themselves about me, repeating themselves, with minor variations, for another golden day. The two vegetarians in our group—Rennie and Lorna—prepare their breakfast oatmeal, a viscous gray slime. I dump two pounds of Buck-sliced bacon into the expedition's wok, to the horror of the vegetarians, and stir it roughly about with a fork. Stir-cooking. The four carnivores look on with hungry eyes. The vegetarians smile in pity. "Pig meat," says Lorna, "for the four fat pork faces." "Eat your pussy food," says Dusty Teale, "and be quiet."

The melody of morning. Black-throated desert sparrows chatter in the willows; *chirr . . . chirr . . . chit chit chit.* The sun comes up, a glaring cymbal, over yonder canyon rim. Quickly the temperature rises five, ten, twenty degrees, at the rate of a degree a minute, from freezing to fifty-two. Or so it feels. We peel off parkas, sweaters, shirts, thermal underwear. Ravens croak, a rock falls, the river flows.

The fluvial life. The alluvial shore. "A river is superior to a lake," writes Henry in his *Journal,* "in its liberating influence. It has motion and indefinite length. . . . With its rapid current it is a slightly fluttering wing. River towns are winged towns."

Down the river. Lorna rows the dory, I row the raft. We are edified by water music from our string trio, a rich enchanting tune out of Peru called "Urubamba." The song goes on and on and never long enough. The Indians must have composed it for a journey down the Amazon.

Fresh slides appear on the mud banks; a beaver plops into the water ahead of us, disappears. The beavers are making a comeback on the Green. Time for D. Julien, Jim Bridger, Joe Meek, Jed Smith, and Jim Beckwourth to reappear. Eternal recurrence, announced Nietzsche. Time for the mountain men to return. The American West has not given us, so far, sufficient men to match our mountains. Or not since the death of Crazy Horse, Sitting Bull, Dull Knife, Red Cloud, Chief Joseph, Little Wolf, Red Shirt, Gall, Geronimo, Cochise, Tenaya (to name but a few), and their comrades. With their defeat died a bold, brave, heroic way of life, one as fine as anything recorded history has to show us. Speaking for myself, I'd sooner have been a liver-eating, savage horseman, riding with Red Cloud, than a slave-owning sophist sipping tempered wine in Periclean Athens. For example. Even Attila the Hun, known locally as the Scourge of God, brought more fresh air and freedom into Europe than the crowd who gave us the syllogism and geometry, Aristotle and his *Categories,* Plato and his *Laws.*

Instead of mountain men we are cursed with a plague of diggers, drillers, borers, grubbers; of asphalt-spreaders, dam-builders, overgrazers, clear-cutters, and strip-miners whose object seems to be to make our mountains match our men—making molehills out of mountains for a race of rodents—for the rat race.

Oh well . . . revenge is on the way. We see it in those high thin clouds far on the northern sky. We feel it in those rumbles of discontent deep in the cupboards of the earth: tectonic crockery trembling on the continental shelves. We hear it down the slipface of the dunes, a blue wind moaning out of nowhere. We smell it on the air: the smell of danger. Death before dishonor? That's right. What else? Liberty or death? Naturally.

When no one else would do it, it was Thoreau, Henry Thoreau the intolerable bore, the mean skulker, the "quaint stump figure of a man," as William Dean Howells saw him, who rang the Concord firebell to summon the villagers to a speech by Emerson attacking slavery. And when John Brown stood on trial for his life, when all America, even the most ardent abolitionists, was denouncing him, it was himself—Henry—who delivered a public address first in Concord, then in Boston, not only defending but praising, even eulogizing, the "madman" of Harpers Ferry.

We go on. Sheer rock—the White Rim—rises from the river's left shore. We pause at noon to fill our water jugs from a series of potholes half filled with last week's rainwater. We drink, and sitting in the sunlight on the pale sandstone, make our lunch—slabs of dark bread, quite authentic, from a bohemian bakery in Moab; a serious hard-core hippie peanut butter, heavy as wet concrete, from some beatnik food coop in Durango, Colorado (where Teale and Corson live); raspberry jam; and wild honey, thick as axle grease, for esophageal lubrication.

"What is your favorite dish?" another guest asked Thoreau as they sat down to a sumptuous Emersonian dinner.

"The nearest," Henry replied.

"At Harvard they teach all branches of learning," said Ralph Waldo.

66

"But none of the roots," said Henry.

Refusing to pay a dollar for his Harvard diploma, he said, "Let every sheep keep its own skin." When objections were raised to his habit of exaggeration, Henry said, "You must speak loud to those who are hard of hearing." Asked to write for the *Ladies' Companion,* he declined on the grounds that he "could not write anything companionable." He defines a pearl as "the hardened tear of a diseased clam, murdered in its old age." On the art of writing he said to a correspondent, "You must work very long to write short sentences." And added that "the one great rule of composition . . . is to speak the truth." Describing the flavor of a certain wild apple, he wrote that it was "sour enough to set a squirrel's teeth on edge, or make a jay scream."

And so on. The man seemingly composed wisecracks and epigrams in his sleep. Even on his deathbed. "Henry, have you made your peace with God?" asked a relative. "I am not aware that we had ever quarreled, Aunt," said Henry. To another visitor, attempting to arouse in him a decent Christian concern with the next world, Henry said, "One world at a time."

One could make a book of Henry's sayings. And call it *Essais. Areopagitica. Walden.*

Many of his friends, neighbors, relatives, and relative friends must have sighed in relief when Henry finally croaked his last, mumbling "moose . . . Indians . . ." and was safely buried under Concord sod. Peace, they thought, at long last. But, to paraphrase the corpse, they had *somewhat hastily* concluded that he was dead.

His passing did not go unnoticed outside of Concord. Thoreau had achieved regional notoriety by 1862. But at the time when the giants of New England literature were thought to be Emerson, Hawthorne, Alcott, Channing, Irving, Longfellow, Dr. Lowell, and Dr. Holmes, Thoreau was but a minor writer. Not even a major minor writer.

Today we see it differently. In the ultimate democracy of time, Henry has outlived his contemporaries. Hawthorne and Emerson are still read, at least in university English departments, and it may be that in a few elementary schools up in Maine and Minnesota children are being compelled to read Longfellow's *Hiawatha* (I doubt it; doubt that they can, even under compulsion), but as for the others they are forgotten by everyone but specialists in American literature. Thoreau, however, becomes more significant with each passing decade. The deeper our United States sinks into industrialism, urbanism, militarism—with the rest of the world doing its best to emulate America—the more poignant, strong, and appealing becomes Thoreau's demand for the right of every man, every woman, every child, every dog, every tree, every snail darter, every lousewort, every living thing, to live its own life in its own way at its own pace in its own square mile of home. Or in its own stretch of river.

Looking at my water-soaked, beer-stained, grease-spotted cheap paperback copy of *Walden,* I see that mine was from the thirty-third printing. And this is only one of at least a dozen current American editions of the book. Walden has been published abroad in every country where English can be read, as in India—God knows they need it there—or can be translated, as in Russia, where they need it even more. The Kremlin's commissars of literature have classified Thoreau as a nineteenth-century social reformer, proving once again that censors can read but seldom understand.

The village crank becomes a world figure. As his own Johnny Appleseed, he sows the seeds of liberty around the planet, even on what looks like the most unpromising soil. Out of Concord, apples of discord. Truth threatens power, now and always.

We walk up a small side canyon toward an area called Soda Springs Basin; the canyon branches and branches again, forming more canyons. The floor of each is flood-leveled sand, the walls perpendicular sandstone. Each canyon resembles a winding corridor in a labyrinth. We listen for the breathing of the Minotaur but find only cottonwoods glowing green and gold against the red rock, rabbitbrush with its mustard-yellow bloom, mule-ear sunflowers facing the sunlight, their coarse petals the color of butter, and the skull and curled horns of a desert bighorn ram, half buried in the auburn sand.

The canyons go on and on, twisting for miles into the plateau beyond. We turn back without reaching Soda Springs. On our return Dusty Teale takes up the bighorn trophy, carries it back to the dory and mounts it on the bow, giving his boat dignity, class, and unearned but warlike glamour.

We camp today at Anderson Bottom, across the river from Unknown Bottom. We find pictographs and petroglyphs here, pictures of deer, bighorns, warriors, and spectral figures representing— who knows—gods, spirits, demons. They do not trouble us. We cook our dinner and sing our songs and go to sleep.

November 9, 1980

Early in the morning I hear coyotes singing again, calling up the sun. There's something about the coyotes that reminds me of Henry. What is it? After a moment the answer comes.

Down near Tucson, Arizona, where I sometimes live—a grim and grimy little-big town, swarming with nervous policemen, dope dealers, resolute rapists, and geriatric bank robbers, but let this pass for the moment—the suburban parts of the city are infested with pet dogs. Every home owner in these precincts believes that he needs whatever burglar protection he can get; and he is correct. Most evenings at twilight the wild coyotes come stealing in from the desert to penetrate the suburbs, raid garbage cans, catch and eat a few cats, dogs, and other domesticated beasts. When this occurs the dogs raise a grim clamor, roaring like maniacs, and launch themselves in hot but tentative pursuit of the coyotes. The coyotes retreat into the brush and cactus, where they stop, facing the town, to wait and sit and laugh at the dogs. They yip, yap, yelp, howl, and holler, teasing the dogs, taunting them, enticing them with the old-time call of the wild. And the dogs stand and tremble, shaking with indecision, furious, hating themselves, tempted to join the coyotes, run off with them into the hills, but—afraid. Afraid to give up the comfort, security, and safety of their housebound existence. Afraid of the unknown and dangerous.

Thoreau was our suburban coyote. Town dwellers have always found him exasperating.

"I have traveled a good deal in Concord; and everywhere, in shops and offices and fields, the inhabitants have appeared to me to be doing penance in a thousand remarkable ways. . . . By a seeming fate, commonly called necessity, they are employed, as it says in an old book, laying up treasures which moth and rust will corrupt and thieves break through and steal. It is a fool's life, as they will find when they get to the end of it, if not before. . . . I sometimes wonder that we can be so frivolous. . . . As if you could kill time without injuring eternity."

Oh, come now, Henry, stop yapping at us. Go make love to a pine tree (all Nature being your bride). Lay off. Leave us alone. But he will not stop.

"The mass of men lead lives of quiet desperation. What is called resignation is confirmed desperation. . . . A stereotyped but unconscious despair is concealed even under what are called the games and amusements of mankind. There is no play in them."

But is it *true* that the mass of men lead lives of quiet desperation? And if so, did Henry escape such desperation himself? And who, if anyone, can answer these questions?

As many have noted, the mass of men—and women—lead lives today of *un*quiet desperation. A frantic busyness ("business") pervades our society wherever we look—in city and country, among young and old and middle-aged, married and unmarried, all races, classes, sexes, in work and play, in religion, the arts, the sciences, and perhaps most conspicuously in the self-conscious cult of meditation, retreat, withdrawal. The symptoms of universal un-ease or dis-ease are apparent on every side. We hear the demand by conventional economists for increased "productivity," for example. Productivity of what? for whose benefit? to what end? by what means and at what cost? Those questions are not considered. We are belabored by the insistence on the part of our politicians, businessmen and military leaders, and the claque of scriveners who serve them, that "growth" and "power" are intrinsic goods, of which we can never have enough, or even too much. As if gigantism were an end in itself. As if a commendable rat were a rat twelve hands high at the shoulders—and still growing. As if we could never have peace on this planet until one state dominates all others.

The secondary symptoms show up in the lives of individuals, the banalities of everyday soap opera; crime, divorce, runaway children, loneliness, alcoholism, mental breakdown. We live in a society where suicide (in its many forms) appears to more and more as a sensible solution; as a viable alternative; as a workable option.

Yes, there are many who seem to be happy in their lives and work. But strange lives, queer work. Space technicians, for example, busily refining a new type of inertial guidance system for an intercontinental ballistic missile bearing hydrogen bombs. Laboratory biologists testing the ability of mice, dogs, chimpanzees to cultivate cancer on a diet of cigarettes and Holsum bread, to propel a treadmill under electric stimuli, to survive zero gravity in a centrifuge. And the indefatigable R. Buckminster Fuller hurtling around the globe by supersonic jet with six wristwatches strapped to each forearm, each watch set to a different time zone. "The world is big," says Fuller, "but it is comprehensible."

And also, to be fair, young dancers in a classroom; an old sculptor hacking in fury at a block of apple wood; a pinto bean farmer in Cortez, Colorado, surveying his fields with satisfaction on a rainy day in July (those rare farmers, whom Thoreau dismissed with such contempt, we now regard with envy); a solitary fly fisherman unzipping his fly on the banks of the Madison River; wet children playing on a shining, sun-dazzled beach.

Compared with ours, Thoreau's was an open, quiet, agrarian society, relatively clean and uncluttered. The factory system was only getting under way in his time, though he took note of it when he remarked that "the shop girls have no privacy, even in their thoughts." In his day England, not America, was "the workhouse of the world." (America now in the process of being succeeded by Japan.) What would Henry think of New England, of the United States, of the Western world, in the year 1980? 1984? 2001? Would he not assert, confidently as before, that the mass of humans continue to lead lives of quiet desperation?

Quiet desperation. The bite of the phrase comes from the unexpected, incongruous juxtaposition of ordinarily antithetical words. The power of it comes from our sense of its illuminating force—"a light which makes the darkness visible." Henry's shocking pronouncement continues to resonate in our minds, with deeper vibrations, 130 years after he made it. He allows for exceptions, indicating the "mass of men," not all men, but as for the truth of his observation no Gallup Poll can tell us; each must look into his own heart and mind and then deny it if he can.

And what about Henry himself? When one of his friends, William Ellery Channing, declared morosely that no man could be happy "under present conditions," Thoreau replied without hesitation, "But I am." He spent nearly a year at his dying and near the end, too weak to write any more, he dictated the following, in answer to a letter from his friend Blake:

"You ask particularly after my health. I *suppose* that I have not many months to live; but of course I know nothing about it. I may add that I am enjoying existence as much as ever, and regret nothing."

When the town jailer, Sam Staples, the same who had locked Thoreau up for a night many years before, and had also become a friend, paid a visit to the dying man, he reported to Emerson: "Never spent an hour with more satisfaction. Never saw a man dying with so much pleasure and peace." A trifle lugubrious, but revealing. Henry's sister Sophia wrote, near his end, "It is not possible to be sad in his presence. No shadow attaches to anything connected with my precious brother. His whole life impresses me as a grand miracle. . . ."

A cheerful stoic all the way, Thoreau refused any drugs to ease the pain or let him sleep; he rejected opiates, according to Channing, "on the ground that he preferred to endure the worst sufferings with a clear mind rather than sink into a narcotic dream." As he would never admit to a vulgar sadness, so he would not allow himself to surrender to mere physical pain.

It must have seemed to Henry during his last year that his life as an author had been a failure. Only two of his books were published during his lifetime and neither received much recognition. His contemporaries, without exception—Emerson included—had consigned him to oblivion, and Henry could not have been unaware of the general opinion. But even in this he refused to acknowledge defeat. Noting the dismal sales of his books, he wrote in his *Journal:* "I believe that the result is more inspiring and better for me than if thousands had bought my wares. It affects my privacy less and leaves me freer."

Emerson declared that Thoreau was a coldly unemotional man, stoical but never cheerful; Emerson had so convinced himself of this that when, in editing some of Thoreau's letters for publication, he came across passages that indicated otherwise, he deleted them. But Ralph Waldo's son, Edward, in his book *Henry Thoreau as Remembered by a Young Friend,* wrote that Henry loved to sing and dance, and was always popular with the children of Concord.

In her *Memories of Hawthorne,* Hawthorne's daughter Rose gives us this picture of Thoreau ice skating, with Emerson and Hawthorne, on the frozen Concord River: "Hawthorne," she writes, "moved like a self-impelled Greek statue, stately and grave" (the marble faun); Emerson "closed the line, evidently too weary to hold himself erect, pitching headforemost . . ."; while Thoreau, circling around them, "performed dithyrambic dances and Bacchic leaps."

But what of the photographs of Henry referred to earlier, the daguerreotype in his thirty-ninth year by B. W. Maxham, made in 1856, and the ambrotype by E. S. Dunshee, made in 1861? Trying to get some sense of the man himself, in himself, which I do not get from his words alone, or from the accounts of Thoreau by others, I find myself looking again and again at these old pictures. Yes, the eyes are unusually large, very sensitive and thoughtful, as is the expression of the whole face. The nose is too long, the chin too small, neither an ornament; the face deeply lined, the brow high, the hair and beard luxuriant. A passable face, if not a handsome one. And it still seems to me that I read in his eyes, in his look, an elemental melancholy. A resigned sadness. But the man was ailing with tuberculosis when the former picture was made, within a year of his death when the second was made. These facts should explain the thoughtful look, justify a certain weariness. In neither picture can we see what might be considered a trace of self-pity—the *vulgar* sadness. And in neither can we perceive the faintest hint of any kind of desperation. Henry may have been lonely; he was never a desperate man.

What does it matter? For us it is Henry's words and ideas that count, or more exactly, the symbiotic and synergistic mutually reinforcing logic of word and idea, and his successful efforts to embody both in symbolic acts. If it were true that he never had a happy moment (I doubt this) in his entire life, he surely had an intense empathy with the sensations of happiness:

". . . I have penetrated to those meadows on the morning of many a first spring day, jumping from hummock to hummock, from willow root to willow root, when the wild river valley and the woods were bathed in so pure and bright a light as would have waked the dead, if they had been slumbering in their graves, as some suppose. There needs no stronger proof of immortality."

The paragraph is from the springtime of Henry's life. *Walden* is a young man's book, most of it written before his thirtieth year. But the infatuation with the sun and sunlight carries on into the premature autumn of his years as well; he never gave them up, never surrendered. Near the end of his life he wrote:

"We walked [jumping has become walking, but the spirit remains the same] in so pure and bright a light, gilding the withered grass and leaves, so softly and serenely bright, I thought I had never bathed in such a golden flood, without a ripple or a murmur to it. The west side of every wood and rising ground gleamed like the boundary of Elysium, and the sun on our backs seemed like a gentle herdsman driving us home at evening.

And concluding: "So we saunter toward the Holy Land, till one day the sun shall shine more brightly than ever he has done, shall perchance shine into our minds and hearts, and light up our whole lives with a great awakening light, as warm and serene and golden as on a bankside in autumn."

November 10, 1980

Onward, into Stillwater Canyon. We have left Labyrinth behind, though how Major Powell distinguished the two is hard to determine. The current is slow, but no slower than before, the canyons as serpentine as ever. In the few straight stretches of water we gain a view of Candlestick Tower, now behind us, and off to the southwest, ahead, the great sandstone monadnock three hundred feet high known as Cleopatra's Chair, "bathed," as Henry would say, "in a golden flood of sunlight."

We row around an anvil-shaped butte called Turk's Head. Hard to see any reason for the name. Is there any reason, out here, for any name? These huge walls and giant towers and vast mazy avenues of stone resist attempts at verbal reduction. The historical view, the geological view, the esthetical view, the rock climber's view, give us only aspects of a massive *presence* that remains fundamentally unknowable. The world is big and it is incomprehensible.

A hot, still morning in Stillwater Canyon. We row and rest and glide, at two miles per hour, between riparian jungles of rusty willow, coppery tamarisk, brown cane, and gold-leaf cottonwoods. On the shaded side the crickets sing their dirgelike monotone. They know, if we don't, that winter is coming.

But today is very warm for mid-November. An Indian-summer day. Looking at the rich brown river, jungle on both banks, I think how splendid it would be, and apposite, to see the rugose snout of an alligator come sliding through the water toward us. We need alligators here. Crocodiles, also. A few brontosauri, pteranodons, and rocs with twenty-five-foot wingspan would not be amiss. How tragic that we humans arrived too late, to the best of our conscious recollection, to have witnessed the fun and frolic of the giant thunder lizards in their time of glory. Why was that great chapter ripped too soon from the Book of Life? I would give ten years off the beginning of my life to see, only once, *Tyrannosaurus rex* come rearing up from the elms of Central Park, a Morgan police horse screaming in its jaws. We can never have enough of nature.

We explore a couple of unnamed side canyons on the right, searching for a natural stone arch I found ten years ago, on a previous river journey. Hallucination Arch, we named it then, a lovely span of two-tone rosy sandstone—not shown on any map—somewhere high in the northern fringes of the Maze. We do not find it this time. We pass without investigating a third unknown canyon; that must have been the right one.

We camp for two nights at the mouth of Jasper Canyon, spend the day between the nights exploring Jasper's higher ramifications, toward the heart of the Maze. If the Maze has a heart. We go on the following day, down the river, and come sailing out one fine afternoon into the confluence of the two great desert streams. The Green meets the Colorado. They do not immediately merge, however, but flow along side by side like traffic lanes on a freeway, the greenish Colorado, the brownish Green, with a thin line of flotsam serving as median.

Henry never was a joiner either.

"Know all men by these presents that I, Henry Thoreau, do not wish to be considered a member of any incorporated body which I have not joined."

A crusty character, Thoreau. An unpeeled man. A man with the bark on him.

We camp today at Spanish Bottom, near the first rapids of Cataract Canyon. Sitting around our fire at sundown, four of us gnawing on spareribs, the other two picking at their pussy food—tofu and spinach leaves and stewed kelp (it looks like the testicles of a sick octopus)—we hear the roar of tons of silty water plunging among the limestone molars of Brown Betty Rapid: teeth set on edge. The thunderous vibrations rise and fall, come and go, with the shifting evening winds.

We spend the next day wandering about the top of the Maze, under the shadows of Lizard Rock, Standing Rock, the Chimney, looking down into five-hundred-foot-deep canyons, into the stems, branches, and limbs of an arboreal system of part-time drainages. It took a liberal allowance of time, indeed, for the rare storms of the canyon country to carve out of solid rock these intricate canyons, each with its unscalable walls, boxlike heads, stomach-turning dropoffs. A man could spend the better part of a life exploring this one area, getting to know, so far as possible, its broad outline and its intimate details. You could make your summer camp on Pete's Mesa, your winter camp down in Ernie's Country, and use Candlestick Spire all year round for a personalized private sundial. And die, when you're ready, with the secret center of the Maze clutched to your bosom. Or, more likely, never found.

Henry spent his life—or earned his life—exploring little more than the area surrounding his hometown of Concord. His jaunts beyond his own territory do not amount to much. He traveled once to Minnesota, seeking health, but that was a failure. He never came west, although, as he says, he preferred walking in a westerly direction. He never saw our Rocky Mountains, or the Grand Canyon, or the

Maze. He never reached the Amazon, Alaska, Antarctica, the Upper Nile, or the Mountains of the Moon. He journeyed once to Staten Island but was not impressed.

Instead, he made a world out of Walden Pond, Concord, and their environs. He walked, he explored, every day and many nights, he learned to know his world as few ever know any world. Once, as he walked in the woods with a friend (Thoreau had many friends, we come to realize, if not one in his lifetime with whom he could truly, deeply share his life; it is we, his readers, over a century later, who must be and are his true companions), the friend expressed his long-felt wish to find an Indian arrowhead. At once Henry stopped, bent down, and picked one up.

November 14, 1980

Today will be our last day on the river. We plan to run the rapids of Cataract Canyon this morning, camp on Lake Powell this afternoon, go on to Hite Marina and back to civilization, such as it is, tomorrow.

I rise early, as usual, and before breakfast go for a walk into the fields of Spanish Bottom. I see two sharp-shinned hawks roosting in a cottonwood. A tree of trembling leaves, pale gold and acid green. The hawks rise at my approach, circle, return to the tree as I go on. Out in the field, one hundred yards away, I see an erect neck, a rodentian head, a pair of muley ears displayed in sharp silhouette against the redrock cliffs. I stop, we stare at each other—the transient human, the ephemeral desert mule deer. Then I notice other deer scattered beyond the first: one, two, three, four, five—nine all told. Two with antlers.

My first thought is *meat*. Unworthy thought—but there they are, waiting, half of them standing broadside to me, their dear beating hearts on level with the top of the sand sage, saltbush, rice grass. Two of them within a hundred yards—easy range for a thirty-thirty. Meat means survival. Survival, by Christ, with honor. With *honor!* When the cities lie at the monster's feet, we shall come here, my friends, my very few friends and I, my sons and my daughter, and we will survive. We shall live.

My second thought is more fitting, for the moment. Leave them in peace. Let them be. Efface yourself, for a change, and let the wild things be.

What would Henry say? Henry said, "There is a period in the history of the individual, as of the race, when the hunters are the 'best men,' as the Algonquins called them. We cannot but pity the boy who has never fired a gun; he is no more humane, while his education has been sadly neglected." But then he goes on to say: "No humane being, past the thoughtless age of boyhood, will wantonly murder any creature which holds its life by the same tenure that he does. The hare in its extremity cries like a child. I warn you, mothers, that my sympathies do not make the usual *philanthropic* distinctions." Is that his last word on the subject? Hardly. Henry had many words for every subject, and no last word for any. He also writes, "But I see that if I were to live in a wilderness, I should become . . . a fisher and hunter in earnest."

So let them be for now. I turn back to camp, making one step. The deer take alarm, finally, and move off at a walk. I watch. Their fear becomes contagious. One begins to run, they all run, bounding away toward the talus slopes of the canyon wall. I watch them leap upward into the rocks, expending energy with optimum ease, going farther and rising higher until they disappear, one by one, somewhere among the boulders and junipers at the foot of the vertical wall.

Back to camp and breakfast. We load the boats, secure the hatches, lash down all baggage, strap on life jackets, face the river and the sun, and growing roar of the rapids. First Brown Betty, then Ben Hur and Capsize Rapids, then the Big Drop and Satan's Gut. Delightful names, and fitting. We feel the familiar rush of adrenaline as it courses through our blood. We've been here before, however, and know that we'll get through. Most likely. The odds are good. Our brave boatman and boatwoman, Dusty and Lorna, ply the oars and steer our fragile craft into the glassy tongue of the first rapid. The brawling waters roar below, rainbows of broken sunlight dance in the spray. We descend.

Henry thou should be with us now.

I look for his name in the water, his face in the airy foam. He must be here. Wherever there are deer and hawks, wherever there is liberty and danger, wherever there is wilderness, wherever there is a living river, Henry Thoreau will find his eternal home.

ANN ZWINGER

Ann Zwinger "grew up on a river. Not a very big river, across the street and down the bank, but it was always there, running downhill to the Wabash and the Ohio and the Mississippi." The river was the White River, flowing through Muncie, Indiana. "When there is a river in your growing up, you probably always hear it."

These words come from the introduction to *Run River Run*, her book about a western river, the Green, for which she won the 1976 John Burroughs Medal. Zwinger was trained as an art historian at Wellesley, Harvard, and Indiana University. She taught art and art history, raised three daughters, made frequent moves with her husband, Herman (who photographs for her books), during his air force career, and finally settled in Colorado Springs in 1960.

The Zwingers bought forty acres of land at 8,300 feet in the nearby

Rockies, and her first book, *Beyond the Aspen Grove* (1970), tells the story of getting to know that land. "I spent a year pruning the paths, just like a city girl. It takes a long time not to prune paths."

She learned that "when you spend a great deal of time outdoors, there comes to be a need to spend it alone. One's life quickly sifts into new patterns that are satisfying and rewarding in a sense that the 'real' world isn't."

From her time alone out there, she has brought back books that combine her artist's eye for color and texture, an ear for rich language, and a passion for the intricate detail of the natural world. She illustrates her books with her own delicate pencil drawings.

Zwinger's other books include *Land Above the Trees: A Guide to American Alpine Tundra* (1972); *Wind in the Rock* (1978), journeys through four canyons leading into Utah's San Juan River; *A Conscious Stillness: Two Naturalists on Thoreau's Rivers* (1982), coauthored with the late Edwin Way Teale; and *A Desert Country Near the Sea* (1983), about the southern tip of Baja California. She has edited two books of letters by pioneer naturalist John Xantus, and her newest work is a major book on the four North American deserts, *The Mysterious Lands* (1989), from which this selection comes.

Zwinger says, "I write just for pure self-indulgence. Because I like to write. Because it gives me an entrée into another world where there is always sunshine. If you pick up the morning paper and go by that, life could be pretty dreary. You just need to wander a while and things make good sense."

"I write for people who have no idea of what's going on out there. I want to say, *look,* this is the best of all possible natural worlds. If we don't pay attention to it, we won't have it, and if we don't have it, we won't have us either."

Cabeza Prieta

Cabeza Prieta means "dark head" in Spanish, and the peak that inspired that name blocks the horizon in front of us, a pearly-tan pyramid draped with a hangman's hood. It is the landmark of the Cabeza Prieta Mountains, a forbidding mountain range rising from an endless, boundless desert beset with a dryness and heat that devours colors and horizons and fades outlines until all that is left is shimmer.

This third week in June I enter the Cabeza Prieta National Wildlife Refuge with Chuck Bowden and Bill Broyles, both of whom have a desert fixation. We have volunteered to take part in a formal yearly desert bighorn sheep count that has been done here since 1961. In 1939 ninety bighorns were estimated to be on the refuge; today the number is approximately three hundred.

Bill, who in his sensible life is an English teacher, drives this road with an insane joy in the required high clearance, four-wheel drive vehicle. We enter the Cabeza from the north, and it seems to me we have come thousands of miles from Refuge Headquarters where we have been briefed on bighorn sheep counting, driving first on highways, then narrow roads, then dirt roads, and now this ridiculous track.

"Road" is a euphemism. When wet, it imbibes rear axles. When dry, it buries you in sand for hours at a time. It has tracks with ruts that snap a wheel, high centered enough to batter a crankcase to oblivion. Washboard is too complimentary a word. We're talking lurching, jolting, tilting, loose rock or deep sand, kidney-busting roads that just don't care whether you travel them or not. Ocotillos, rude toll takers, thrust their thorny hands into the open window with intent to rob. On this questionable road, not only does one eat dust, and smell an overloaded and overheated engine, but also has to endure the squeals of vegetation dragging its fingernails along the sides of the truck, thorny plants as ill-tempered as the road itself.

Desert bighorns roam the mountain range of the Cabeza in male and female bands (young rams are included in both) throughout most of the year, getting enough water through eating green vegetation. They combine and come in to waterholes only during the driest period of the year. The commingling announces the onset of the herd's breeding season, correlated to producing lambs at the most propitious time, a timing that varies in each desert bighorn herd according to the particular mountain range climate in which it lives.

We lurch and bounce into the small east-west valley that holds the tank where I will be stationed. The valley is narrow, rising to a notch between two high, rocky ridges to the west. Without gentling bajadas, the mountains bolt out of the floor, rubbled and pitted, short and steep, boldly tilted, roughed out with a chisel but never worked down or refined.

Bill offloads three jerry cans of water, enough for my six days here, a folding chair and a cot; he and Chuck will go on to their own stations twenty miles away. I heave my duffle down, eager for them to be gone, anxious to set up desert housekeeping, ready to begin these days of immaculate solitude. The truck disappears in a cloud of dust and I stand blissfully alone. For a moment I breathe in the desert isolation. Then I unfold my chair, set up the tripod, and mount and focus the spotting telescope on the far slope where I hope the sheep will be, wire the thermometer at eye level, and survey my kingdom.

The blind is semi-hidden beneath an ancient mesquite tree. Open in the rear and with a window facing the natural water pocket of the tank, it is made of four steel posts faced with saguaro ribs. The plywood roof has wallboard nailed underneath and, in the space between, a heaping of mesquite twigs and pods hints at a woodrat nest, a suspicion soon confirmed when the occupant itself sashays off to forage.

The *tinaja* itself is invisible from the blind. It is tucked to the right against a vertical wall, two hundred yards up the draw. Huge boulders flank it on the left; one, fifteen feet high, provides a provident perch for the several dozen birds that frequent it. I watch two Gambel's quail bobble down the hillside to the tank, and locate the pool by watching where they disappear. Tomorrow I shall walk up to record its water level.

Promptly at seven o'clock bats begin cruising my campsite, looping lower than I'm used to seeing bats fly, whishing right past my ear, winnowing the air for insects with great efficiency. After they leave, I get bitten. The light fades so gently that my eyes are adjusted to starlight by 9:30 when I go to bed. The air is a comfortable 86 degrees F. The quiet is soporific.

An infinitesimal breeze feathers my ankles and face all night long, the gentlest whisper of air in a great friendliness of night.

Gazing up through the lacy canopy of mesquite leaves, I try only to doze, unwilling to miss anything. I do not succeed.

In the midst of a sound sleep, a sudden gust of wind looses an avalanche of mesquite twigs and pods and I shoot awake. The moon fingerpaints the sky with clouds. The Milky Way forms a great handle to the basket of the earth in which I lie. It is deliciously cool, enough so to pull up my sleeping bag. When I awake again, a big old saguaro hoists a near-full moon on its shoulder. By 4:30 in the morning, when sheep watchers are expected to be sentient and at their task, pearly gray light seeps into the valley.

Mesquite pods litter my sleeping bag and shoes. My cup holds three like swizzle sticks. These mesquites have longer, slimmer fruits, smaller leaves and leaflets, and a larger number of leaflets than those in damper climes. Unlike other Pea Family pods, these do not split open but retain the seeds imbedded in a hard covering, and an animal who eats one must eat pod and all. So important were they to Seri Indians that they had names for eight different stages of fruit development. The Seris even sifted through woodrat nests several months after harvest to recover the mesquite seeds; their high protein and carbohydrate content made them a significant addition to their diet. Amargosa and Pinacateno Indians developed a special gyratory crusher for grinding them. All I can say from experience is that the fruits are as hard as rocks, dry as bones, fall like lead weights, and truly do taste sweet.

I settle into my watcher's chair in which I will spend so much time I will come to feel as if it is annealed to my behind. Immediately in front of the blind wends a dry, sandy stream bed about eight feet wide, in the middle of which grows a twenty-five foot saguaro with one massive arm raised in greeting. Some fruit remains on the cactus, looking like red blossoms—indeed, an early botanist mistook them for that—but most have fallen to the sand and seethe with ants. George Thurber, botanist for the Mexican Boundary Survey in the early 1850s, who first collected saguaro seeds, followed the Indian practice of rolling the coarse fruit pulp into balls for storage; the seeds he thus preserved later germinated in East Coast botanical gardens. A mature cactus can produce more than two hundred fruits, each with some thousand seeds. Of these, perhaps half a dozen will be left after birds and rodents and ants have devoured them. Not only are they nutritious but they fullfil the water requirements of most animals.

Behind the blind a low divide centers the valley between two narrow, quirky sandy draws. The soil of the small divide, without benefit of shading mesquite or protective creosote bush, is cement hard, paved with rough granite chips from the flanking and flaking mountains. Anthills stud it, a cholla or two, a few bristly borages, and many spiny-herbs, but mostly it is bare.

On the crest of the rise is a small hill of harvester ants whose twelve-inch wide stream of foragers was busy long before I found them this morning. With the temperature well above their threshold of 50 degrees F, they likely were out at first light. They swarm on a small gilia with cottony heads, harvest cryptantha seeds, explore peppergrass, and cart back not only seeds but leaf scraps and bracts, anything detachable. Thus laden they totter back to the nest's dime-sized opening, rimmed with a foot-wide circle of salt-and-pepper debris. There they disappear underground to feed the larvae that serve as the colony's collective stomach. Although the adults gather the food, chew it and prepare it, it is the larvae that digest it and regurgitate part of it to the adults, and the colony cannot survive without their services.

By nine o'clock, when it is 90 degrees, not an ant is to be seen.

The white-winged dove contingent begins calling when the day warms, a repetitive call, not the familiar measured "who cooks for who" but an erratic, petulant "don't tell me, don't *tell* me!" White-wings continue to call across the canyon, antiphonal with the mourning doves in stereophonic sound.

The air between the blind and the water tank begins to quiver. More doves arrive on the big rock overlooking the water. Throughout the day they gather, sometimes twenty or so in plenary session, sometimes playing musical chairs as they drop down to drink and return. But always in full sun.

Judging from their numbers, white-winged doves are exceedingly successful desert dwellers. The white-wing's clutch size is always two, and population increase depends upon how many clutches are laid and reared. They are capable of producing throughout the year, apparently oblivious to the usual environmental spurs, such as day length and rain, that govern many birds. Very resistant to dehydration, they can lose up to twenty percent of their body weight, and go four to five days without water. With access to water, they can tank up within five minutes. The doves' mobility makes it possible for them to water long distances away from where they nest.

White-wings throw together a haphazard nest that provides little in the way of insulation or protection from sun when there are eggs in the nest. A "brood patch," a small area of bare skin on the breast, serves as a heat sink for the eggs that prevents them from overheating. Given that most desert birds maintain elevated body temperatures, white-winged doves have lower body temperatures and are able to maintain it during incubation. Males have an even lower body temperature during the day than females and do most of the egg-tending during the highest heat periods. In the Sonoran Desert in Arizona, even in direct sun, brooding birds never leave the nest unattended.

When they take off as a flock, their white wing patches shine while their dun-colored bodies fade into the background, and from a distance they look like a flock of small white birds. But I did not come to watch birds. I came to watch desert bighorns. And not one has shown itself.

In the noon stillness, the heat presses down, lies as heavy as a mohair wool blanket on my head, on my shoulders. The air around the rocks near the tank wavers with banners of heat. I feel stooped and bent with heat as I leave the shade of the blind to walk up to the tank. I need to record the tank's water level so I can see how much it evaporates during the week (it goes down 3¼ inches). Bur sage is so dry that its leaves disintegrate as I brush by them. Big clumps of saltbush outline the wash, many of them festooned with vining milkweed, twined and twisted in great heaps like a giant-sized helping of vermicelli. Seed pods load catclaw, how ripe the seeds are depending upon where the bush is set. By the blind the seeds are still green and held tightly in the pod; lower, they are drying to a salmon color, and in the hollow at the foot of the draw, they are dark brown and shiny, rattling in the twisted pods.

A four-foot high elephant tree has found a purchase in the rocky slope along the trail, leaves just perceptible, ready to unfold the moment it rains. The papery-barked trunks, fat for water storage, give off a familiar turpentine odor. This is as far north as they are found. Low holly-leaved bur sage marches up the slope to the ridge crest, endemic to the lower Sonoran. Its scalloped leaves mark different years' growth in their different colors, the oldest faded nearly pure white. Blooming over for the year, only the nettlelike seed heads remain on the bushes.

The granite surrounding the water tank is pale gray, crystalline, flecked with sparkling black mica and roughened with large quartz crystals. A silt-holding wall has been built above the pool, and some cementing at the lower end provides an apron where animals can safely approach the water. Higher water levels limn the rocky sides with gray lines, and an inch of vivid green algae circles the present pool. It swarms with bees. I hesitate a long minute, then make my measurement as quickly as possible because in recent years I've become allergic to bee stings.

Anaphylactic shock, due to bee or insect stings, is not common, but on a percentage basis, it kills more people than venomous snake bite does. While it would take over a hundred stings to affect a non-sensitized person, just one sting can cause anaphylatic shock in one who is sensitized. Mellitin, a basic protein composed of thirteen different amino acids, is the chief toxin; it affects the permeability of skin capillaries, instigates a drop in blood pressure followed by a swift rise, and damages nerves and muscle tissue. With no medical help within a hundred miles, one becomes exceedingly conservative.

On the way back to the blind, primed to thinking about venomous bites, I gasp when I round a boulder and five feet in front of me, on a large flat rock, is a Gila monster. Handsomely marked, it is fearsome looking because it is such a large lizard, well bigger than my forearm, heavy of body and tail, built like a tank. Warning colors of black and yellow vividly mark its beaded skin.

It is justifiably feared because of its virulent bite. After it locks onto its victim with its jaws, it chews. While doing so, it injects venom from glands set just under the skin of the lower jaw. Each tooth has sharp flanges flanking the grooves through which the venom flows, making the bite very painful. In small animals death comes from respiratory paralysis. Although the bite brings severe pain and swelling, quickly followed by nausea and weakness, it is not usually fatal to man.

Restricted to the Sonoran Desert in the United States, Gila monsters are generally active at night. They eat the eggs of reptiles and birds, and the young of small rodents. I cannot imagine why it is out at this time of day. I back off with more speed than grace and leave it to its meditation.

An innocent little four-inch, zebra-tailed lizard, perhaps more tan-and-white than zebra-striped, scampers across the sand, its ringed tail curled up over its back like a pug dog. The chevron pattern on its back and legs blends in with the flickering shadows and becomes invisible when it skitters under an acacia. It is one of the fastest lizards around. In comparison to its quick legs, mine feel made of lead.

I spend well over an hour out of the blind, most of it by necessity in the sun. The temperature is 100 degrees F when I get back at noon, despite a high overcast. Not only does air heat envelop me but heat billows up from the ground and radiates from everything around me: the ceiling of the blind, the sand in the wash, the desert pavement.

I gulp down a half bottle of fruit juice, made highly conscious of the necessity to drink lots of liquids by my packet of reading material with articles on death by thirst in gory detail, including "tumid tongue and livid lips," "unclean blow-flies" to gather on the eyes and ears of my "already festering carcass" pierced again and again by the cruel spines of cholla. My pulse rate is up to 90. My face feels flushed. Small blood vessels at the skin's surface dilate in order to radiate heat back to the environment because skin temperature is generally cooler (around 92 degrees F) than the surrounding hot air. The dilation of skin vessels also ensures an ample supply of blood to activate the two million sweat glands on the body because, when air temperature reaches between 86 and 92 degrees F, the only way the body can lose heat is by sweating. No amount of training, no amount of acclimation, alters the amount of water that is needed to replace water lost in perspiration. Because the body can neither store water nor reduce its need, water supply is critical.

Sweating allows man to exist in the desert. Sweating is controlled by centers in the hypothalamus which have set points keyed to temperature. Overheating can be prevented by the loss of one cup of sweat per hour, but it's easy to lose more without sensing it because in the desert, sweat evaporates the minute it reaches the skin surface. But there is a limit to how much cooling can be provided by sweating before the loss of constituents of the blood—chloride, sodium, potassium, lactic acid—begins which can bring on muscle cramping and severe electrolyte imbalance.

Acclimatization helps by increasing the perspiration rate and beginning sweating at a slightly lower temperature. Circulating blood increases in volume, urinary sodium and chloride are better retained, and salt concentrations in perspiration decrease. But no matter how much one is acclimated, water remains critical to well-being. Paradoxically, even if ample water is available, it is impossible to drink and recover the amount lost because thirst is satiated before one's water intake equals water loss. Possibly this is a natural governor that prevents seriously upsetting the concentration of salt in the blood. Complete restoration usually comes when water is taken with food, and food replaces the lost salts. While it is almost impossible for a man to replace in one draught the full amount of water lost, animals do so by drinking greater quantities in shorter periods of time—a mourning dove can drink ten times (per body volume) what a man can drink in the same amount of time, and bighorn sheep can gulp down enough water at one time to recover their original weight.

If water loss from sweating cannot be replaced, sweat glands extract water from blood plasma, and body temperature rises. The loss of plasma water increases the viscosity of the blood, placing an additional stress on the heart, requiring more work to pump thickened blood through arteries and veins, also raising the pulse rate. Even though the heart beats faster, the amount of blood pumped out per minute remains nearly the same. In turn, retarded blood flow fosters a continuing rise in the inner core of the body; pulse rate and rectal temperature increase, breathing quickens. (Such a reduced blood volume does not affect the heart rate in dehydrated bighorn sheep.)

At two-percent loss of body weight due to loss of body water, thirst for some may already be fierce, accompanied by anorexia and flushed skin and increased pulse rate. Small increments of debit have large symptoms: at four percent the mouth and throat go dry; by eight percent the tongue feels swollen, salivary functions cease and speech becomes difficult. After ten percent the ability to cooperate, or even to operate, is gone. At twelve-percent loss, circulation is so impaired that an explosive heat rise deep in the body is imminent, and deep body temperature rises dangerously fast. Death follows, although lethal limits may be as high as eighteen- to twenty-percent loss of body weight. If liquid is available, recovery can be nearly complete an hour after drinking.

At 104 degrees F, one must evaporate an equivalent of 1.5 percent of one's body weight per hour to maintain a constant body temperature. I do a quick calculation: at 110 pounds 1.5 percent of my body weight is 1.65 pounds. Since a pint's a pound the world around, to replace what I lost means at least a quart of liquid since, in addition, the body produces about eighty calories of heat per hour through metabolic activity. Dissipating this through sweat requires another five ounces of water per hour. In other words, an hour's wandering has started me toward a two-percent deficit.

I look with new respect at the bighorn sheep than can lose twenty percent of its body weight (or a thirty-percent loss of total body water), or the white-winged dove that can lose twenty-five percent, with no ill effects and even drink salt water, or the quail who can survive fifty percent loss of body weight without succumbing.

I had anticipated some anorexia. Normal enough. I had also anticipated some of the psychological by-products of dehydration such as lethargy or depression. Lethargy just may be the better part of valor in this heat, and as for depression, I have done nothing but walk around with a grin on my face the whole time. Like Carl Lumholtz, traveling the Cabeza three-quarters of a century ago, I find it "radiant with good cheer."

Nevertheless I take another slug of juice. It insureth my well-being.

The sheep do not come in until the second day, as if they were waiting to be sure their intruder was from a friendly planet. Periodically scanning the mountain slope with binoculars, it is some moments before I realize that the "rocks" are moving. I nearly tip over the spotting scope in my delight and eagerness to get it focused. They come downslope with a deliberate, measured tread, pausing occasionally to stand and look about or pull off an acacia twig. There are either eight or nine, and they change places just often enough to make them exasperatingly difficult to count.

My first impression is of smallness and grace. Less than four feet high at the shoulder, they move easily over what I know to be a rocky, ruggedly steep and treacherous slope. With their smooth coats and long legs they bear little resemblance to domestic sheep to which they are *not* related. The genus *Ovis* originated in Eurasia more than two and a half million years ago, migrated, as man did, across the Bering Land Bridge, and became isolated during glacial periods into different groups. Today desert bighorns inhabit the arid mountain ranges of the southwest deserts.

Although there is no pushing or shoving, there appears to be a hierarchy at the water tank. A large, mature ewe waters first, at two-minute and then four-minute draughts, and then stands aside. She has shed most of her winter coat but a continuous remnant cloaks her shoulders like a shawl, making her easy to identify when she appears again. Bighorn shedding begins at the rump and moves forward. All the other sheep have less old pelage clinging, suggesting that this ewe may be an older animal. She is accompanied by two lambs with their mothers that never do go down to drink. The ability of desert bighorns to go without water exceeds even that of the camel. Their extensive rumen complex—the first stomach of ruminant animals—holds and supplies enough water to support their needs. They produce a concentrated urine and to some extent, resorb moisture from their feces. They can drink rapidly enough to get back into water balance within five minutes.

The next morning the large ewe again waters first, taking two long drinks before she leaves. Other sheep move in, either young ewes and/or yearling rams—their size and horn shape is so close that I cannot differentiate at this distance. Mature rams are easy to identify as the horns are much more developed and the four-year mark is usually a very prominent and readily visible groove on the horn, especially visible from the back. From that line one counts rings back toward the head for a fairly accurate age estimate as each ring marks the cessation of growth for one year. The sheep remain at the tank, either drinking or standing around, for half an hour, and then slowly amble up the hill.

After they leave a single large ram comes to drink. His horns have almost a full curl with the ends broomed off, leaving the tips worn and blunt (brooming results from rubbing the horns against rocks and dirt). The ram takes almost ten minutes to come down the hillside; after reading about their prowess in leaping rocks and sheer cliffs, I am struck by his extreme deliberateness, which I take also as a measure of his serenity. When motivated, a bighorn sheep covers ground with astonishing leaps and bounds, fast and fleet. I dutifully record that he drinks for two minutes, then six, then two more minutes before picking his way back up the hill.

Meanwhile the earlier crew reaches the crest of the ridge to the east and disappears over the rim. The ram stops twenty feet or so below the top, where a ewe and her lamb lie hidden in a shadowy niche. He nuzzles the lamb, which tags along after him over the hill.

The ewe remains, entering a deep overhang I have not noticed before, disappearing from view in the shadow.

I am inordinately pleased. I feel as if the sheep have honored me with their presence and I sit here smiling as I write up my observations.

I had three concerns before I came on this bighorn sheep count: that I would be uneasy, that time would hang heavy, that I could not endure the heat. Instead I have felt at home, there have not been enough hours in the day, and the heat has become a bearable, even if not an always welcome, companion. Joseph Wood Krutch's words come to mind: "Not to have known—as most men have not—either the mountain or the desert is not to have known one's self. Not to have known one's self is to have known no one."

At noontime I am concentrating so hard on taking notes that when a cicada lets off a five-second burst like a bandsaw going through metal, I jump. When there has been no activity at the tank for over an hour I opt for a can of tuna sprinkled with the juice of half a lemon.

No sooner do I open the can and get my fork out than the doves explode from the rock on which they've congregated. A red-tailed hawk bullets straight at me, talons extended, tail spread. No sound, no screaming. The doves evaporate. The red-tail makes no hit. It wheels, spirals upward, and stoops again. Again no luck. It makes no third try. Its disappearance is followed by a great shocked silence. It is half an hour before the doves venture, one by one, back to their sentinel rock.

I see only this one hawk stoop. The only time that the red-tails are quiet is on the attack. Long before they come into water I hear their eerie "KEEEeeeer KEEeeer" that ricochets off the sky itself. One afternoon a red-tail sits on the steep cliff to the right; a couple of feet below it perch some house finches; a black-throated sparrow searches the bush beside it; a pair of ash-throated flycatchers rest above it; and across the tank, a batch of doves roost peacefully on their rock. All birds must be vulnerable to attack at the waterhole and many avoid too much exposure by being able to drink very quickly, or coming in early and late when raptors are not hunting. Yet there are also these moments when the lion and the lamb lie down together.

I put aside my field glasses and lift my fork. This time a turkey vulture lands on the big rock. The rock is nearly vertical on the side overlooking the tank and this is where it chooses to descend to water. It gets about a quarter of the way down, contorted in a most awkward position, big feet splayed out on the rock, tail pushed up behind at a painful angle, as it looks intently down at the water, a hilarious study in reluctance. Gingerly it inches down (they have no claw-grasping capability as birds of prey like eagles and hawks do) until it can hold no longer and crash-lands onto the apron to drink. The closeness of the rocks around the pool disallows it maneuvering space because of its broad wing span, and so it edges as close as it can to insure dropping upon the only place where it can stand and drink.

Four species of this misanthropic-looking bird existed in the Pleistocene, and this creature looks like one of the originals. On the ground, vultures are hunched and awkward bundles of feathers, but in the air, where I watch them during much of the day, they are magnificent graceful soarers. They ascend with the updrafts coming off the hot desert floor, floating and lifting to cooler air where visibility is superb—at 500 feet the visible horizon is 27 miles away, and at 2,000 feet, 55 miles. Such big raptors get as much water as they need from the carrion they consume. At this dry time of year, if the carrion supply is not sufficient, they must come in to water.

One more try on the tuna fish. A Yuma antelope squirrel scuttles down the wash to nibble on saguaro seeds. Tail held high over its back for shade, a single white stripe on each flank, its quick, jerky movements typical of the ever-watchful. It stands slightly hunkered up on its hind legs to eat, but its rear end never touches the hot sand. It spends little time feeding and soon tucks back into the brush near the blind where it undoubtedly has a burrow. There it can spread-eagle itself on the cool soil and unload its body heat, before taking on the desert again.

The next day, emboldened by curiosity, it hops up on my iceless ice chest inside the blind and puts its head in my empty cup which immediately tips over, sending the squirrel flying.

By four o'clock the tank is in shadow and the blind is in sun. The resident robber fly alights in front of me, makes a slapdash attempt at a wandering fly, misses, and returns to watching. Flies, belonging to many genera, are very widely dispersed in the desert, more prevalent than any other insect order, and of these, bee flies and robber flies are the most numerous. The huge eyes of robber flies give them peripheral vision; with streamlined stilletolike bodies that allow swift flight, and needlelike beaks, they are efficent predators on the desert wafters and drifters. I've watched this one impale a smaller fly in flight, almost too quickly for the eye to follow, then alight to suck out the juices. Today it seems scarcely to care.

A cloud cover that has kept the temperature relatively low all day disappears and the sun has an unobstructed shot at my back. Until the sun drops behind the ridge, I am unqualifiedly miserable.

In the evening a caterwauling of Gambel's quail issues from the mesquite trees where they have been up in the branches. Their fussing is interspersed with a short, soapsudsy cluck, embellished by a silvery "tink" at the end. The first night I was here they scolded and fumed about the stranger in their midst. The second night they gossiped and fussed, and the third night I awoke to find one within a few feet of the cot.

They visit the big mesquites only in the evening, always after sunset, and they are always noisy. I would think a predator could hear them a mile off. Their drinking patterns have evolved to avoid raptors; they commonly come in to water twice a day, one period beginning at dawn, the other ending at dusk. Birds of prey tend to arrive around noon, so the quail's watering time does not overlap.

Quail have been termed "annual" birds because of their variable yearly populations. In the Sonoran Desert, the number of young quail per adult found in the fall correlates to rainfall during the previous December-April; in the Mojave, the same high correlation exists between young and October-March precipitation (a relationship that exists in other desert animals, among them bighorn sheep). Their reproductive activity begins before green vegetation becomes a part of their diet, the amount of which would give them clues to the amount of nourishment available, and hence the clutch size that could survive. Although quail do not breed at all in exceptionally dry years, they are one of the most prolific of birds.

Painted wooden birds, charming wind-up toys with staccato movements, officiously giving each other instructions, they bustle among the creosote bushes. As I watch them puttering about, I recall the Kawaiisu Indian story about the tear marks on Quail's face because her young died, one after the other, because she made her cradles out of sandbar willow—a wood which the Indians therefore do not use for cradles.

The day dims and I stretch out to count the stars, framed in a triangle of mesquite branches. Content, I realize I have reached, as Sigurd Olson wrote, "the point where days are governed by daylight and dark, rather than by schedules, where one eats if hungry and sleeps when tired, and becomes completely immersed in the ancient rhythms, then one begins to live." Yes.

JOHN
MCPHEE

Photograph © Stephen Trimble

John McPhee was born in Princeton Hospital in 1931, attended Princeton University, where his father was a staff doctor, and now writes and teaches there as Ferris Professor of Journalism. In his office is a photograph of him surrounded by his four daughters—a swirl of female energy. He says, "My daughters tell me I'm the most provincial thing they ever heard of in their lives. Princeton is sort of a fixed foot; New Jersey is an interesting place. One could also say that if you are from New Jersey you might have a strong desire to go somewhere else—see Alaska, spend time in Maine, in the Grand Canyon. I do get out of here, as my work testifies."

Indeed it does. McPhee is a staff writer for the *New Yorker*. His "pieces of writing," as he always calls them—models of careful observation and masterful structure—first appear there, then subsequently in book form. In subject matter they embrace oranges, tennis players, canoemaking, the Swiss Army, and nuclear physicists. He says he writes without regard to whether the piece will be read in the magazine or as a book. At the same time, he is aware that "the physical form in which something is printed affects the taste of the thing. I mean, you wouldn't drink a martini in a paper cup."

His office contains a word processor, which David Love's mother, "years after her life had ended," convinced him to acquire. He was using parts of her pioneer journal in his most recent book, *Rising from the Plains*, and couldn't bear to retype them thirty times. Now, when he writes, "I draft on this machine, but after that I still lie down on the couch over there with a clipboard and the manuscript pages and a pencil. And I prop my chin up on a pillow and I fiddle with the pencil."

Rising from the Plains is McPhee's nineteenth book. He knew early on that he wanted to write for the *New Yorker*, and submitted pieces to them for fourteen years before they accepted one. After his profile of basketball star Bill Bradley appeared, McPhee became a *New Yorker* staff writer in 1964. Landscape appears conspicuously in many of his books, including *The Pine Barrens* (1968); *Encounters with the Archdruid* (1972), a profile of conservationist David Brower and three of his adversaries; *Coming Into the Country* (1977), a many sided look at Alaska; and a series on geology and geologists—*Basin and Range* (1981), *In Suspect Terrain* (1983), and *Rising from the Plains* (1986)—with the overall title "Annals of the Former World."

This excerpt from *Basin and Range* comes from near the beginning of the "Annals." The series of books, says McPhee, "just happened to me, that's all. I thought I was going to be doing some interesting little sketch of a geologist, and soon be into some other subject." Three books later, there are "Annals" still to come.

Basin and Range

I used to sit in class and listen to the terms come floating down the room like paper airplanes. Geology was called a descriptive science, and with its pitted outwash plains and drowned rivers, its hanging tributaries and starved coastlines, it was nothing if not descriptive. It was a fountain of metaphor—of isostatic adjustments and degraded channels, of angular unconformities and shifting divides, of rootless mountains and bitter lakes. Streams eroded headward, digging from two sides into mountain or hill, avidly struggling toward each other until the divide between them broke down, and the two rivers that did the breaking now became confluent (one yielding to the other, giving up its direction of flow and going the opposite way) to become a single stream. Stream capture. In the Sierra Nevada, the Yuba had captured the Bear. The Macho member of a formation in New Mexico was derived in large part from the solution and collapse of another formation. There was fatigued rock and incompetent rock and inequigranular fabric in rock. If you bent or folded rock, the inside of the curve was in a state of compression, the outside of the curve was under great tension, and somewhere in the middle was the surface of no strain. Thrust fault, reverse fault, normal fault—the two sides were active in every fault. The inclination of a slope on which boulders would stay put was the angle of repose. There seemed, indeed, to be more than a little of the humanities in this subject. Geologists communicated in English; and they could name things in a manner that sent shivers through the bones. They had roof pendants in their discordant batholiths, mosaic conglomerates in desert pavement. There was ultrabasic, deepocean, mottled green-and-black rock—or serpentine. There was the slip face of the barchan dune. In 1841, a paleontologist had decided that the big creatures of the Mesozoic were "fearfully great lizards," and had therefore named them dinosaurs. There were festooned crossbeds and limestone sinks, pillow lavas and petrified trees, incised meanders and defeated streams. There were dike swarms and slickensides, explosion pits, volcanic bombs. Pulsating glaciers. Hogbacks. Radiolarian ooze. There was almost enough resonance in some terms to stir the adolescent groin. The swelling up of mountains was described as an orogeny. Ontogeny, phylogeny, orogeny—accent syllable two. The Antler Orogeny, the Avalonian Orogeny, the Taconic, Acadian, Alleghenian Orogenies. The Laramide Orogeny. The center of the United States had had a dull geologic history—nothing much being accumulated, nothing much being eroded away. It was just sitting there conservatively. The East had once been radical—had been unstable, reformist, revolutionary, in the Paleozoic pulses of three or four orogenies. Now, for the last hundred and fifty million years, the East had been stable and conservative. The far-out stuff was in the Far West of the country—wild, weirdsma, a leather-jacket geology in mirrored shades, with its welded tuffs and Franciscan mélange (internally deformed, complex beyond analysis), its strike-slip faults and falling buildings, its boiling springs and fresh volcanics, its extensional disassembling of the earth.

There was, to be sure, another side of the page—full of geological language of the sort that would have attracted Gilbert and Sullivan. Rock that stayed put was called autochthonous, and if it had moved it was allochthonous. "Normal" meant "at right angles." "Normal" also meant a fault with a depressed hanging wall. There was a Green River Basin in Wyoming that was not to be confused with the Green River Basin in Wyoming. One was topographical and was *on* Wyoming. The other was structural and was *under* Wyoming. The Great Basin, which is centered in Utah and Nevada, was not to be confused with the Basin and Range, which is centered in Utah and Nevada. The Great Basin was topographical, and extraordinary in the world as a vastness of land that had no drainage to the sea. The Basin and Range was a realm of related mountains that coincided with the Great Basin, spilling over slightly to the north and considerably to the south. To anyone with a smoothly functioning bifocal mind, there was no lack of clarity about Iowa in the Pennsylvanian, Missouri in the Mississippian, Nevada in Nebraskan, Indiana in Illinoian, Vermont in Kansan, Texas in Wisconsinan time. Meteoric water, with study, turned out to be rain. It ran downhill in consequent, subsequent, obsequent, resequent, and not a few insequent streams.

As years went by, such verbal deposits would thicken. Someone developed enough effrontery to call a piece of our earth an epieugeosyncline. There were those who said interfluve when they meant between two streams, and a perfectly good word like mesopotamian would do. A cactolith, according to the American Geological Institute's *Glossary of Geology and Related Sciences,* was "a quasi-horizontal chonolith composed of anastomosing ductoliths, whose distal ends curl like a harpolith, thin like a sphenolith, or bulge discordantly like an akmolith or ethmolith." The same class of people who called one rock serpentine called another jacupirangite. Clinoptilolite, eclogite, migmatite, tincalconite, szaibelyite, pumpellyite. Meyerhofferite. The same class of people who called one rock paracelsian called another despujolsite.

Metakirchheimerite, phlogopite, katzenbuckelite, mboziite, noselite, neigh-borite, samsonite, pigeonite, muskoxite, pabstite, aenigmatite. Joesmithite. With the X-ray diffractometer and the X-ray fluorescence spectrometer, which came into general use in geology laboratories in the late nineteen-fifties, and then with the electron probe (around 1970), geologists obtained ever closer examinations of the components of rock. What they had long seen through magnifying lenses as specimens held in the hand—or in thin slices under microscopes—did not always register identically in the eyes of these machines. Andesite, for example, had been given its name for being the predominant rock of the high mountains of South America. According to the machines, there is surprisingly little andesite in the Andes. The Sierra Nevada is renowned throughout the world for its relatively young and absolutely beautiful granite. There is precious little granite in the Sierra. Yosemite Falls, Half Dome, El Capitan—for the most part the "granite" of the Sierra is granodiorite. It has always been difficult enough to hold in the mind that a magma which hardens in the earth as granite will—if it should flow out upon the earth—harden as rhyolite, that what hardens within the earth as diorite will harden upon the earth as andesite, that what hardens within the earth as gabbro will harden upon the earth as basalt, the difference from pair to pair being a matter of chemical com-position and the differences within each pair being a matter of texture and of crystalline form, with the darker rock at the gabbro end and the lighter rock the granite. All of that—not to mention such wee appendixes as the fact that diabase is a special texture of gabbro—was difficult enough for the layman to remember before the diffractometers and the spectrom-eters and the electron probes came along to present their multiplex cavils. What had previously been described as the granite of the world turned out to be a large family of rock that included granodiorite, monzonite, syenite, adamellite, trondhjemite, alaskite, and a modest amount of true granite. A great deal of rhyolite, under scrutiny, became dacite, rhyoda-cite, quartz latite. Andesite was found to contain enough silica, potassium, sodium, and aluminum to be the fraternal twin of granodiorite. These points are pretty fine. The home terms still apply. The enthusiasm geolo-gists show for adding new words to their conversation is, if anything, exceeded by their affection for the old. They are not about to drop gran-ite. They say granodiorite when they are in church and granite the rest of the week.

When I was seventeen and staring up the skirts of Eastern valleys, I was taught the rudiments of what is now referred to as the Old Geology. The New Geology is the package phrase for the effects of the revolution that occurred in earth science in the nineteen-sixties, when geologists clambered onto seafloor spreading, when people began to discuss continents in terms of their velocities, and when the interac-tions of some twenty parts of the globe became known as plate tectonics.

There were few hints of all that when I was seventeen, and now, a shake later, middle-aged and fading, I wanted to learn some geology again, to feel the difference between the Old and the New, to sense if possible how the science had settled down a decade after its great upheaval, but less in megapictures than in day-to-day contact with country rock, seeing what had not changed as well as what had changed. The thought occurred to me that if you were to walk a series of roadcuts with a geologist something illuminating would in all likelihood occur. This was long before I met Karen Kleinspehn, or, for that matter, David Love, of the United States Geological Survey, or Anita Harris, also of the Survey, or Eldridge Moores, of the University of California at Davis, all of whom would eventually take me with them through various stretches of the continent. What I did first off was what anyone would do. I called my local geologist. I live in Princeton, New Jersey, and the man I got in touch with was Kenneth Deffeyes, a senior professor who teaches introductory geology at Princeton University. It is an assignment that is angled wide. Students who have little aptitude for the sciences are required to take a course or two in the sciences en route to some cerebral Valhalla dangled high by the designers of curriculum. Deffeyes' course is one that such students are drawn to select. He calls it Earth and Its Resources. They call it Rocks for Jocks.

Deffeyes is a big man with a tenured waistline. His hair flies behind him like Ludwig van Beethoven's. He lectures in sneakers. His voice is syllabic, elocutionary, operatic. He has been described by a colleague as "an intellectual roving shortstop, with more ideas per square metre than anyone else in the department—they just tumble out." His surname rhymes with "the maze." He has been a geological engineer, a chemical oceanographer, a sedimentary petrologist. As he lectures, his eyes search the hall. He is careful to be clear but also to bring forth the full promise of his topic, for he knows that while the odd jock and the pale poet are the white of his target the bull's-eye is the future geologist. Undergraduates do not come to Princeton intending to study geology. When freshmen fill out cards stating their three principal interests, no one includes rocks. Those who will make the subject their field of major study become interested after they arrive. It is up to Deffeyes to interest them—and not a few of them—or his department goes into a subduction zone. So his eyes search the hall. People out of his course have been drafted by the Kansas City Kings and have set records in distance running. They have also become professors of geological geophysics at Caltech and of petrology at Harvard.

Deffeyes' own research has gone from Basin and Range sediments to the floor of the deep sea to unimaginable events in the mantle, but his enthusiasms are catholic and he appears to be less attached to any one part of the story than to the entire narrative of geology in its four-dimensional recapitulations of space and time. His goals as a teacher are ambitious to the point of irrationality: At the very least, he seems to expect a hundred mint geologists to emerge from his course—expects perhaps to turn on his television and see a certified igneous petrographer up front with the starting Kings. I came to know Deffeyes when I wondered how gold gets into mountains. I knew that most old-time hard-rock prospectors had little to go on but an association of gold with quartz. And I knew the erosional details of how gold comes out of mountains and into the rubble of streams. What I wanted to learn was what put the gold in the mountains in the first place. I asked a historical geologist and a geomorphologist. They both recommended Deffeyes. He explained that gold is not merely rare. It can be said to love itself. It is, with platinum, the noblest of the noble metals—those which resist combination with other elements. Gold wants to be free. In cool crust rock, it generally is free. At very high temperatures, however, it will go into compounds; and the gold that is among the magmatic fluids in certain pockets of interior earth may be combined, for example, with chlorine. Gold chloride is "modestly" soluble, and will dissolve in water that comes down and circulates in the magma. The water picks up many other elements, too: potassium, sodium, silicon. Heated, the solution rises into fissures in hard crust rock, where the cooling gold breaks away from the chlorine and—in specks, in flakes, in nuggets even larger than the eggs of geese—falls out of the water as metal. Silicon precipitates, too, filling up the fissures and enveloping the gold with veins of silicon dioxide, which is quartz.

When I asked Deffeyes what one might expect from a close inspection of roadcuts, he said they were windows into the world as it was in other times. We made plans to take samples of highway rock. I suggested going north up some new interstate to see what the blasting had disclosed. He said if you go north, in most places on this continent, the geology does not greatly vary. You should proceed in the direction of the continent itself. Go west. I had been thinking of a weekend trip to Whiteface Mountain, or something like it, but now, suddenly, a vaulting alternative came to mind. What about Interstate 80, I asked him. It goes the distance. How would it be? "Absorbing," he said. And he mused aloud: After 80 crosses the Border Fault, it pussyfoots along on morainal till that levelled up the fingers of the foldbelt hills. It does a similar dance with glacial debris in parts of Pennsylvania. It needs no assistance on the craton. It climbs a ramp to the Rockies and a fault-block staircase up the front of the Sierra. It is geologically shrewd. It was the route of animal migrations, and of human history that followed. It avoids melodrama, avoids the Grand Canyons, the Jackson Holes, the geologic operas of the country, but it would surely be a sound experience of the big picture, of the history, the construction, the components of the continent. And in all likelihood it would display in its roadcuts rock from every epoch and era.

In seasons that followed, I would go back and forth across the interstate like some sort of shuttle working out on a loom, accompanying geologists on purposes of their own or being accompanied by them from cut to cut and coast to coast. At any location on earth, as the rock record goes down into time and out into earlier geographies it touches upon tens of hundreds of stories, wherein the face of the earth often changed, changed utterly, and changed again, like the face of a crackling fire. The rock beside the road exposes one or two levels of the column of time and generally implies what went on immediately below and what occurred (or never occurred) above. To tell all the stories would be to tell pretty much the whole of geology in many volumes across a fifty-foot shelf, a task for which I am in every conceivable way unqualified. I am a layman who has travelled for a couple of years with a small core sampling of academic and government geologists ranging in experience from a graduate student to an authentic *éminence grise,* and what I intend to do now is to distill the trips of those years. I wish to make no attempt to speak for all geology or even to sweep in a great many facts that came along. I want to choose some things that interested me and through them to suggest the general history of the continent by describing events and landscapes that geologists see written in rocks.

To poke around in a preliminary way, Deffeyes and I went up to the Palisades Sill, where I was to return with Karen Kleinspehn, borrowed some diabase with a ten-pound sledge, and then began to travel westward, traversing the Hackensack Valley. It was morning. Small airplanes engorged with businessmen were settling into Teterboro. Deffeyes pointed out that if this were near the end of Wisconsinan time, when the ice was in retreat, those airplanes would have been settling down through several hundred feet of water, with the runway at the bottom of a lake. Glacial Lake Hackensack was the size of Lake Geneva and was host to many islands. It had the Palisades Sill for an eastern shore line, and on the west the lava hill that is now known as the First Watchung Mountain. The glacier had stopped at Perth Amboy, leaving its moraine there to block the foot of the lake, which the glacier fed with meltwater as it retreated to the north. Nearly two hundred million years earlier, the runway would have been laid out on a baking red flat beside the first, cooling Watchung—glowing from cracks, from lava fountains, but generally black as carbon. Basalt flows don't light up the sky. Three hundred million years before that, the airplanes would have been settling down toward the same site through water—in this instance, salt water—on the eastern shelf of a broad low continent, where an almost pure limestone was forming, because virtually nothing from the worn-away continent was

99

eroding into the shallow sea. Three random moments from the upper ninth of time.

In Paterson, I-80 chops the Watchung lava. Walking the cut from end to end, Deffeyes picked up some peripheral shale—Triassic red shale. He put it in his mouth and chewed it. "If it's gritty it's a silt bed, and if it's creamy it's a shale," he said. "This is creamy. Try it." I would not have thought to put it in coffee. In the blocky basaltic wall of the road, there were many small pockets, caves the size of peas, caves the size of lemons. As magma approaches the surface of the earth, it is so perfused with gases that it fizzes like ginger ale. In cooling basalt, gas bubbles remain, and form these minicaves. For a century and more, nothing much fills them. Slowly, though, over a minimum of about a million years, they can fill with zeolite crystals. Until well after the Second World War, not a whole lot was known about the potential uses of zeolite crystals. Nor was it known where they could be found in abundance. Deffeyes did important early work in the field. His doctoral dissertation, which dealt with two basins and two ranges in Nevada, included an appendix that started the zeolite industry. Certain zeolites (there are about thirty kinds) have become the predominant catalysts in use in oil refineries, doing a job that is otherwise assigned to platinum. Now, in Paterson, Deffeyes searched the roadcut vugs (as the minute caves are actually called) looking for zeolites. Some vugs were large enough to suggest the holes that lobsters hide in. They did indeed contain a number of white fibrous zeolite crystals—smooth and soapy, of a type that resembled talc or asbestos— but the cut had been almost entirely cleaned out by professional and amateur collectors, undeterred by the lethal traffic not many inches away. Nearly all the vugs were now as empty as they had been in their first hundred years. In the shale beyond the lava we saw the burrows of Triassic creatures. An ambulance from Totowa flew by with its siren wailing.

We moved on a few miles into the Great Piece Meadows of the Passaic River Valley, flat as a lake floor, poorly drained land. A meadow in New Jersey is any wet spongy acreage where you don't sink in above your chin. Great Piece Meadows, Troy Meadows, Black Meadows, the Great Swamp—Whippany, Parsippany, Madison, and Morristown are strewn among the reeds. The whole region, very evidently, was the bottom of a lake, for a lake itself is by definition a sign of poor drainage, an aneurysm in a river, a highly temporary feature on the land. Some lakes dry up. Others disappear after the outlet stream, cutting back into the outlet, empties the water. This one—Glacial Lake Passaic— vanished about ten thousand years ago after the retreating glacier exposed what is now the Passaic Valley. The lake drained gradually into the new Passaic River, which fell a hundred feet into Glacial Lake Hackensack, and, en route, went over a waterfall that would one day in effect found the city of Paterson by turning its first mill wheel. At the time of its greatest extent, Lake Passaic was two hundred feet deep, thirty miles long, and ten miles wide, and seems to have been a scene of great beauty. Its margins are still decorated with sandpits and offshore bars, wavecut cliffs and stream deltas, set in suburban towns. The lake's west shore was the worn-low escarpment of the Border Fault, and its most arresting feature was a hook-shaped basaltic peninsula that is now known to geologists as a part of the Third Watchung Lava Flow and to the people of New Jersey as Hook Mountain.

Deffeyes became excited as we approached Hook Mountain. The interstate had blasted into one toe of the former peninsula, exposing its interior to view. Deffeyes said, "Maybe someone will have left some zeolites here. I want them so bad I can taste them." He jumped the curb with his high-slung Geology Department vehicle, got out his hammers, and walked the cut. It was steep and competent, with brown oxides of iron over the felt-textured black basalt, and in it were tens of thousands of tiny vugs, a high percentage of them filled with pearl-lustred crystals of zeolite. To take a close look, he opened his hand lens—a small-diameter, ten-power Hastings triplet. "You can do a nice act in a jewelry store," he suggested. "You whip this thing out and you say the price is too high. These are beautiful crystals. Beautiful crystals imply slow growth. You don't get in a hurry and make something that nice." He picked up the sledge and pounded the cut, necessarily smashing many crystals as he broke their matrix free. "These crystals are like Vietnamese villages," he went on. "You have to destroy them in order to preserve them. They contain aluminum, silicon, calcium, sodium, and an incredible amount of imprisoned water. 'Zeolite' means 'the stone that boils.' If you take one small zeolite crystal, of scarcely more than a pinhead's diameter, and heat it until the water has come out, the crystal will have an internal surface area equivalent to a bedspread. Zeolites are often used to separate one kind of molecule from another. They can, for example, sort out molecules for detergents, choosing the ones that are biodegradable. They love water. In refrigerators, they are used to adsorb water that accidentally gets into the Freon. They could be used in automobile gas tanks to adsorb water. A zeolite called clinoptilolite is the strongest adsorber of strontium and cesium from radioactive wastes. The clinoptilolite will adsorb a great deal of lethal material, which you can then store in a small space. When William Wyler made *The Big Country,* there was a climactic chase scene in which the bad guy was shot and came clattering down a canyon wall in what appeared to be a shower of clinoptilolite. Geologists were on the phone to Wyler at once. 'Loved your movie. Where was that canyon?' There are a lot of zeolites in the Alps, in Nova Scotia, and in North Table Mountain in Colorado. When I was at the School of Mines, I used to go up to North Table Mountain just to wham around. Some of the best zeolites in the world are in this part of New Jersey."

There were oaks and maples on top of Hook Mountain, and, in the wall of the roadcut, basal rosettes of woolly mullein, growing in the rock. The Romans drenched stalks of mullein with suet and used them for funeral torches. American Indians taught the early pioneers to use the long flannel leaves of this plant as innersoles. Only three miles west of us was the Border Fault, where the basin had touched the range, where the stubby remnants of the fault scarp are now under glacial debris. Deffeyes said that the displacement along the fault—the eventual difference between two points that had been adjacent when the faulting began—exceeded fifteen thousand feet. Of course, this happened over several millions of years, and the mountains fronting the basin were all the while eroding, so they were never anything like fifteen thousand feet high. Generally, though, in the late Triassic, there would have been about a mile of difference, a mile of relief, between basin and range. In flash floods, boulders came raining off the mountains and piled in fans at the edge of the basin, ultimately to be filled in with sands and muds and to form conglomerate, New Jersey's so-called Hammer Creek Conglomerate—multicircled, polka-dotted headcheese rock, sometimes known as puddingstone. Here where the basin met the range, the sediments piled up so much that after all of the erosion of two hundred million years what remains is three miles thick. "I was in a bar once in Austin, Nevada," Deffeyes said, "and there was a sudden torrential downpour. The bartender began nailing plywood over the door. I wondered why he was doing that, until boulders came tumbling down the main street of the town. When you start pulling a continent apart, you have a lot of consequences of the same event. Faulting produced this basin. Sediments filled it in. Pull things apart and you produce a surface vacancy, which is faulting, and a subsurface vacancy, which causes upwelling of hot mantle that intrudes as sills or comes out as lava flows. In the Old Geology, you might have seen a sill within the country rock and said, 'Ah, the sill came much later.' With the New Geology, you see that all this was happening more or less at one time. The continent was splitting apart and the ultimate event was the opening of the Atlantic. If you look at the fold-belt in northwest Africa, you see the other side of the New Jersey story. The folding there is of the same age as the Appalachians, and the subsequent faulting is Triassic. Put the two continents together on a map and you will see what I mean. Fault blocks like this one are still in evidence, but discontinuously, from the Connecticut Valley to South Carolina. They are all parts of the suite that opened the Atlantic seaway. The story is very similar in the Great Basin—in the West, in the Basin and Range. The earth is splitting apart there, quite possibly opening a seaway. It is not something that happened a couple of hundred million years ago. It only began in the Miocene, and it is going on today. What we are looking at here in New Jersey is not just some little geologic feature, like a zeolite crystal.

This is the opening of the Atlantic. If you want to see happening right now what happened here two hundred million years ago, you can see it all in Nevada."

Basin. Fault. Range. Basin. Fault. Range. A mile of relief between basin and range. Stillwater Range. Pleasant Valley. Tobin Range. Jersey Valley. Sonoma Range. Pumpernickel Valley. Shoshone Range. Reese River Valley. Pequop Mountains. Steptoe Valley. Ondographic rhythms of the Basin and Range. We are maybe forty miles off the interstate, in the Pleasant Valley basin, looking up at the Tobin Range. At the nine-thousand-foot level, there is a stratum of cloud against the shoulders of the mountains, hanging like a ring of Saturn. The summit of Mt. Tobin stands clear, above the cloud. When we crossed the range, we came through a ranch on the ridgeline where sheep were fenced around a running brook and bales of hay were bright green. Junipers in the mountains were thickly hung with berries, and the air was unadulterated gin. This country from afar is synopsized and dismissed as "desert"—the home of the coyote and the pocket mouse, the side-blotched lizard and the vagrant shrew, the MX rocket and the pallid bat. There are minks and river otters in the Basin and Range. There are deer and antelope, porcupines and cougars, pelicans, cormorants, and common loons. There are Bonaparte's gulls and marbled godwits, American coots and Virginia rails. Pheasants. Grouse. Sandhill cranes. Ferruginous hawks and flammulated owls. Snow geese. This Nevada terrain is not corrugated, like the folded Appalachians, like a tubal air mattress, like a rippled potato chip. This is not—in that compressive manner—a ridge-and-valley situation. Each range here is like a warship standing on its own, and the Great Basin is an ocean of loose sediment with these mountain ranges standing in it as if they were members of a fleet without precedent, assembled at Guam to assault Japan. Some of the ranges are forty miles long, others a hundred, a hundred and fifty. They point generally north. The basins that separate them—ten and fifteen miles wide—will run on for fifty, a hundred, two hundred and fifty miles with lone, daisy-petalled windmills standing over sage and wild rye. Animals tend to be content with their home ranges and not to venture out across the big dry valleys. "Imagine a chipmunk hiking across one of these basins," Deffeyes remarks. "The faunas in the high ranges here are quite distinct from one to another. Animals are isolated like Darwin's finches in the Galápagos. These ranges are truly islands."

Supreme over all is silence. Discounting the cry of the occasional bird, the wailing of a pack of coyotes, silence—a great spatial silence—is pure in the Basin and Range. It is a soundless immensity with mountains in it. You stand, as we do now, and look up at a high mountain front, and turn your head and look fifty miles down the valley, and there is utter silence. It is the silence of the winter forests of the Yukon, here carried high to the ridgelines of the ranges. As the physicist Freeman Dyson has written in *Disturbing the Universe,* "It is a soul-shattering silence. You hold your breath and hear absolutely nothing. No rustling of leaves in the wind, no rumbling of distant traffic, no chatter of birds or insects or children. You are alone with God in that silence. There in the white flat silence I began for the first time to feel a slight sense of shame for what we were proposing to do. Did we really intend to invade this silence with our trucks and bulldozers and after a few years leave it a radioactive junkyard?"

What Deffeyes finds pleasant here in Pleasant Valley is the aromatic sage. Deffeyes grew up all over the West, his father a petroleum engineer, and he says without apparent irony that the smell of sagebrush is one of two odors that will unfailingly bring upon him an attack of nostalgia, the other being the scent of an oil refinery. Flash floods have caused boulders the size of human heads to come tumbling off the range. With alluvial materials of finer size, they have piled up in fans at the edge of the basin. ("The cloudburst is the dominant sculptor here.") The fans are unconsolidated. In time to come, they will pile up to such enormous thicknesses that they will sink deep and be heated and compressed to form conglomerate. Erosion, which provides the material to build the fans, is tearing down the mountains even as they rise. Mountains are not somehow created whole and subsequently worn away. They wear down as they come up, and these mountains have been rising and eroding in fairly even ratio for millions of years—rising and shedding sediment steadily through time, always the same, never the same, like row upon row of fountains. In the southern part of the province, in the Mojave, the ranges have stopped rising and are gradually wearing away. The Shadow Mountains. The Dead Mountains, Old Dad Mountains, Cowhole Mountains, Bullion, Mule, and Chocolate Mountains. They are inselberge now, buried ever deeper in their own waste. For the most part, though, the ranges are rising, and there can be no doubt of it here, hundreds of miles north of the Mojave, for we are looking at a new seismic scar that runs as far as we can see. It runs along the foot of the mountains, along the fault where the basin meets the range. From out in the valley, it looks like a long, buff-painted, essentially horizontal stripe. Up close, it is a gap in the vegetation, where plants growing side by side were suddenly separated by several metres, where, one October evening, the basin and the range—Pleasant Valley, Tobin Range—moved, all in an instant, apart. They jumped sixteen feet. The erosion rate at which the mountains were coming down was an inch a century. So in the mountains' contest with erosion they gained in one moment about twenty thousand years. These mountains do not rise like bread. They sit still for a long time and build up tension, and then suddenly jump. Passively, they are eroded for millennia, and they they jump again. They have been doing this for about eight million years. This fault, which jumped in 1915, opened like a zipper far up the valley, and, exploding into the silence, tore along the mountain base for upward of twenty miles with a sound that suggested a runaway locomotive.

"This is the sort of place where you really do not put a nuclear plant," says Deffeyes. "There was other action in the neighborhood at the same time—in the Stillwater Range, the Sonoma Range, Pumpernickel Valley. Actually, this is not a particularly spectacular scarp. The lesson is that the whole thing—the whole Basin and Range, or most of it—is alive. The earth is moving. The faults are moving. There are hot springs all over the province. There are young volcanic rocks. Fault scars everywhere. The world is splitting open and coming apart. You see a sudden break in the sage like this and it says to you that a fault is there and a fault block is coming up. This is a gorgeous, fresh, young, active fault scarp. It's growing. The range is lifting up. This Nevada topography is what you see *during* mountain building. There are no foothills. It is all too young. It is live country. This is the tectonic, active, spreading, mountain-building world. To a nongeologist, it's just ranges, ranges, ranges."

Most mountain ranges around the world are the result of compression, of segments of the earth's crust being brought together, bent, mashed, thrust and folded, squeezed up into the sky—the Himalaya, the Appalachians, the Alps, the Urals, the Andes. The ranges of the Basin and Range came up another way. The crust—in this region between the Rockies and the Sierra—is spreading out, being stretched, being thinned, being literally pulled to pieces. The sites of Reno and Salt Lake City, on opposite sides of the province, have moved apart fifty miles. The crust of the Great Basin has broken into blocks. The blocks are not, except for simplicity's sake, analogous to dominoes. They are irregular in shape. They more truly suggest stretch marks. Which they are. They trend north-south because the direction of the stretching is east-west. The breaks, or faults, between them are not vertical but dive into the earth at roughly sixty-degree angles, and this, from the outset, affected the centers of gravity of the great blocks in a way that caused them to tilt. Classically, the high edge of one touched the low edge of another and formed a kind of trough, or basin. The high edge—sculpted, eroded, serrated by weather—turned into mountains. The detritus of the mountains rolled into the basin. The basin filled with water—at first, it was fresh blue water—and accepted layer upon layer of sediment from the mountains, accumulating weight, and thus unbalancing the block even further. Its tilt became more pronounced. In the manner of a seesaw, the high, mountain side of the block went higher and the low, basin side went lower until the block as a whole reached a state of precarious and temporary truce with God, physics, and mechanical and chemical erosion, not to mention, far below, the agitated mantle, which was running a temperature hotter than normal, and was, almost surely, controlling the action. Basin and range. Integral fault blocks: low side the basin, high side the range. For five hundred miles they nudged one another across the province of the Basin and Range. With extra faulting, and whatnot, they took care of their own irregularities. Some had their high sides on the west, some on the east. The escarpment of the Wasatch Mountains—easternmost expression of this immense suite of mountains—faced west. The Sierra—the western-most, the highest, the predominant range, with Donner Pass only half-way up it—presented its escarpment to the east. As the developing Sierra made its skyward climb—as it went on up past ten and twelve and four-teen thousand feet—it became so predominant that it cut off the incom-ing Pacific rain, cast a rain shadow (as the phenomenon is called) over lush, warm, Floridian and verdant Nevada. Cut it off and kept it dry.

We move on (we're in a pickup) into dusk—north up Pleasant Valley, with its single telephone line on sticks too skinny to qualify as poles. The big flanking ranges are in alpenglow. Into the cold clear sky come the ranking stars. Jackrabbits appear, and crisscross the road. We pass the darkening shapes of cattle. An eerie trail of vapor traverses the basin, sent up by a clear, hot stream. It is only a couple of feet wide, but it is running swiftly and has multiple sets of hot white rapids. In the source springs, there is a thumping sound of boiling and rage. Beside the springs are lucid green pools, rimmed with accumulated travertine, like the travertine walls of Lincoln Center, the travertine pools of Havasu Canyon, but these pools are too hot to touch. Fall in there and you are Brunswick stew. "This is a direct result of the crustal spreading," Deffeyes says. "It brings hot mantle up near the surface. There is probably a frac-ture here, through which the water is coming up to this row of springs. The water is rich in dissolved minerals. Hot springs like these are the source of vein-type ore deposits. It's the same story that I told you about the hydrothermal transport of gold. When rainwater gets down into hot rock, it brings up what it happens to find there—silver, tungsten, copper, gold. An ore-deposit map and a hot-springs map will look much the same. Seis-mic waves move slowly through hot rock. The hotter the rock, the slower the waves. Nowhere in the continental United States do seismic waves move more slowly than they do beneath the Basin and Range. So we're not woofing when we say there's hot mantle down there. We've meas-ured the heat."

The basin-range fault blocks in a sense are float-ing on the mantle. In fact, the earth's crust everywhere in a sense is float-ing on the mantle. Add weight to the crust and it rides deeper, remove cargo and it rides higher, exactly like a vessel at a pier. Slowly disassem-ble the Rocky Mountains and carry the material in small fragments to the Mississippi Delta. The delta builds down. It presses ever deeper on the mantle. Its depth at the moment exceeds twenty-five thousand feet. The heat and the pressure are so great down there that the silt is turning into siltstone, the sand into sandstone, the mud into shale. For another example, the last Pleistocene ice sheet loaded two miles of ice onto Scotland, and that dunked Scotland in the mantle. After the ice melted, Scotland came up again, lifting its beaches high into the air. Isostatic adjustment. Let go a block of wood that you hold underwater and it adjusts itself to the surface isostatically. A frog sits on the wood. It goes down. He vomits. It goes up a little. He jumps. It adjusts. Wherever landscape is eroded away, what remains will rise in adjustment. Older rock is lifted to view. When, for whatever reason, crust becomes thicker, it adjusts downward. All of this—with the central image of the basin-range fault blocks floating in the mantle—may suggest that the mantle is molten, which it is not. The mantle is solid. Only in certain pockets near the surface does it turn into magma and squirt upward. The temperature of the mantle varies widely, as would the temperature of anything that is two thousand miles thick. Under the craton, it is described as chilled. By surface standards, though, it is generally white hot, everywhere around the world—white hot and solid but magisterially viscous, permitting the crust above it to "float." Deffeyes was in his bathtub one Saturday afternoon thinking about the viscosity of the mantle. Suddenly he stood up and reached for a towel. "Piano wire!" he said to himself, and he dressed quickly and went to the library to look up a book on piano tuning and to calculate the viscosity of the wire. Just what he guessed—10^{22} poises. Piano wire. Look under the hood of a well-tuned Steinway and you are looking at strings that could float a small continent. They are rigid, but ever so slowly they will sag, will slacken, will deform and give way, with the exact viscosity of the earth's mantle. "And that," says Deffeyes, "is what keeps the piano tuner in business." More miles, and there appears ahead of us something like a Christmas tree alone in the night. It is Winnemucca, there being no other possibility. Neon looks good in Nevada. The tawdriness is refined out of it in so much wide black space. We drive on and on toward the glow of colors. It is still far away and it has not increased in size. We pass nothing. Deffeyes says, "On these roads, it's ten to the minus five that anyone will come along." The better part of an hour later, we come to the beginnings of the casino-flashing town. The news this year is that dollar slot machines are outdrawing nickel slot machines for the first time, ever.

II

OTHER NATIONS

"We need another and a wiser and perhaps a more mystical concept of animals. . . . In a world older and more complete than ours they move finished and complete, gifted with extensions of the senses we have lost or never attained, living by voices we shall never hear. They are not brethren, they are not underlings; they are other nations, caught with ourselves in the net of life and time, fellow prisoners of the splendour and travail of the earth."

HENRY BESTON, *THE OUTERMOST HOUSE,* 1928

EDWARD
HOAGLAND

Photograph © Thomas Victor

Edward Hoagland divides his time between New York City, where he was born in 1932, and a small farmhouse without electricity in Barton, Vermont. His essays come from both worlds, from the city and from the woods, with asides along the way into circuses, turtles, British Columbia, literary criticism, the Sudan, and anything and anyplace else that happens to catch his interest.

Hoagland's first novel, *Cat Man* (1956), earned him a Houghton Mifflin Literary Fellowship as he was graduating from Harvard. He has continued to write novels, most recently, *Seven Rivers West* (1986). But he has said that it was his divorce from his first wife in 1964, which left him "floundering," that turned him toward journal-writing and essays, now his

110

primary voice. His essays are collected in *The Courage of Turtles* (1971); *Walking the Dead Diamond River* (1973), from which this piece comes; *Red Wolves and Black Bears* (1976); and *The Tugman's Passage* (1982). Landscape dominates his two books of travel, as well: *Notes from the Century Before: A Journal from British Columbia* (1969) and *African Calliope: A Journey to the Sudan* (1979).

He writes about the vanishing grizzly bears and red wolves, not so much to save them as to note that their existence matters: "It's as though the last bit of ocean were about to become more dry land, planted and paved. The loss would be not to us who have already sailed it, who have no wish to be seamen, and who can always go back and relive in our minds what we've experienced. The loss is to people unborn who might have turned into seamen, or who might have seen it and loved it as we, alive now and not seamen, have seen it and loved it."

Hoagland has written of the "vulnerable" essay, "Fiction can be hallucinatory if it wishes, and journalism impassive, and so each continues through thick and thin, but essays presuppose a certain standard of education in the reader, a world ruled by some sort of order . . . where people seek not fragmentation but a common bond." He gives order to his own world in written words. Hoagland stutters, and he says, "being in these vocal handcuffs made me a devoted writer at twenty. I worked like a dog, choosing each word."

Although Hoagland clearly values his choice to live in two places as well as his freedom to write about most anything, still he wonders, ". . . where is *home*? Is home going to be only our hats? Can we possibly function like that? Marriage, death and fear and joy under our hats? That's where the question really rests, for architects and pharmacists and you and me."

111

Hailing the Elusory
Mountain Lion

The swan song sounded by the wilderness grows fainter, ever more constricted, until only sharp ears can catch it at all. It fades to a nearly inaudible level, and yet there never is going to be any one time when we can say right *now* it is gone. Wolves meet their maker in wholesale lots, but coyotes infiltrate eastward, northward, southeastward. Woodland caribou and bighorn sheep are vanishing fast, but moose have expanded their range in some areas.

Mountain lions used to have practically the run of the Western Hemisphere, and they still do occur from Cape Horn to the Big Muddy River at the boundary of the Yukon and on the coasts of both oceans, so that they are the most versatile land mammal in the New World, probably taking in more latitudes than any other four-footed wild creature anywhere. There are perhaps only four to six thousand left in the United States, though there is no place that they didn't once go, eating deer, elk, pikas, porcupines, grasshoppers, and dead fish on the beach. They were called mountain lions in the Rockies, pumas (originally an Incan word) in the Southwestern states, cougars (a naturalist's corruption of an Amazonian Indian word) in the Northwest, panthers in the traditionalist East—"painters" in dialect-proud New England—or catamounts. The Dutchmen of New Netherland called them tigers, red tigers, deer tigers, and the Spaniards *leones* or *leopardos.* They liked to eat horses—wolves preferred beef and black bears favored pork—but as adversaries of mankind they were overshadowed at first because bears appeared more formidable and wolves in their howling packs were more flamboyant and more damaging financially. Yet this panoply of names is itself quite a tribute, and somehow the legends about "panthers" have lingered longer than bear or wolf tales, helped by the animal's own limber, far-traveling stealth and as a carry-over from the immense mythic force of the great cats of the Old World. Though only Florida among the Eastern states is known for certain to have any left, no wild knot of mountains or swamp is without rumors of panthers; nowadays people delight in these, keeping their eyes peeled. It's wishful, and the wandering, secretive nature of the beast ensures that even Eastern panthers will not soon be certifiably extinct. An informal census among experts in 1963 indicated that an island of twenty-five or more may have survived in the New Brunswick-Maine-Quebec region, and Louisiana may still have a handful, and perhaps eight live isolated in the Black Hills of South Dakota, and the Oklahoma panhandle may have a small colony—all outside the established range in Florida, Texas, and the Far West. As with the blue whale, who will be able to say when they have been eliminated?

"Mexican lion" is another name for mountain lions in the border states—a name that might imply a meager second-best rating there yet ties to the majestic African beasts. Lions are at least twice as big as mountain lions, measuring by weight, though they are nearly the same in length because of the mountain lion's superb long tail. Both animals sometimes pair up affectionately with mates and hunt in tandem, but mountain lions go winding through life in ones or twos, whereas the lion is a harem-keeper, harem-dweller, the males eventually becoming stay-at-homes, heavy figureheads. Lions enjoy the grassy flatlands, forested along the streams, and they stay put, engrossed in communal events—roaring, grunting, growling with a racket like the noise of gears being stripped—unless the game moves on. They sun themselves, preside over the numerous kibbutz young, sneeze from the dust, and bask in dreams, occasionally waking up to issue reverberating, guttural pronouncements which serve notice that they are now awake.

Mountain lions spirit themselves away in saw-toothed canyons and on escarpments instead, and when conversing with their mates they coo like pigeons, sob like women, emit a flat slight shriek, a popping bubbling growl, or mew, or yowl. They growl and suddenly caterwaul into falsetto—the famous scarifying, metallic scream functioning as a kind of hunting cry close up, to terrorize and start the game. They ramble as much as twenty-five miles in a night, maintaining a large loop of territory which they cover every week or two. It's a solitary, busy life, involving a survey of several valleys, many deer herds. Like tigers and leopards, mountain lions are not sociably inclined and don't converse at length with the whole waiting world, but they are even less noisy; they seem to speak most eloquently with their feet. Where a tiger would roar, a mountain lion screams like a castrato. Where a mountain lion hisses, a leopard would snarl like a truck stuck in snow.

Leopards are the best counterpart to mountain lions in physique and in the tenor of their lives. Supple, fierce creatures, skilled at concealment but with great self-assurance and drive, leopards are bolder when facing human beings than the American cats. Basically they are hot-land beasts and not such remarkable travelers individually, though as a race they once inhabited the broad Eurasian land mass all the way from Great Britain to Malaysia, as well as Africa. As late as the 1960s, a few were said to be still holding out on the shore of the Mediterranean at Mount Mycale, Turkey. (During a forest fire twenty years ago a yearling swam the narrow straits to the Greek island Samos and holed up in a cave, where he was duly killed—perhaps the last leopard ever to set foot in Europe on his own.) Leopards are thicker and shorter than adult mountain lions and seem to lead an athlete's indolent, incurious life much of the time, testing their perfected bodies by clawing tree trunks, chewing on old skulls, executing acrobatic leaps, and then rousing themselves to the semiweekly antelope kill. Built with supreme hardness and economy, they make little allowance for man—they don't see him as different. They relish the flesh of his dogs, and they run up a tree when hunted and then sometimes spring down, as heavy as a chunk of iron wrapped in a flag. With stunning, gorgeous coats, their tight, dervish faces carved in a snarl, they head for the hereafter as if it were just one more extra-emphatic leap—as impersonal in death as the crack of the rifle was.

The American leopard, the jaguar, is a powerfully built, serious fellow, who, before white men arrived, wandered as far north as the Carolinas, but his best home is the humid basin of the Amazon. Mountain lions penetrate these ultimate jungles too, but rather thinly, thriving better in the cooler, drier climate of the untenanted pampas and on the mountain slopes. They are blessed with a pleasant but undazzling coat, tan except for a white belly, mouth and throat, and some black behind the ears, on the tip of the tail and at the sides of the nose, and so they are hunted as symbols, not for their fur. The cubs are spotted, leopardlike, much as lion cubs are. If all of the big cats developed from a common ancestry, the mountain lion's specialization has been unpresumptuous—away from bulk and savagery to traveling light. Toward deer, their prey, they may be as ferocious as leopards, but not toward chance acquaintances such as man. They sometimes break their necks, their jaws, their teeth, springing against the necks of quarry they have crept close to—a fate in part resulting from the circumstance that they can't ferret out the weaker individuals in a herd by the device of a long chase, the way wolves do; they have to take the luck of the draw. None of the cats possess enough lung capacity for grueling runs. They depend upon shock tactics, bursts of speed, sledge-hammer leaps, strong collarbones for hitting power, and shearing dentition, whereas wolves employ all the advantages of time in killing their quarry, as well as the numbers and gaiety of the pack, biting the beast's nose and rump—the technique of a thousand cuts—lapping the bloody snow. Wolves sometimes even have a cheering section of flapping ravens accompanying them, eager to scavenge after the brawl.

It's risky business for the mountain lion, staking the strength and impact of his neck against the strength of the prey animal's neck. Necessarily, he is concentrated and fierce; yet legends exist that mountain lions have irritably defended men and women lost in the wilderness against marauding jaguars, who are no friends of theirs, and (with a good deal more supporting evidence) that they are susceptible to an odd kind of fascination with human beings. Sometimes they will tentatively seek an association, hanging about a campground or following a hiker out of curiosity, perhaps, circling around and bounding up on a ledge above to watch him pass. This mild modesty has helped preserve them from extinction. If they have been unable to make any adjustments to the advent of man, they haven't suicidally opposed him either, as the buffalo wolves and grizzlies did. In fact, at close quarters they seem bewildered. When treed, they don't breathe a hundred-proof ferocity but puzzle over what to do. They're too light-bodied to bear down on the hunter and kill him easily, even if they should attack—a course they seem to have no inclination for. In this century in the United States only one person, a child of thirteen, has been killed by a mountain lion; that was in 1924. And they're informal animals. Lolling in an informal sprawl on a high limb, they can't seem to summon any Enobarbas-like front of resistance for long. Daring men occasionally climb up and toss lassos about a cat and haul him down, strangling him by pulling from two directions, while the lion, mortified, appalled, never does muster his fighting aplomb. Although he could fight off a pack of wolves, he hasn't worked out a posture to assume toward man and his dogs. Impotently, he stiffens, as the dinosaurs must have when the atmosphere grew cold.

Someday hunting big game may come to be regarded as a form of vandalism, and the remaining big creatures of the wilderness will skulk through restricted reserves wearing radio transmitters and numbered collars, or bearing stripes of dye, as many elephants already do, to aid the busy biologists who track them from the air. Like a vanishing race of trolls, more report and memory than a reality, they will inhabit children's books and nostalgic articles, a special glamour attaching to those, like mountain lions, that are geographically incalculable and may still be sighted away from the preserves. Already we've become enthusiasts. We want game about us—at least at a summer house; it's part of privileged living. There is a precious privacy about seeing wildlife, too. Like meeting a fantastically dressed mute on the road, the fact that no words are exchanged and that *he's* not going to give an account makes the experience light-hearted; it's wholly ours. Besides, if anything out of the ordinary happened, we know we can't expect to be believed, and since it's rather fun to be disbelieved—fishermen know this—the privacy is even more complete. Deer, otter, foxes are messengers from another condition of life, another mentality, and bring us tidings of places where we don't go.

Ten years ago at Vavenby, a sawmill town on the North Thompson River in British Columbia, a frolicsome mountain

lion used to appear at dusk every ten days or so in a bluegrass field alongside the river. Deer congregated there, the river was silky and swift, cooling the summer air, and it was a festive spot for a lion to be. She was
thought to be a female, and reputedly left tracks around an enormous territory to the north and east—Raft Mountain, Battle Mountain, the Trophy
Range, the Murtle River, and Mahood Lake—territory on an upended,
pelagic scale, much of it scarcely accessible to a man by trail, where the
tiger lilies grew four feet tall. She would materialize in this field among
the deer five minutes before dark, as if checking in again, a habit that may
have resulted in her death eventually, though for the present the farmer
who observed her visits was keeping his mouth shut about it. This was
pioneer country; there were people alive who could remember the time
when poisoning the carcass of a cow would net a man a pile of dead
predators—a family of mountain lions to bounty, maybe half a dozen
wolves, and both black bears and grizzlies. The Indians considered lion
meat a delicacy, but they had clans which drew their origins at the Creation from ancestral mountain lions, or wolves or bears, so these massacres
amazed them. They thought the outright bounty hunters were crazy men.

Even before Columbus, mountain lions were
probably not distributed in saturation numbers anywhere, as wolves may
have been. Except for the family unit—a female with her half-grown
cubs—each lion seems to occupy its own spread of territory, not as a result
of fights with intruders but because the young transients share the same
instinct for solitude and soon sheer off to find vacant mountains and valleys. A mature lion kills only one deer every week or two, according to
a study by Maurice Hornocker in Idaho, and therefore is not really a notable factor in controlling the local deer population. Rather, it keeps watch
contentedly as that population grows, sometimes benefitting the herds
by scaring them onto new wintering grounds that are not overbrowsed,
and by its very presence warding off other lions.

This thin distribution, coupled with the moun-
tain lion's taciturn habits, make sighting one a matter of luck, even for
game officials located in likely country. One warden in Colorado I talked
to had indeed seen a pair of them fraternizing during the breeding season. He was driving a jeep over an abandoned mining road, and he passed
two brown animals sitting peaceably in the grass, their heads close together.
For a moment he thought they were coyotes and kept driving, when all
of a sudden the picture registered that they were *cougars!* He braked and
backed up, but of course they were gone. He was an old-timer, a man
who had crawled inside bear dens to pull out the cubs, and knew where
to find clusters of buffalo skulls in the recesses of the Rockies where the
last bands had hidden; yet this cryptic instant when he was turning his
jeep round a curve was the only glimpse—unprovable—that he ever got
of a mountain lion.

Such glimpses usually are cryptic. During a summer I spent in Wyoming in my boyhood, I managed to see two coyotes, but both occasions were so fleeting that it required an act of faith on my part afterward to feel sure I had seen them. One of the animals vanished between rolls of ground; the other, in rougher, stonier, wooded country, cast his startled gray face in my direction and simply was gone. Hunching, he swerved for cover, and the brush closed over him. I used to climb to a vantage point above a high basin at twilight and watch the mule deer steal into the meadows to feed. The grass grew higher than their stomachs, the steep forest was close at hand, and they were as small and fragile-looking as filaments at that distance, quite human in coloring, gait and form. It was possible to visualize them as a naked Indian hunting party a hundred years before—or not to believe in their existence at all, either as Indians or deer. Minute, aphid-sized, they stepped so carefully in emerging, hundreds of feet below, that, straining my eyes, I needed to tell myself constantly that they were deer; my imagination, left to its own devices with the dusk settling down, would have made of them a dozen other creatures.

Recently, walking at night on the woods road that passes my house in Vermont, I heard footsteps in the leaves and windfalls. I waited, listening—they sounded too heavy to be anything less than a man, a large deer or a bear. A man wouldn't have been in the woods so late, my dog stood respectfully silent and still, and they did seem to shuffle portentously. Sure enough, after pausing at the edge of the road, a fully grown bear appeared, visible only in dimmest outline, staring in my direction for four or five seconds. The darkness lent a faintly red tinge to his coat; he was well built. Then, turning, he ambled off, almost immediately lost to view, though I heard the noise of his passage, interrupted by several pauses. It was all as concise as a vision, and since I had wanted to see a bear close to my own house, being a person who likes to live in a melting pot, whether in the city or country, and since it was too dark to pick out his tracks, I was grateful when the dog inquisitively urinated along the bear's path, thereby confirming that at least I had witnessed *something*. The dog seemed unsurprised, however, as if the scent were not all that remarkable, and, sure enough, the next week in the car I encountered a yearling bear in daylight two miles downhill, and a cub a month later. My farmer neighbors were politely skeptical of my accounts, having themselves caught sight of only perhaps a couple of bears in all their lives.

So it's with sympathy as well as an awareness of the tricks that enthusiasm and nightfall may play that I have been going to nearby towns seeking out people who have claimed at one time or another to have seen a mountain lion. The experts of the state—game wardens, taxidermists, and most accomplished hunters—emphatically discount the claims, but the believers are unshaken. They include some

117

summer people who were enjoying a drink on the back terrace when the apparition of a great-tailed cat moved out along the fringe of the woods on a deer path; a boy who was hunting with his .22 years ago near the village dump and saw the animal across a gully and fired blindly, then ran away and brought back a search party, which found a tuft of toast-colored fur; and a state forestry employee, a sober woodsman, who caught the cat in his headlights while driving through Victory Bog in the wildest corner of the Northeast Kingdom. Gordon Hickok, who works for a furniture factory and has shot one or two mountain lions on hunting trips in the West, saw one cross U.S. 5 at a place called Auger Hole near Mount Hor. He tracked it with dogs a short distance, finding a fawn with its head gnawed off. A high-school English teacher reported seeing a mountain lion cross another road, near Runaway Pond, but the hunters who quickly went out decided that the prints were those of a big bobcat, splayed impressively in the mud and snow. Fifteen years ago a watchman in the fire tower on top of Bald Mountain had left grain scattered in the grooves of a flat rock under the tower to feed several deer. One night, looking down just as the dusk turned murky, he saw two slim long-tailed lions creep out of the border of spruce and inspect the rock, sniffing deer droppings and dried deer saliva. The next night, when he was in his cabin, the dog barked and, looking out the window, again he saw the vague shape of a lion just vanishing.

A dozen loggers and woodsmen told me such stories. In the Adirondacks I've also heard some persuasive avowals—one by an old dog-sled driver and trapper, a French Canadian; another by the owner of a tourist zoo, who was exhibiting a Western cougar. In Vermont perhaps the most eager rumor buffs are some of the farmers. After all, now that packaged semen has replaced the awesome farm bull and so many procedures have been mechanized, who wants to lose *all* the adventure of farming? Until recently the last mountain lion known to have been killed in the Northeast was recorded in 1881 in Barnard, Vermont. However, it has been learned that probably another one was shot from a tree in 1931 in Mundleville, New Brunswick, and still another trapped seven years later in Somerset County in Maine. Bruce S. Wright, director of the Northeastern Wildlife Station (which is operated at the University of New Brunswick with international funding), is convinced that though they are exceedingly rare, mountain lions are still part of the fauna of the region; in fact, he has plaster casts of tracks to prove it, as well as a compilation of hundreds of reported sightings. Some people may have mistaken a golden retriever for a lion, or may have intended to foment a hoax, but all in all the evidence does seem promising. Indeed, after almost twenty years of search and study, Wright himself finally saw one.

The way these sightings crop up in groups has often been pooh-poohed as greenhorn fare or as a sympathetic hysteria among neighbors, but it is just as easily explained by the habit mountain lions have of establishing a territory that they scout through at intervals, visiting an auspicious deer-ridden swamp or remote ledgy mountain. Even at such a site a successful hunt could not be mounted without trained dogs, and if the population of the big cats was extremely sparse, requiring of them long journeys during the mating season, and yet with plenty of deer all over, they might not stay for long. One or two hundred miles is no obstacle to a Western cougar. The cat might inhabit a mountain ridge one year, and then never again.

Fifteen years ago, Francis Perry, who is an ebullient muffin of a man, a farmer all his life in Brownington, Vermont, saw a mountain lion "larger and taller than a collie, and grayish yellow" (he had seen them in circuses). Having set a trap for a woodchuck, he was on his way to visit the spot when he came over a rise and, at a distance of fifty yards, saw the beast engaged in eating the dead woodchuck. It bounded off, but Perry set four light fox traps for it around the wood-chuck. Apparently, a night or two later the cat returned and got caught in three of these, but they couldn't hold it; it pulled free, leaving the marks of a struggle. Noel Perry, his brother, remembers how scared Francis looked when he came home from the first episode. Noel himself saw the cat (which may have meant that Brownington Swamp was one of its haunts that summer), once when it crossed a cow pasture on another farm the brothers owned, and once when it fled past his rabbit dogs through under-brush while he was training them—he thought for a second that its big streaking form was one of the dogs. A neighbor, Robert Chase, also saw the animal that year. Then again last summer, for the first time in fifteen years, Noel Perry saw a track as big as a bear's but round like a mountain lion's, and Robert's brother, Larry Chase, saw the actual cat several times one summer evening, playing a chummy hide-and-seek with him in the fields.

Elmer and Elizabeth Ambler are in their forties, populists politically, and have bought a farm in Glover to live the good life, though he is a truck driver in Massachusetts on weekdays and must drive hard in order to be home when he can. He's bald, with large eye-brows, handsome teeth and a low forehead, but altogether a strong-looking, clear, humane face. He is an informational kind of man who will give you the history of various breeds of cattle or a talk about taxation in a slow and musical voice, and both he and his wife, a purposeful, self-sufficient redhead, are fascinated by the possibility that they live in wilder-ness. Beavers inhabit the river that flows past their house. The Amblers say that on Black Mountain nearby hunters "disappear" from time to time, and bears frequent the berry patches in their back field—they see them, their visitors see them, people on the road see them, their German shepherds meet them and run back drooling with fright. They've stocked

119

their farm with horned Herefords instead of the polled variety so that the creatures can "defend themselves." Ambler is intrigued by the thought that apart from the danger of bears, someday "a cat" might prey on one of his cows. Last year, looking out the back window, his wife saw through binoculars an animal with a flowing tail and "a cat's gallop" following a line of trees where the deer go, several hundred yards uphill behind the house. Later, Ambler went up on snowshoes and found tracks as big as their shepherds'; the dogs obligingly ran alongside. He saw walking tracks, leaping tracks and deer tracks marked with blood going toward higher ground. He wonders whether the cat will ever attack him. There are plenty of bobcats around, but they both say they know the difference. The splendid, nervous *tail* is what people must have identified in order to claim they have seen a mountain lion.

I, too, cherish the notion that I may have seen a lion. Mine was crouched on an overlook above a grass-grown, steeply pitched wash in the Alberta Rockies—a much more likely setting than anywhere in New England. It was late afternoon on my last day at Maligne Lake, where I had been staying with my father at a national-park chalet. I was twenty; I could walk forever or could climb endlessly in a sanguine scramble, going out every day as far as my legs carried me, swinging around for home before the sun went down. Earlier, in the valley of the Athabasca, I had found several winter-starved or wolf-killed deer, well picked and scattered, and an area with many elk antlers strewn on the ground where the herds had wintered safely, dropping their antlers but not their bones. Here, much higher up, in the bright plenitude of the summer, I had watched two wolves and a stately bull moose in one mountain basin, and had been up on the caribou barrens on the ridge west of the lake and brought back the talons of a hawk I'd found dead on the ground. Whenever I was watching game, a sort of stopwatch in me started running. These were moments of intense importance and intimacy, of new intimations and aptitudes. Time had a jam-packed character, as it does during a mile run.

I was good at moving quietly through the woods and at spotting game, and was appropriately exuberant. The finest, longest day of my stay was the last. Going east, climbing through a luxuriant terrain of up-and-down boulders, brief brilliant glades, sudden potholes fifty feet deep—a forest of moss-hung lodgepole pines and firs and spare, gaunt spruce with the black lower branches broken off—I came upon the remains of a young bear, which had been torn up and shredded. Perhaps wolves had cornered it during some imprudent excursion in the early spring. (Bears often wake up while the snow is still deep, dig themselves out and rummage around in the neighborhood sleepily for a day or two before bedding down again under a fallen tree.) I took the skull along so that I could extract the teeth when I got hold of some tools. Discoveries like this represent a superfluity of wildlife and show how many beasts there are scouting about.

I went higher. The marmots whistled familially; the tall trees wilted to stubs of themselves. A pretty stream led down a defile from a series of openings in front of the ultimate barrier of a vast mountain wall which I had been looking at from a distance each day on my outings. It wasn't too steep to be climbed, but it was a barrier because my energies were not sufficient to scale it and bring me back the same night. Besides, it stretched so majestically, surflike above the lesser ridges, that I liked to think of it as the Continental Divide.

On my left as I went up this wash was an abrupt, grassy slope that enjoyed a southern exposure and was sunny and wind-blown all winter, which kept it fairly free of snow. The ranger at the lake had told me it served as a wintering ground for a few bighorn sheep and for a band of mountain goats, three of which were in sight. As I approached laboriously, these white, pointy-horned fellows drifted up over a rise, managing to combine their retreat with some nippy good grazing as they went, not to give any pursuer the impression that they had been pushed into flight. I took my time too, climbing to locate the spring in a precipitous cleft of rock where the band did most of its drinking, and finding the shallow, high-ceilinged cave where the goats had sheltered from storms, presumably for generations. The floor was layered with rubbery droppings, tramped down and sprinkled with tufts of shed fur, and the back wall was checkered with footholds where the goats liked to clamber and perch. Here and there was a horn lying loose—a memento for me to add to my collection from an old individual that had died a natural death, secure in the band's winter stronghold. A bold, thriving family of pack rats emerged to observe me. They lived mainly on the nutritives in the droppings, and were used to the goats' tolerance; they seemed astonished when I tossed a stone.

I kept scrabbling along the side of the slope to a section of outcroppings where the going was harder. After perhaps half an hour, crawling around a corner, I found myself faced with a bighorn ram who was taking his ease on several square yards of bare earth between large rocks, a little above the level of my head. Just as surprised as I, he stood up. He must have construed the sounds of my advance to be those of another sheep or goat. His horns had made a complete curl and then some; they were thick, massive and bunched together like a high Roman helmet, and he himself was muscly and military, with a grave-looking nose. A squared-off, middle-aged, trophy-type ram, full of imposing professionalism, he was at the stage of life when rams sometimes stop herding and live as rogues.

He turned and tried a couple of possible exits from the pocket where I had found him, but the ground was badly pitched and would require a reeling gait and loss of dignity. Since we were within a national park and obviously I was unarmed, he simply was not inclined to put himself to so much trouble. He stood fifteen or twenty feet above me pushing his tongue out through his teeth, shaking his head slightly and dipping it into charging position as I moved closer by a step or two, raising my hand slowly toward him in what I proposed as a friendly greeting. The day had been a banner one since the beginning, so while I recognized immediately that this meeting would be a valued memory, I felt as natural in his company as if he were a friend of mine reincarnated in a shag suit. I saw also that he was going to knock me for a loop, head over heels down the steep slope, if I sidled nearer, because he did not by any means feel as expansive and exuberant at our encounter as I did. That was the chief difference between us. I was talking to him with easy gladness, and beaming; he was not. He was unsettled and on his mettle, waiting for me to move along, the way a bighorn sheep waits for a predator to move on in wildlife movies when each would be evenly matched in a contest of strength and position. Although his warlike nose and high bone helmet, blocky and beautiful as weaponry, kept me from giving in to my sense that we were brothers, I knew I could stand there for a long while. His coat was a down-to-earth brown, edgy with muscle, his head was that of an unsmiling veteran standing to arms, and despite my reluctance to treat him as some sort of boxed-in prize, I might have stayed on for half the afternoon if I hadn't realized that I had other sights to see. It was not a day to dawdle.

I trudged up the wash and continued until, past tree line, the terrain widened and flattened in front of the preliminary ridge that formed an obstacle before the great roaring, silent, surflike mountain wall that I liked to think of as the Continental Divide, although it wasn't. A cirque separated the preliminary ridge from the ultimate divide, which I still hoped to climb to and look over. The opening into this was roomy enough, except for being littered with enormous boulders, and I began trying to make my way across them. Each was boat-sized and rested upon underboulders; it was like running in place. After tussling with this landscape for an hour or two, I was limp and sweating, pinching my cramped legs. The sun had gone so low that I knew I would be finding my way home by moonlight in any case, and I could see into the cirque, which was big and symmetrical and presented a view of sheer barbarism; everywhere were these cruel boat-sized boulders.

Giving up and descending to the goats' draw again, I had a drink from the stream and bathed before climbing farther downward. The grass was green, sweet-smelling, and I felt safely close to life after that sea of dead boulders. I knew I would never be physically younger or in finer country; even then the wilderness was singing its swan song. I had no other challenges in mind, and though very tired, I liked looking up at the routes where I'd climbed. The trio of goats had not returned, but I could see their wintering cave and the cleft in the rocks where the spring was. Curiously the bighorn ram had not left; he had only withdrawn upward, shifting away from the outcroppings to an open sweep of space where every avenue of escape was available. He was lying on a carpet of grass and, lonely pirate that he was, had his head turned in my direction.

It was from this same wash that looking up, I spotted the animal I took to be a mountain lion. He was skulking among some outcroppings at a point lower on the mountainside than the ledges where the ram originally had been. A pair of hawks or eagles were swooping at him by turns, as if he were close to a nest. The slant between us was steep, but the light of evening was still more than adequate. I did not really see the wonderful tail—that special medallion—nor was he particularly big for a lion. He was gloriously catlike and slinky, however, and so indifferent to the swooping birds as to seem oblivious of them. There are plenty of creatures he wasn't: he wasn't a marmot, a goat or other grass-eater, a badger, a wolf or coyote or fisher. He *may* have been a big bobcat or a wolverine, although he looked ideally lion-colored. He had a cat's strong collarbone structure for hitting, powerful haunches for vaulting, and the almost mystically small head mountain lions possess, with the gooseberry eyes. Anyway, I believed him to be a mountain lion, and standing quietly I watched him as he inspected in leisurely fashion the ledge that he was on and the one under him savory with every trace of goat— frosty-colored with the white hairs they'd shed. The sight was so dramatic that it seemed to be happening close to me, though in fact he and the hawks or eagles, whatever they were, were miniaturized by distance.

If I'd kept motionless, eventually I could have seen whether he had the proper tail, but such scientific question had no weight next to my need to essay some kind of communication with him. It had been exactly the same when I'd watched the two wolves playing together a couple of days before. They were above me, absorbed in their game of noses-and-paws. I had recognized that I might never witness such a scene again, yet I couldn't hold myself in. Instead of talking and raising my arm to them, as I had with the ram, I'd shuffled forward impetuously as if to say *Here I am!* Now, with the lion, I tried hard to dampen my impulse and restrain myself as long as I could. Then I stepped toward him, just barely squelching a cry in my throat but lifting my hand—as clumsy as anyone is who is trying to attract attention.

At that, of course, he swerved aside instantly and was gone. Even the two birds vanished. Foolish, triumphant and disappointed, I hiked on down into the lower forests, gargantuanly tangled, another life zone—not one which would exclude a lion but one where he would not be seen. I'd got my second wind and walked lightly and softly, letting the silvery darkness settle around me. The blowdowns were as black as whales; my feet sank in the moss. Clearly this was as crowded a day as I would ever have, and I knew my real problem would not be to make myself believed but rather to make myself understood at all, simply in reporting the story, and that I must at least keep the memory straight for myself. I was so happy that I was unerring in distinguishing the deer trails going my way. The forest's night beauty was supreme in its promise, and I didn't hurry.

JOHN
MADSON

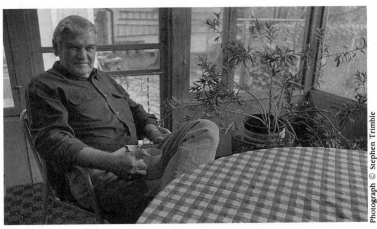

John Madson was trained in wildlife biology at Iowa State University, and did graduate work in fisheries biology, but he has been working as a writer ever since. Born in 1923, he grew up along a little prairie river in Iowa—the south fork of the Skunk—hunting rabbits for a quarter apiece and tramping the prairie.

After World War II, Madson edited the Iowa Conservation Commission magazine, wrote features for the Des Moines *Register*, and subsequently spent twenty years as half of the Conservation Department of the Winchester Western division of the Olin Corporation. He and his partner had only one responsibility: "to promote professional game management and wildlife biology." Madson did this by writing books about most of the major American game species.

In 1979, when his free-lancing began to demand more time than his job would allow, he quit to write full-time on his own. His books include *Stories From Under the Sky* (1961), *Out Home* (1979), *Where the Sky*

126

Began: Land of the Tallgrass Prairie (1982), and *Up on the River* (1985), about the upper Mississippi. His articles appear frequently in *Smithsonian, National Geographic* and *Audubon*, where this piece first appeared. Madson is now at work on a book about the Missouri River—the stream that flows near his home in Godfrey, Illinois. From there, he and his wife, Dycie, who illustrates his books, range outward across the Midwest and West, hiking, drawing, "loafing," and learning.

Madson writes with a salty voice, full of local stories and characters. He says, "Unable to really portray what I see in words, I am more inclined to portray what I feel—a very profound emotional response to certain landscapes, to certain creatures that inhabit those landscapes, which are part of the spirit of place. I can think of no more interesting intellectual exercise than to try to crystallize some of those emotions in writing."

He believes that "one of the prime motives for going outdoors is to hunt for something, whether it's the perfect photographic exposure, mushrooms, birdwatching, or meat for food. Active hunting as a predator closes this magic circle of man, animal, and land. When you are hunting an animal, you are seeing that animal at its best." But Madson distinguishes sharply between hunting and killing, saying that he would be just as happy if he stalked an antelope for five or six miles, and finally "came up on that last pile of rocks, as close as I could get (300 yards) and brought up that empty Winchester and went click on that buck."

Madson values "writing from the gut, writing from personal experience." At the same time he notes "poetry can enhance fact. It can deepen fact, it can color fact." Today, he remains "just as excited as I was when I sold my first story. I can't think of a better recommendation for an occupation than that."

Life on the Back Side
of the Moon

I t's an improbable kind of place, looking like the set of a science-fiction movie with cardboard-and-plaster backdrops left by some slightly mad producer. All that's needed are little green bipeds peering around the corners at tourists.

But the Big Badlands of western South Dakota are real enough. Too real, for those who are vaguely disturbed by the raw, scalded look of these brooding landscapes with their naked sediments in banded layers of white, yellow, red, buff, and green. It is a tangled fastness of fluted spires, ridges, sheer walls, and deep ravines left by thirty-five million years of erosion. A ravaged land whose only vitality is in the elements that scourge it—or so it would seem.

Oh, the impression of death and desolation is valid enough; the Badlands are a vast charnel house of extinct animals, and their ancient bones are exhumed by every rainfall that washes the clays and mudstones. But there's life among the wrecked battlements, and the vital spark is all the brighter for burning among monuments of a dead world. And even though every kind of Badlands creature today can be found elsewhere in the West, nowhere else do they stand in more vivid contrast with such backdrops, or are they set off to better advantage.

Especially in late autumn, when the spark burns brightest.

The day was running out, with shadows deepening in the arroyos and the late sunlight flat and orange against the pinnacles at the head of Sage Creek Basin.

For over an hour I had been trying to find a way up to the rim and out of there, and now I was scrambling down again, knowing I'd have to return by the way I'd come in, which was a very bad way. I stopped on a crumbling ledge to rest and cuss, looking out over the tortured floor of the basin toward the heights a mile to the southeast. A faint movement up there. High up, coming into the late October sunlight across a narrow, grassy table just below the crenulated skyline.

All day I had packed the 32X spotting 'scope without using it, and since early afternoon I had been wishing it was an

extra canteen. Now, feeling wiser than a treeful of owls, I set up the telescope and focused on the distant ridge. The moving speck resolved into
a young bighorn ram whose attention was fixed on something that moved
at the edge of the 'scope's field. A ewe, grazing along the tableland fifty
yards from the ram. He walked a few paces, watching her intently, then
paused and looked back. A second ram appeared from somewhere below
the far edge and came over to join the first. Each carried half-curve horns
and still had some of the slimness of adolescence. I reckoned them to be
in their fourth autumn—old enough to be interested in comely ewes but
young enough to be discreet in the presence of any senior rams. They
stood together, longingly watching the ewe, and then—as if on signal—
both looked back down the trail.

He materialized instantly; one moment there was
nothing behind the two young rams, and then the shadow behind a low
spire seemed to intensify and coalesce, and a great dark ram walked slowly
onto the stage. The ewe was almost fawn-colored in that light; the two
young rams were darker, with cream-colored rump patches. The new-
comer was darkest of all, almost chocolate, appearing to be twice as heavy
as either of the youngsters—a barrel-bodied old herd-sire whose massive
horns swept back and out, curving in a full curl with broomed tips ending on the level of his amber eyes. I had been looking for him most of
the week, the principal actor in this theater of the high Badlands, and now
the old ham had finally appeared with his supporting juveniles and ingenue
waiting onstage, and his entrance couldn't have been more spectacular.

He stalked past the two young rams, which fell
into single file behind him. The ewe gave no sign of having seen them
and grazed on peacefully as the old herd-master slowly approached, stealing toward the ewe like a bird dog closing on a covey of quail that has
started to move, head lowered and neck stretched to catch her scent. The
two striplings were close behind, taking care not to come between their
chief and his lady.

The old ram made no attempt to mount the ewe;
he simply bird-dogged her across the grassy tableland, keenly interested
but also showing that neither he nor the ewe was quite ready for the
autumn mating games and jousts of bighorn sheep. He never lost that taut
alertness, and she never displayed the slightest interest. This prenuptial
parade made a full circuit of the ridge before the ewe finally kicked up
her heels and vanished into a ravine. The three rams walked over to a
point and stood there in the waning light, the two younger animals facing
the old one almost nose to nose. I've never been one to anthropomor-
phize, but I could almost hear the dialogue:

"Well, it's Saturday night. What's up?"

"Oh, I don't know. You kids want to do
something?"

"Sure. What do you want to do?"

"Beats me. Anything you'd like to do?"

And so on.

The tableau lasted for twenty minutes in the lowering light. Some high cloud cover was moving in from the northwest, promising snow, and the last of the sun burnished the rich sable of the big ram as he stood with head raised while the young rams attended him, facing him as if awaiting the word. Then one of them broke away and disappeared over the far edge, followed by the other. The old herd-master walked slowly after them and went out of sight, and I figured that was the last of it. But a minute later he reappeared on a grassy table far below, splendid head held high, in an airy, weightless trot that took him into the gathering darkness.

I did the same, but less gracefully and at considerable cost to my outfit, scrambling, sliding, and tearing out the seat of my patched Levis. When I finally got to the road it was full dark, which was okay, considering the state of my pants. Dust-dry, scuffed and spavined, but still exulting in the sight of those splendid animals and secure in the hope that there'd be lambs next spring in hidden nurseries high in the pinnacles.

The little herd of bighorns in South Dakota's Badlands National Park needs all the lambs it can get. The herd hasn't done all that well since wild sheep were stocked there twenty years ago.

Twenty-two bighorns were brought into the old Badlands National Monument and held in an enclosed pasture for several years, and after half the sheep died of disease in 1967 the survivors were set free. Today, the National Park Service puts the herd at about forty animals, although there are probably more than that; about a year ago wildlife photographer Ron Spomer was in the highest, most rugged part of the park's north unit and counted forty-five bighorns in one band. No one can be sure of the real total. Small groups of the sheep travel widely. Some have been reported as far south as Stronghold Table in the southern unit of the park and as far east as the rough breaks near the town of Kadoka—an extreme spread of at least sixty miles. Most of these far-ranging sheep appear to be young rams, the "social castrates" of the main population, but one bighorn seen near Kadoka was an older animal. As far as anyone knows there are three dominant rams in the herd—two with at least a three-quarter curl of horn, and the big full-curl ram.

The Badlands' bighorn population appears stable. Too stable. National Park Service biologist Hank McCutchen feels the sheep aren't expanding as they should, and the cause is hard to nail down. Food is in good supply. Hank has found plenty of "ice cream" forage species that the sheep have hardly used. Is there any evidence of climbing injuries on the treacherous Badlands formations? No. He's found nothing to indicate that. Something, though, is wrong with lamb survival. Hank has seen flocks that might include eight or ten lambs but only a few yearlings. It appears that many lambs aren't getting past their first winter. Disease, possibly. He has found a low incidence of lungworm in the droppings of both adults and lambs, and suspects that lungworm larvae are being transmitted through the placentas and infecting lambs before birth.

Another problem is suitable water supply. There is less water in the Badlands than there once was; overgrazing and farming have sapped the old aquifers, and many of the original springs have dried up or been silted in. If existing waterholes are very far from escape terrain (and a couple of hundred yards can be too far) and there's cover capable of concealing coyotes, predation on young sheep can be serious. There is no firm evidence of eagle predation. There is a recent case of a mature ram being shot by a poacher, but the general behavior of the sheep doesn't show the spookiness usually reflected in a hunted population. Adding it all up, McCutchen feels that the major limiting factor of bighorns in Badlands National Park is probably one of the distribution and condition of watering places.

These sheep are of Rocky Mountain stock, replacing the extinct Audubon bighorn—the subspecies that may have been the largest of all our North American wild sheep. Audubon reported a ram that weighed 344 pounds, of which horns and skull alone weighed 44.5 pounds. Variously called "Audubon bighorn," "Badlands bighorn," and "Black Hills bighorn," the type locality for this sheep was generally in the Badlands between the Cheyenne and White rivers.

It's been said that the Audubon bighorn's last stand was in the Black Hills, where it was wiped out around 1895. However some records go well beyond that. A Badlands hunter named Charley Jones shot a ram on Sheep Mountain Table just south of the little town of Scenic in 1903. Some bighorns still ranged on and around Sheep Mountain Table in 1908 and 1909 but were last recorded there in 1910. However, there's an old photo of a professional wolf hunter standing before his tent with a typical Badlands "wall" in the background. Captioned "fresh meat in camp" it shows the head, foreparts, and dressed hindquarters of a bighorn sheep hanging from the meat pole. This may have been the animal killed about 1918 between Big and Little Corral draws, just west of Sheep Mountain Table.

That was about it. But in 1926, several miles southwest of Camp Crook, South Dakota, near the breaks of the Little Missouri, a lone bighorn ram was shot. And while it's possible that ram may have wandered up the Little Missouri from the North Dakota Badlands, the chances of its having come from bighorn country farther west are considered remote. Veteran South Dakota game biologists strongly suspect that this ram—which was known locally and diligently hunted by ranchers for several years—was the last of its race. *Sic transit auduboni.*

The big stuff went early. Bison once used the Badlands for summer range, but a severe three-year drought that began in 1861 put a temporary end to that. By the time the drought finally broke, the bison had been pushed farther west by hunting pressure, and they never did return in any real numbers.

For a skilled professional of the 1870s, market hunting for the Black Hills gold camps could be more profitable than mining. And compared with other early western mining regions, the Black Hills and adjacent Badlands must have provided relatively easy picking. The valleys of the White and Cheyenne rivers were among the best hunting grounds, and the deer, elk, bighorns, and antelope of the Badlands were hard hit, early on.

Elk were abundant in the Badlands until about 1877, but during that summer hunters for the mining camps killed large numbers. When the surviving elk migrated just east of Rapid City that fall, miners and townspeople "turned out en masse and slaughtered hundreds." Their herds depleted and their old migration pattern shattered, the Badlands elk never really returned.

Black bears and grizzlies were mostly gone by the turn of the century, and wolves weren't far behind. The big plains "loafers" were anathema to stockmen, who paid bounties of $5 to $20 for each old wolf and $3 for pups, and sometimes gave free room and board to the wolf hunter. A big spread with a predation problem might even hire a wolfer by the month, paying $50 per month and board. On top of that, the State of South Dakota paid bounty of $5 for each adult wolf and $2 for a pup. What with one thing and another, a good wolfer might make better wages than a top cowhand.

The grizzly bear and gray wolf never really returned to the Badlands—and for a long time not much of anything else was there either. A 1919 survey revealed little wildlife in the Badlands and concluded, "The entire region seems void of all wild animal life."

When I first saw the South Dakota Badlands in 1939, the year they were named a national monument, wildlife was still at a low ebb. There were a few deer in the wilder corners but no bighorns or bison and not many pronghorn antelope. With national monument status and the emergence of modern wildlife conservation, though, things began to pick up. Antelope were trickling back into the Badlands during the early 1940s but were still rare. Into the 1950s and 1960s pronghorns continued to build, drifting freely in and out of the Badlands and steadily increasing. In 1963 they were joined by their old plains partners when a herd of fifty-three bison was stocked in the western part of the monument. The buffs settled in, found it to their liking, and prospered. Within five years the original herd had increased threefold, putting new life into the Tyree and Sage Creek basins. Every now and then they put some spice into my life, too.

A "table" in Badlands parlance is simply a flat-topped mesa with vertical sides. Some are hundreds of acres; others are only a few feet across. Most are perfectly flat and covered with grass. On overnight trips into the western basins of the Badlands I coveted certain little tables that were maybe thirty feet in diameter with sheer sides. In buffalo country this makes for a feeling of security and a sound night's sleep. At least, it used to. Several winters ago I backpacked alone into the Tyree Basin during a period of relatively deep snow. I set up camp on a Badlands table no larger than my living room at home. It was five feet high with no breaks in its sheer sides—an unassailable bastion. I'd seen fresh buffalo trails on the way in, but they'd never follow me up on that table, right? Wrong.

Deep-winter camping alone is a great way to catch up on sleep. There's no one to talk to, and I can't say I enjoy reading in a subzero tent. So that evening I dined early, was in my sleeping bag by 6:00 P.M., and slept the clock around. Stepping out into a dazzling sunrise I found a neat line of buffalo tracks around my tent. Sometime during the night a curious buff had found me there, scaled the vertical bank, looked things over, and politely exited over the far side without waking me. I still camp on those little tables because they're floor-flat and comfortable, but I do so with no illusions.

When I travel up the south side of the Sage Creek Basin at the edge of the main buffalo range I always have binoculars, for I like to stop frequently and check the landscape for signs of company. This south ridge is a favorite hangout for solitary buffalo bulls that may be half hidden by a fold of land or a clump of trees. I like to know when they're on my line of march. It's rude to startle them, and I'm the soul of courtesy to a cranky old buffalo bull.

I was returning to camp along that upland one fine afternoon, checking the route ahead as I usually do, and saw a lone buffalo skylined about a mile ahead. I noted the location and altered course to give him a wide berth. But there were two deep drainages to cross, and the second one forced me off my course. As I topped out I was distracted by a gleaming slope of polished chalcedony pebbles and didn't have buffalo in mind.

Passing within fifty feet of some junipers, I heard a deep groan. In the shadow between two of the trees stood a huge, black frowning stormcloud of a bull buffalo. Again, that deep groan. His head was slowly going down and his tail was slowly coming up, and he was beginning to look downright uncharitable. He was 2,000 pounds of growing unhappiness, but he wasn't any unhappier than I was. The nearest cover was another clump of junipers about 400 yards down the open ridge. It was one of my longer walks. There was no point in running; that might have triggered him, and there's no outrunning a buffalo. When last seen he was still frowning down the hillside at me and rumbling curses. A bland little adventure, looking back at it. I guess you had to be there.

There are several hundred buffalo in the Badlands. The number varies. A few years back, when there were about 450, they began breaking out of the park boundaries into pastures and wheatlands north and west of the park. This incursion was frowned upon by ranchers. The herd was summarily reduced; the surplus was shipped down to the Pine Ridge Indian Reservation for the Oglala Sioux herd. And twenty-two miles of high, strong, reasonably buffalo-tight fence was put up along the north and west sides of the Sage Creek Basin. In the fall of 1983 park officials estimated the bison count to be 370, mostly cows with about fifty calves and fifty mature bulls. Their general health is very good, and the herd went into the winter of 1983 with modest numbers and an immoderate amount of good forage. But water is always something of a range problem in the Badlands, and some of the old catch basins at the head of Sage Creek are being blasted out, deepened, and their capacity doubled.

Mixed in with the bison is an ephemeral population of pronghorn antelope that range through much of the Badlands as single animals, small bunches, or herds of a hundred or more. Highly mobile, they come and go. Mule deer are much the same, although they tend to be more sedentary than pronghorns. Last fall, as I skirted the north side of the Conata Basin, I happened to notice some clumps of sumac several hundred yards away in a field of deep grass—a singular thing, seeing as how sumac didn't belong there. I climbed a little formation and glassed the situation, and deer began to sprout out of the grass. Four were very large bucks whose antlers accounted for the "sumac." There were eight others, too—an even dozen mule deer bucks having a bachelor party before the rise of the rutting moon.

Antelope and bison are two sides of the same coin. They are grasslanders that flourish together. Mule deer are another matter. They may not do well in plains country with high bison populations that often forage and bed down in the brushy draws on which muleys depend for browse and cover. This probably had a lot to do with the early distribution of deer on the Great Plains. But in the big basins on the west side of Badlands National Park the two species have struck an ecological bargain—and on a day when I've counted two hundred bison I have also seen as many as twenty mule deer.

134

Badlands deer and antelope appear to be in excellent physical shape—thanks in no small part to steady pressure by predators. With wolves gone there's not much that can work on the buffalo, but the Badlands have coyotes aplenty, and the little songdogs prey on young antelope and deer when they can. The Badlands are one of the few places I've heard coyotes singing in the middle of the day, and their sign can be found almost anywhere up in the big formations that a person can climb. I've yet to see bobcats in the Badlands, although I've seen their tracks. But in recent years there have been several reports of Little Bob's big cousin. Three mountain lion cubs were seen playing on one of the main roads in 1963, and the U. S. Fish and Wildlife Service has reported an adult cougar east of the park in the Kadoka area.

There's no one I'd rather be with in the deep Badlands than Bill Lone Hill. A former tribal policeman on the huge Pine Ridge Indian Reservation, and a park ranger for fifteen years, Bill knows the Badlands National Park like his own backyard—which it is, come to think of it.

One evening when he and his family lived near the White River Visitor Center in the south unit, Bill heard a woman crying for help from beside the road. He looked around, calling out but getting no reply. He searched again next day, helped by a nearby road crew, but they found nothing. He had nearly forgotten the incident several days later when he was hiking along the White River north of his house. Coming back by a different route he paused on some high ground above the river. Down below, walking along the bank and unaware of Bill's presence, was a full-grown mountain lion. Later that day some boys came tearing up to Bill's headquarters saying they had seen a "huge cat" while swimming in the river. They had dressed en route, which isn't easy on motor bikes.

If there's a winged counterpart of the mountain lion, it's the golden eagle. This magnificent raptor frequents the Badlands, nesting and hunting there, and investing the stark landscapes with a mysticism that isn't lost on the Sioux.

One day Bill Lone Hill and I were deep in the "baddest of the badlands" between the village of Red Shirt and Stronghold Table when a golden eagle soared over us, the white rondels of its underwings marking it as a youngster. While we watched it off into the north, Bill told of a funeral he'd attended not long before. A middle-aged man, a devotee of the Old Ways, had insisted on dancing in an annual pow-wow in spite of a heart condition and his doctor's warnings. But the ceremony meant too much to him to give up—and in his final dance he had made but one circle when he fell dead.

Dying as he did during the ceremonial, he was given a traditional Sioux funeral that embodied many of the old customs. There was a four-day waiting period between the time of death and the funeral, with burial at the dancing grounds, and there was a basket of gift turquoise and silver to which the people helped themselves. It was spring, and during the funeral a pair of golden eagles swung high overhead. That was impressive enough, but even more stirring was the sound of the eagles screaming—something few of the people had ever heard, and which was regarded as highly significant. A person may live for years in golden eagle country and never hear one scream—but then, it isn't every day that such a dancer passes on.

Golden eagle predation on young bighorns, antelope, and deer fawns is probably nominal at most, although you'll hear heated claims to the contrary through much of the West. Little is known of eagle predation on big game in the Badlands, although the huge birds are capable of preying successfully on fawns and antelope kids. Author Joy Hauk tells of a Badlands eagle that took on a grown coyote and actually lifted the songdog off the ground. Both fell to earth and faced each other in a standoff until the coyote's nerve failed and he lit out for a nearby fence where he cowered behind a post until the eagle had gone. You can't really blame that coyote; a free golden eagle is an awesome thing. I was walking along the north rim of Sage Creek Basin one wild spring day, a few hundred yards from a large prairie dog town, when I flushed a golden eagle that had been perched just over the edge out of the wind. The great bird turned downwind and passed close by on wings spanning almost eight feet, and if there'd been a fencepost to hide behind I'd have damn well gone for it.

Much as I admire such majesty, though, my favorite Badlands birds are far more modest. In the deepest part of winter along the windswept roads, clouds of snow buntings whirl into the air— their shadings of white and light brown a perfect match to the landscape. In high summer, long after the buntings have gone to the Arctic, the Badlands are set with sapphires. I've never been in the main campground near park headquarters during the warm months when there weren't Rocky Mountain bluebirds in attendance. A half-dozen or more may be in sight at one time, and on sunny days they are that fine shade of deep blue that you'll see in Badlands sunsets, just above the bands of salmon and gold.

The abiding horror of many Badlands visitors is the prairie rattlesnake, which probably keeps some people on the roads and off the trails. The irony of this is that the Big Badlands, by and large, are a good place to get away from rattlesnakes. I've seen fewer rattlers within the park than in South Dakota range country as a whole. Most of the Badlands is just too barren for rattlesnakes, with neither cover nor food. The few rattlers I've seen in the park were either in prairie dog towns or heavy grass, but never in the bare formations themselves. Anyway, the threat of rattlesnakes shouldn't keep anyone out of those fine backlands. The more you're in rattler country the more you accept snakes as a fact of life—like cactus and gyp water. At the risk of being struck, stuck or fluxed, you live with such things but don't take liberties with them.

Prairie rattlers have a reputation for being on the hot side as rattlesnakes go, and it's true that they can be touchy. But the only really het-up prairie rattler I've seen was the one that Bob "He-Dog" Henderson grabbed barehanded as it was escaping down a prairie dog hole. He-Dog hauled it out by the tail and tossed it aside. Hot? That snake was plumb incandescent.

Sometimes, however, the forbearance of a prairie rattler surpasseth all understanding. Like one that a friend of ours brought into Scenic.

The social center of that little Badlands village was the infamous Longhorn Saloon, which was owned by "Halley" Merrill who looked like Buffalo Bill's uncle. The real tough uncle. There were bullet holes in the ceiling and sawdust on the floor—to soak up the blood, some said. Anyway, my friend Jim Brandenburg brought this big prairie rattler to town one summer afternoon and turned it loose in front of the Longhorn. The cowboys tried to stir it up, but it was a sleepy sort of day and the snake wasn't mad at anybody. In the prevailing spirit of conviviality, someone offered to buy the rattler a drink. They took the snake inside and put it on the bar and tried to interest it in some beer. The rattlesnake couldn't be corrupted. It wouldn't drink and it wouldn't fight, so it plainly didn't belong in the Longhorn. Jim turned it loose just outside of town, figuring it might start a whole line of highly moral rattlesnakes.

"And we can but hope," he told the waddies back at the saloon, "that the example shown by the noble buzztail will not be lost on certain unrefined, bowlegged, White River brush-poppers I could name."

The backlands. Those parts away from roads, out back of beyond, behind the postcard pictures in those special reaches of the Badlands that our friend Curt Twedt calls "the Goodlands."

There are many ways to go, from the short walks and well-marked longer paths (mostly in the eastern part of the park) to the trails that grow dimmer as they fade into wilderness, pointing to places where you make your own tracks and follow no one's lead. There's a way for everyone; the main thing is to get out of your car and away from the parking lot if you can. The Park Service schedules guided "nature walks" during the tourist season, and the ones I've taken have been excellent. They usually last several hours, leaving shortly after dawn to avoid midday heat and often leading in, through, and over some fine Badlands. If you are gimpy, though, ask about these hikes in advance. They have their rough spots.

For a few, the best trails of all are the skeined game trails that wander up and over the "tables" and through the passes, into lost canyons and up onto cedared shelves.

The best times for such doings are in spring and fall when temperatures are moderate, tourists are fewer, and the midday light is better for photography than in full summer. I've packed into the Badlands in all seasons and known July days of 104 degrees F and February nights of –15 degrees F and have enjoyed it all, though I'll freely admit that late October is hard to beat. High summer presents problems. Getting into some of the best backlands may require a long day—and that means at least a gallon of water. If you plan to be active in early afternoon when the sun is thrown back by those beige walls and the breathless canyons swim with heat, consider taking two gallons per person per day. There's no drinkable water back there unless you pack a purification outfit and share a waterhole with buffalo.

This matter of water is why I like to leave camp very early and come in at sundown in midsummer—spending the middle of the day shut down on some high shelf under a canopy of cedars. Might as well. Most critters are shut down, too, and the light is bad for pictures. So relax in a pool of deep shade, looking out into the white blaze of a Badlands afternoon and listening to pure silence. Take off your boots and dry your socks. Have an apple. Sooner or later, something will develop. As it did one day when I was sharing a fine loafing-place with some pale Badlands chipmunks and happened to glance at a patch of bare ground only a few feet away. The place was an on-site museum of Oligocene fossils. Scattered over several square yards were fragments of ancient turtle shell that looked like shards of terra-cotta pottery. A closer look revealed bone fragments and teeth of an oreodont—a sheep-sized mammal that grazed here in herds thirty million years ago. Nearby was part of a lower jaw, teeth in place, tinted a shade of cinnamon and clearly defined in its matrix of light Badlands clay. A private exhibit and a transient one, for the fossils were already deteriorating with exposure.

I leave such things in place, rarely even touching them anymore. Not because I don't hanker after curios, for I do, nor just because it's unlawful to take fossils from federal lands, which it is. But an Oligocene mammal bone in place has a quality that it loses when disinterred. Out of place its aura of great antiquity begins to fade, and with it vanishes that faint perception of a remote and unpeopled world. Any mystery or message is lost. The fossil becomes just a lump of minerals. So I leave the turtle shells and oreodont teeth as I find them, interred in

138

their monuments of richly tinted alluvia and volcanic ash. I can offer them nothing better.

Much as I like prowling around in these high formations—and especially along the trackless eaves that overlook the Conata and Sage Creek basins—I'm often nervous about it. Not of falling from a high wall, necessarily, but of getting stuck in a hole.

Badlands sediments erode in odd ways, with rainfall draining almost unchecked to cut tunnels and caves, deep vertical trenches, and treacherous sinkholes. One summer afternoon last year I was climbing along Deer Haven at the head of the Conata Basin. This is a "slump" where high ground has collapsed to form a rough, semicircular shelf forested with juniper, and although I'm familiar with most of it, I had never been over on a grassy eastern part. I worked over that way, farther than I'd ever gone, tramping blithely across an open plateau. Then I began noticing odd holes almost hidden in the heavy grass. Some were like animal burrows a foot or two in diameter; just beyond were pits twenty feet across, constricting funnel-like in their depths. I was at the edge of the worst before I ever saw it—a hidden sinkhole only four feet in diameter but perhaps ten times that deep. A fearsome place and a terrible thing for a lone hiker to step into. Cliffs, and the possibility of falls, I can face. But I can't handle the thought of stepping into one of those pits while alone and being wedged deep out of sight and hearing. Next time up there, I'm packing a climber's ax.

Getting back into the stark, flayed mazes of the Big Badlands is mandatory for anyone loving good boondocks, but most of the wildlife action occurs throughout the grasslands of the big tables and the main basins. The heart of this is an eight-by-twelve-mile designated wilderness that embraces the Sage Creek and Tyree basins and is best entered from the primitive campground at its northwest corner. If you keep to high ground south of Sage Creek Basin it's reasonably good walking without having to fight past some of those god-awful Badlands arroyos. And while I don't usually keep careful head counts, in the course of one long day above Sage Creek I've seen buffalo, antelope, mule deer, sharptailed grouse, a prairie falcon, a golden eagle, a rattlesnake, bluebirds, turkey buzzards, magpies, a sprinkling of prairie dogs, a coyote, and a badger on his way home after a hard night. During that particularly fine September day there were no other hikers. At any time of year I seldom meet anyone in the remote Badlands. Not that I crave company back there, but something about this bothers me, and it's a note to close on.

The people I do meet are almost invariably on the sunny side of thirty, full of prance and the strong juices of youth. Now and then I'll meet a youngster of forty or so, but it's been years since I've seen anyone my own age in the backcountry, and I sometimes wonder about the boys and girls I used to play with. Where are they, now that our hair is white? I've only to go to the main campground to find out. They'll be there overnight on their way to Mount Rushmore, in air-conditioned mobile homes looking at pictures of grandchildren and "Love Boat" reruns. All those years while I've been out playing they have been growing up, Growing Substantial, and growing old.

Our ways must have parted forty-five years ago when I got into the Big Badlands, down under the Wall, for the first time. It was a stunning adventure that left me with a clear choice: to get on with the business of growing up and put such places behind me, or to remain fourteen years old and never really leave them at all.

So. See you in the backlands. I'll be the silver-tipped kid with ragged pants and an extra candy bar. Stop and talk. I've learned some good places.

DAVID
QUAMMEN

Born in 1948 in Cincinnati, Ohio, David Quammen studied Faulkner at Yale and Oxford; that particular passion remains with him, as one of the pieces here ("Chambers of Memory") testifies. His mentor in college was Robert Penn Warren, "but I didn't take his writing course, I took his criticism course. He taught me how to read, how to see what writers were doing."

Quammen published a novel in 1970, then moved west to Montana in 1973; he now lives in Bozeman. He spent some time studying aquatic entomology at the university in Missoula—"and I began to think I would really like to do a series of essays for somebody on the creatures that most people consider vermin."

Before that began to happen, Quammen worked as a bartender and a fishing guide. It was after a congenial day of fishing and a steak dinner at a ranch house with John Rasmus, the editor of *Outside* magazine, that Quammen was able to convince Rasmus to take a look at a piece considering the question

"What's good about mosquitoes, if anything?" Rasmus' favorable reaction led to his taking over the "Natural Acts" column in the magazine in 1981, and he has been *Outside*'s monthly natural history essayist ever since. A first "Natural Acts" collection became a book in 1985; another *(The Flight of the Iguana)* was published in 1988. Three of Quammen's columns won the 1987 National Magazine Award for Essays and Criticism. With the support of a Guggenheim Fellowship, he is working on a book about island biogeography.

Quammen estimates that the magazine column takes about one-fourth of his time. He has published two more novels, the first one "a spy novel," with a lot of Edward Teller and Robert Oppenheimer in it, and the second, "a novel with some espionage, marbled with natural history and Eugene Marais." When he finished the latter book, *The Soul of Viktor Tronko,* in 1986, he knew what he wanted. "I want to go outside and travel, and walk around in some swamps and some jungles, and talk to some scientists—do something that's not so godawful lonely as writing a novel and that's not so byzantine complicated as writing something that has the CIA in its plot."

Quammen's humor, what he calls "smart-aleckyness," loops through his pieces as well as his conversation. He gleefully quotes his wife, Kris, on word processors: "they are to writing what the Hammond organ is to music." But he has begun to use one, "trying to see if it's possible to get the same level of care, the same level of polish."

Quammen describes himself as "a science journalist—a writer first, and a naturalist, if at all, second. I didn't grow up reading Aldo Leopold and Eiseley and those folks. I eventually got around to reading them. But I don't read very much of the kind of thing that I write. I like to hike, I like to fish, but that doesn't mean I want to read about those things. I'd rather go out and do it or, if I'm home, I'd rather read Raymond Carver."

Chambers of Memory

The past is not dead, Faulkner told us, it is not even past. Or words to that effect. I'm quoting from memory. Memory believes before knowing remembers, Faulkner told us, another way of making the same point. It was his central and abiding theme: The past is not dead, is not gone, cannot ever be completely escaped or erased or forgotten; the past *is*. This of course was the heresy of a self-educated Mississippi crank, an axiom more Confederate than American, but probably (and despite the fact that Freud agreed) he was right. William Faulkner himself has been in his grave 25 years now, and *he* certainly isn't dead or gone. I got to thinking about—got to remembering—his words this week as I considered the chambered nautilus, an animal that carries its own past evermore forward through life and history, sealed off behind a wall of pearl.

For a living nautilus, the past literally provides balance and buoyancy. And the animal stays in touch with that past, remotely, inextricably, through a long tubular organ known as the siphuncle. This nautiloid siphuncle is a conduit of blood and memory.

The nautilus itself is a staggeringly ancient beast, a marine creature that has remained almost unchanged over 450 million years, since before life on earth had even climbed out of the oceans. Five species within the genus *Nautilus* survive today, last remnants of a line—the chambered cephalopods—that once produced 10,000 different species and dominated the ocean environment as emphatically—and at the same time—as dinosaurs dominated the land. They seem to have been the first successful marine predators, preceding the primitive fishes, preceding the seagoing reptiles, preceding such other cephalopod predators as the squid and the octopus. Like the squid and the octopus, though, nautiloids were soft-bodied animals with multiple tentacles. Long before backbones and toothy jaws and scales came into fashion, the nautiloids and their kin had solved the problems connected with making a living as carnivores in the ocean depths. They did it by secreting ingenious shells.

The shells were most typically spiral in shape, flattened on both sides and coiling gracefully outward from an axis, very much like the shells of surviving *Nautilus* species. The animals secreted these structures progressively as they themselves grew—adding wider extensions to the spiral, vacating ever outward as they needed more elbow room, living always in only the outermost chamber and closing each earlier chamber behind them with a calcareous wall. Of course the shells afforded protection for the early nautiloids, but they also did something more. Like the wings of the first true bird, those chambered shells allowed nautiloids to rise up off the substrate, defying gravity.

Controlled buoyancy may seem a modest feat to us, by hindsight, but the extent to which it expanded nautiloid horizons would be hard to overestimate.

The shells of living *Nautilus* species serve the same function today. Actually they perform less like a bird's wing than like a hot-air balloon. The successive chamber walls are known technically as septa, and behind every septum is a space filled with either liquid or gas, or with some balance of both—depending on whether the individual nautilus is seeking to rise or descend through the levels of water. Behind every septum is a phase of the animal's history.

But the seal isn't absolute. At the center of every septum is a small hole. Only the siphuncle penetrates backward in space and time.

In July of 1962, William Faulkner died suddenly, under somewhat mysterious circumstances involving whiskey and a steep flight of stairs, possibly also a heart attack or a stroke. He was buried quickly and rather quietly (for a Nobel Prize winner) beneath blasting summer heat in the town of Oxford, Mississippi, where he had lived his life. Roughly a year later, an article titled "The Death of William Faulkner" appeared in *The Saturday Evening Post.* I remember seeing it. There was a photograph of a fierce-looking little man with a stiff mustache and the eyes and nose of a peregrine falcon: this person Faulkner, evidently, of whom I knew nothing. I was 15. In the background of the photo, I recall, was a house that looked imposing behind its tall Greek Revival columns. I read the Ogden Nash in that issue of the *Post,* and probably much of the rest, and glanced at the cartoons, but I ignored the article about the dead man, whom I understood only vaguely to be some sort of notorious curmudgeon, maybe a segregationist governor. I was chiefly preoccupied at the time with football and bass fishing and the banjo, and no one had yet so much as forced me to read even "Barn Burning" or "The Bear," thank goodness. This article was by someone named Hughes Rudd, which fact I took note of not at all.

Six years later much had changed, and my life was spiraling around William Faulkner the way a miller spirals around a summer lantern. I had consumed all the novels and was well launched on an obsession that would go beyond appreciative rereading, beyond critical pedantry (and a graduate thesis describing "Centripetal and Contrapuntal Structural Patterns in William Faulkner's Major Novels"), beyond the demented self-assigned task of trying to translate *Absalom, Absalom!* into a film script; beyond all that, I say, and on into cultic veneration. Junior in college now. Late one evening I got a call from a buddy who said I should hustle right down to a certain bar and meet this wild man from CBS News, who was waving tumblers of Cutty Sark through the air and talking about my favorite subject, Faulkner. The man's name was Hughes Rudd.

It turned out that this Mr. Rudd was now a correspondent for CBS, and that he was leaving soon for Oxford, Mississippi, where he would film a documentary about Faulkner and Faulkner's country. Within the elapsed time of one Cutty Sark, I found myself hired. When I protested earnestly that I hadn't come to ask for a job, that I was just genuinely interested, that I didn't want to bust his budget, Hughes Rudd said: "Don't be a fool, David. CBS has money falling off it like dry leaves." And he fluttered his fingers through the air. "Dry leaves."

A few days later I was AWOL from college, camped at a motel in Oxford along with Hughes and a film crew. For two weeks there I lived the exalted life of an editorial consultant and gofer. I stood by while Hughes interviewed the chairman of the English department of Ole Miss, a blindingly tedious man who considered himself chief curator of Faulkner's memory, but whom Faulkner certainly would have loathed and Hughes had no patience for either. I stood by while Hughes talked with Faulkner's old hunting chums and the blacks who had tended his horses. I stood by throughout long days of filming inside the small ramshackle house on the south edge of town, beyond the driveway chain and the huge magnolias, the same house that from a distance looked so imposing behind its Greek Revival columns. And sometimes, though not for many hours, I stayed at the motel to work on a script. Hughes wasn't really shooting this program according to any script. He was shooting from his gut; he was shooting from his memory of what certain phrases, certain scenes, certain novels could mean to a person's life. Clearly Hughes himself was still a writer at heart, despite the CBS business, and he cared about Faulkner in a way that no mere TV commentator ever would. Hughes Rudd was in those days (and remained) an anomaly in the sleek world of network news—a jowly man with a bloodshot glare and a fast, sardonic wit who stubbornly worked the fringes of American culture, the flea circuses and hog-calling fests and tattooists' conventions by which he could illuminate, with his dour deadpan, the important truth that life itself was a benign but very ridiculous practical joke. And the notion of putting a thoughtful contemplation of Faulkner and Faulkner's Mississippi before a prime-time CBS audience was itself so mischievously improbable that it suited Hughes perfectly. Meanwhile there was no need for a script. Faulkner himself had written the script; it was *his* words we would use. For now, Hughes just wanted to capture the flavor of Faulkner's place—which happened to be rural and small-town Mississippi. Get a shot of those mules. Get a shot of that kudzu. Get a shot of this derelict plantation house with the roof fallen in and the columns awry and the thistles growing up between the porch planks. I stood by.

We ate steaks every night and I learned to drink and Hughes and I talked about Faulkner. We talked about writing, the craft Hughes had abandoned and I was just hoping to begin. We talked about sleekness versus fringes. It emerged that my real job on this assignment was to keep Hughes good company and to prevent him from getting himself into trouble—so Hughes said—when the Cutty Sark piled up over his eyebrows. I was fired and rehired three times within the first week. One harrowing evening at dinner I sat between Hughes and Faulkner's sister-in-law, a middle-aged woman still seething with outrage over something Hughes had written in the *Post* piece six years earlier, and I was obliged to referee. Then it was over. I went back to college and heard six months later that the whole project had been scrapped because CBS News was a half million dry leaves over budget.

I visited Hughes once in New York, where he took me out to a dauntingly fancy lunch. Then for 16 years I didn't see him again, except occasionally on the tube. Mississippi had been wild good fun, but for me it seemed a closed compartment of the past.

Beneath a soft, fleshy hood that covers what might loosely be called its face, the nautilus has jaws like a parrot. With those jaws it crunches the carapaces of crabs and lobsters; in captivity it will feed like a geek on raw chicken. It gapes at the world through a pair of large, empty eyes, almost as blank as Orphan Annie's. It has more arms than ten octopuses. It propels itself horizontally by jet power, farting water out through a flexible funnel and steering on sound Newtonian principles. A creature, then, of many bizarre charms—but none of these is so notable as its buoyancy system. The nautiloid buoyancy system is responsible for a seemingly paradoxical feat: allowing the animal to grow bigger and heftier without changing its weight.

The nautilus does grow more *massive* as it matures. But weight is a measure of gravitational force, not of mass, and for a nautilus that gravitational force depends on its own average density, relative to the density of seawater. The spiral shell is made of very dense material, and tends therefore to sink. The buoyancy system compensates, enabling the nautilus to levitate. The shell is constantly being enlarged, constantly getting heavier, and so every five or six weeks the nautilus slides its soft body toward the opening and secretes another septum behind, sealing off another abandoned chamber. Then it begins pumping heavy fluid—the ballast—out of that chamber, by means of its siphuncle, and refilling the same space with buoyant gas.

To rise toward the ocean surface, it replaces more fluid with gas. To descend again, it exchanges gas for fluid. None of these transfers is accomplished quickly. To build a new septum and empty the space behind it requires time and energy. To refill a chamber with ballast once it has been emptied was thought by scientists, until recently, to be impossible. Most of its life a nautilus spends in the middle depths, along the steep seaward slopes of drowned reefs in the Pacific, between 1,000 and 2,000 feet down, scavenging there along muddy ledges. Or simply hovering.

The pressure at those depths is enormous. A human body would be squashed. It might seem logical that such pressure would also drive seawater back in through the siphuncle, flooding all chambers of a nautilus shell despite any resistance the animal might try to offer.

This does not happen. A nautilus is immune to nostalgia.

For me there have been a number of compartmented phases, and a number of lucky fortuities. I was fortunate enough to escape the future that seemed to await me, quite manifestly, as an assistant professor of English whose turf was Faulkner. I was fortunate enough to get untangled from the ivy and sneak off without punching the dots for a doctorate. I evaded the looming Volvo and the corduroy jacket with leather elbows and the unfunny early marriage. Any complacency sensed here should be understood in the context that I managed to make my share of other good roaring mistakes instead. But at least those weren't manifest in advance. I spent three years at menial labor while writing a novel about the death of Faulkner, a novel that no one at the time or for years afterward wanted to publish, and now I don't either. But it took much time and energy to seal that one behind a wall. Then for reasons of rent and groceries I drifted into being a science journalist, sort of. By what seemed an accident of disposition I found myself working the fringes, the flea circuses and hog-calling fests of the realm of natural history. Probably if you had asked me about influences, though, I would have mumbled the names Ardrey, McPhee, Gould.

Like a reformed alcoholic I can brag that it's now been seven years, a record, since I last reread *Absalom, Absalom!* Life is short and 11 times probably is enough.

A few months ago I got a letter from Hughes Rudd. He had seen a small bit of my science writing in *The Washington Post* and called the editors for my address. He was now with ABC, Hughes told me, though planning to retire soon and move to France with his wife, Ann. He was glad to hear I had fetched up in the northern Rockies, of which he had fond memories from many years back, when he had rattled through Montana and Wyoming in his old Chevy pickup. "No gunrack, but a WWII Walther P-38 in the glove compartment. I used to go out in the semidesert around Rock Springs, Wyoming, and fire it at rocks, pretending they were newspaper publishers or book publishers." It was vintage Hughes, the whole letter, and it made me smile widely. I had forgotten over the years how much I cherished this man. I had forgotten how much I had learned from him. Even owning a television might have been justified, possibly, on the merits of Hughes Rudd alone.

"I took Ann to Oxford four or five years ago," wrote Hughes, "so she could see the Faulkner house, and the ole burg didn't look the same at all: they were remodeling the courthouse! Insanity. When they get started, they don't leave a fellow nothing, as Hemingway said somewhere." Hughes knew that I knew that in many ways that old courthouse, with its Confederate monument facing stubbornly south, had been a symbol of Faulkner's world and Faulkner's central idea. But of course the civic beavers who remodel historic courthouses can't obliterate the essence of recollection, and in even his cynical moments Hughes would admit it. "I've lost track of all the Faulkner relatives and friends I knew in those days," he added.

People come into our lives and then they go out again. The entropy law, as applied to human relations. Sometimes in their passing, though, they register an unimagined and far-reaching influence, as I suspect Hughes Rudd did upon me. There is no scientific way to discern such effects, but memory believes before knowing remembers. And the past lives coiled within the present, beyond sight, beyond revocation, lifting us up or weighing us down, sealed away—almost completely—behind walls of pearl.

The Miracle of the Geese

Listen: *uh-whongk, uh-whongk, uh-whongk, uh-whongk,* and then you are wide awake, and you smile up at the ceiling as the calls fade off to the north, and already they are gone. Silence again, 3 A.M., the hiss of March winds. A thought crosses your mind before you roll over and, contentedly, resume sleeping. The thought is: "Thank God I live here, right here exactly, in their path. Thank God for those birds." The honk of wild Canada geese passing overhead in the night is a sound to freshen the human soul. The question is why.

What makes the voice of that species so stirring, so mysteriously authoritative, to the ears of our own species? It is more than a matter of beauty. It is more than the majesty of unspoiled nature. Listen again, to America's wisest poet:

> Long ago, in Kentucky, I, a boy, stood
> By a dirt road, in first dark, and heard
> The great geese hoot northward.
>
> I could not see them, there being no moon
> And the stars sparse. I heard them.
>
> I did not know what was happening in my heart.
>
> It was the season before the elderberry blooms,
> Therefore they were going north.

The boy in Kentucky, 70 years ago, was Robert Penn Warren. The lines are from a long meditation on John James Audubon, the American wilderness, and the nature of love and knowledge. What was happening that evening in young Warren's heart? What is it that happens in yours or in mine when we hear the same sound today? Presumptuously, I propose a theory.

Wild geese, not angels, are the images of humanity's own highest self. They show us the apogee of our own potential. They live by the same principles that we, too often, only espouse. They embody liberty, grace, and devotion, combining those three contradictory virtues with a seamless elegance that leaves us shamed and inspired.

150

When they pass overhead, honking so musically, we are treated to (and accused by) a glimpse of the same sort of sublime creaturehood that we want badly to see in ourselves.

The particularities that support this notion are many, and you can find them in any study of the animal's biology and behavior. Here I want to consider just one. Geese mate monogamously, and for life.

Some thinkers would have us believe that monogamy and (still more extreme) fidelity are masochistic inventions of human culture, artificial limitations inspired by superstition and religion, and running counter to all natural imperatives of biology. Doesn't biology dictate that males of any given species should try to perpetrate their sperm as broadly as possible upon the female population? Doesn't evolution require that a female advance her own genes by selecting the strongest and smartest mate when she is first ready to breed, and by then selecting another, if possible even stronger and smarter, the next time around in her cycle? Doesn't the Darwinian dynamic—the relentless competition for reproductive success and survival—entail an equally relentless flirtatiousness among all animal species, an unending lookout for the prospect of a new and better mate? Well, no, not always. The evolutionary struggle, it turns out, is somewhat more complicated than a singles' bar. Among geese, there is an ecological mandate for fidelity.

Geese live a lofty but difficult life, facing the problems of starvation and predation in forms that are acutely particular to them, traveling long distances every year between their wintering ranges and their breeding grounds, struggling each summer to hatch and raise and educate a brood of goslings. Amid these travails, they just can't afford to philander. They need one another there, male and female, each its chosen mate, at all times. They have committed themselves, by physiology and anatomy, to a life of mutual reliance in permanent twosomes.

Curiously, this commitment seems to derive straight from the two other characteristics that we humans most admire in them: their noble size, and their impressive migration.

Of the world's 15 species of wild geese, all are confined to the Northern Hemisphere, and most populations of those species make formidable annual migrations (hundreds or thousands of miles, and in one species, over the Himalayas), traveling northward in spring to breed, south again in the fall. Many fly all the way to the Arctic. Canada geese (also known as Canadian honkers and, scientifically, *Branta canadensis)* are the largest and most familiar on this continent, with populations that follow flyways along the East Coast and the West Coast and several other north-south routes in between. The particular population I happen to know spend their winters in Arizona and thereabouts, feeding busily to build up fat reserves that will be needed later, and then fly up the Rockies to a certain braided stretch of the Madison River here in Montana, where dozens of tiny islands separated by narrow channels give them a choice of ideal nesting sites. As with any animal migration, there are routine costs and acute dangers that must be overcome by these journeying geese. Why the various populations migrate at all is not completely explicable, but flying up into the far northern summer does bring them to fresh food supplies (they eat grasses and other vegetation), to areas relatively empty of man and other predators, and to thawing wetlands that suit their needs for nesting. Also, the type of food they find by chasing spring northward is especially nutritious, since the young plant shoots they favor tend to hold high concentrations of protein and nitrogen. This last fact is quite important. Nutritionally, geese have to play every angle they can.

Their digestive system is damnably inefficient. Unlike other grazing herbivores, geese have no capacity to digest cellulose, which accounts for a large portion of plant tissue and holds the cellular juices (rich in sugar and protein) locked within cell walls. Most grazing animals digest cellulose with the help of bacteria that live in their guts. Along the upper digestive tract of a cow, for instance, is an extra stomach known as the rumen, wherein reside the cellulose-gobbling bacteria. But a Hereford is not obliged to cope with the delicate physics of flight. A goose, already large and heavy for a bird, and destined for long-distance flights, cannot tolerate the extra weight of a rumen. So a goose has no bacteria to help it with cellulose, and therefore no high rate of gastric efficiency. Much of the potential nutriment consumed passes straight through the bird unutilized.

And it passes quickly. A huge, full belly is also forbidden by aerodynamics, and so the whole alimentary process is accelerated. A meal of grass travels the length of a goose gut, top to bottom, in only about two hours. During such a short time, and again with no chemical breakdown of cellulose, digestion is necessarily partial. In consequence, a goose needs to be constantly filling itself and, also constantly, emptying. Hence the expression: "Loose as a goose."

Concerning the storage of fat reserves, it's the same problem once more. Some smaller birds that migrate to the Arctic can lay on as much as another full body-weight before they head north to breed. But the laws of scale as applied to aerodynamics make that sort of provident fattening, for a goose, impossible. Already so large, a goose (despite its great strength) cannot let itself get much bulkier. Anything more than a 25 percent weight gain, during the winter months of serious feeding, will jeopardize its ability to get airborne.

Each of these limitations makes the life of a goose a little harder, the margin of survival a little slimmer. Combined, they dictate that a goose needs to spend every possible moment stuffing its gullet. Hours are precious, benefit is low, and a goose must gorge steadily just to stay even, or to get slightly ahead for those lean and costly breeding months to come. This may be, in fact, another reason why geese migrate north in summer: Under the long northern hours of daylight, they can spend more time feeding.

And all the nutritional hardships of goosehood affect not just survival, but also procreation. A female goose will give away most of her own precious energy reserves—in the form of embryos and yolks—with the laying of a clutch of eggs. Then for almost a month she must incubate those eggs, scarcely leaving the nest to grab a daily meal. She will regain some weight, but not much, since most of her time is spent sitting, supplying the motherly heat. Occasionally, a female will even starve right on the nest. If she dies, the eggs will not hatch or the hatchlings will die also. The gander will be left a childless widower. Obviously, malnourishment of momma is bad for all concerned.

But time spent courting, each year, is time stolen from eating. Energy spent on sexual coquetry is energy that might have gone into eggs, or made the difference in the survival of a female exhausted after laying. Geese have neither time nor energy to spare, so they take a long view. They commit themselves to endurance, to each other, to the future—and not to maximizing their sexual options.

Romance in the life of a goose begins during the second or third winter. A male and a female find one another on the wintering grounds or just after migration north; overtures are made, a modest neckstretching dance is performed, then reciprocated, and before long an understanding has been struck. This arrangement is accomplished almost hastily, but destined to last. The new pair honk in mutual acknowledgment, then fly off to establish a nest, which will be only their first of many. Ducks, closely related to geese in the family Anatidae, go about things differently. Ducks tend toward elaborate courtship displays and garish plumage among the males, both of which serve toward distinguishing males of a given species from females, and males of one species from males of another, all to facilitate the pairing process, which in ducks is repeated each year. Not the geese. Choosing one mate for life and remaining faithful, geese have no need for such fancy displays or flashy dimorphic costumes. They put their resources to other uses. They spurn narcissism and fickleness and that annual flirtatious skirmishing, in favor of economy and a dignified single-mindedness.

It takes the best efforts of two geese, working full time, to hatch and rear a brood of goslings. The female lays a clutch of eggs (five is the average number), and then sits on them throughout the chilling days and nights of April and early May. The male meanwhile stands watch. If a predator approaches, he warns her by honking, and together they fly to safety—or else she hunkers low on the nest and he takes flight alone, noisily, attempting to lead the danger away. After hatching, he is still there, standing sentinel so that the mother and young can devote their undivided attention to feeding—and now *he* loses weight, while his mate restores herself and the goslings grow. For two months the goslings are flightless, but good swimmers, and during that period they all travel their river or lake as a tight family group, the mother paddling in front, then a file of goslings, the father following close behind. In autumn they fly south again, still together, an inseparable family. Evidently the migration routes are not programmed instinctively, but must be taught by the parents to each new generation. On the winter feeding grounds they *still* stay together, with the father again mainly responsible for keeping watch while the others feed. Finally, after a full year, either the youngsters leave voluntarily or the male drives them away. He and his mate are now once again devoting themselves to each other, and to the prospect of a new brood. They fly north as a pair of old lovers, to Manitoba or British Columbia or Newfoundland, or to the channels of the Madison River.

Each year in May I take a boat down through the Madison channels and visit, as unobtrusively as possible, those nesting geese. In good years, when deadlines are distant and life is sane, I get down there more than once, catching the geese at various stages of their cycle. The first time I ever saw a brood of newborn goslings was also the first year I made the float in a kayak; there were two of us, a woman and I, neither of us very familiar with kayak technique nor with each other. As we paddled past a small island, five balls of yellow fluff exploded hysterically out of a nest, straight into the river, and were swept immediately downstream. In the shock of the moment my friend's kayak went upside down; I was no help whatsoever, and it seemed suddenly that five young geese and a tall 28-year-old woman might all drown at once. But everyone survived. Now the woman and I are married and at least some of those five goslings, with decent luck, flew back this spring to become grandparents. The years intervening passed very quickly.

Monogamy and lifelong devotion, like an annual flight to the Arctic and back, are not things accomplished easily. Saint Paul told the Corinthians that "love does not come to an end," but the divorce rate among your friends and mine proves that Saint Paul was either a liar or a fool. Given all trends and pressures to the contrary, I was glad to find an ecological mandate for permanent partnership among animals so estimable as *Branta canadensis*. I was equally glad, recently, to come upon an interesting quote from the French novelist Marguerite Duras:

Fidelity, enforced and unto death, is the price you pay for the kind of love you never want to give up, for someone you want to hold forever, tighter and tighter, whether he's close or far away, someone who becomes dearer to you the more you've sacrificed for his sake. This sacrificial relationship is precisely the one that exists in the Christian church between pain and absolution. It can survive outside the church, but it retains its ecclesiastical form. There can be no more violent, and beautiful, strategy than this for seizing time, for restoring eternity to life.

Geese figured it out for themselves. They know something about violence and beauty and time. Listen: *uh-whongk, uh-whongk, uh-whongk, uh-whongk, uh-whongk* . . .

JOHN HAY

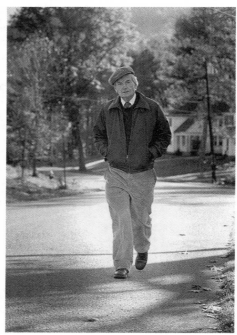

Photograph © Stephen Trimble

"I feel comfortable walking around in the woods, shuffling through the leaves, because I feel I'm inside of something. I've got roots," says John Hay. He has deep roots in New England, but the place that he has made his own, Cape Cod, did not enter his life until his thirties. He was born in 1915 in Ipswich, Massachusetts. His father was curator of archaeology at the American Museum of Natural History in New York City; his grandfather (also John Hay) was personal secretary to Abraham Lincoln and secretary of state under Teddy Roosevelt. (The current John Hay hastens to say that he never knew his grandfather, who died in 1905).

Hay spent his boyhood summers at Lake Sunapee, New Hampshire, where, he says, he developed an "invaluable sense of space." He spent two months with the poet Conrad Aiken on Cape Cod just before being drafted into World War II, and moved to the Cape after the war. Educated at Harvard, he "wanted to write poetry first," publishing his first volume of poems in 1947. Living on the Cape led to his natural history books.

Each spring, he discovered, the alewives returned from the sea to choke the local brook with spawning fish. Hay found this so exciting that he was moved to write a book about it, the classic *The Run* (1959). Many other titles followed, including *The Great Beach* (1963), for which he won the John Burroughs Medal; *In Defense of Nature* (1970); *The Undiscovered Country* (1982), from which these two essays come; and *The Immortal Wilderness* (1987).

Hay is not only a writer and naturalist, but an avowed conservationist; his ethics are as strong as his feel for language. For more than twenty years he was president of the board of the Cape Cod Museum of Natural History. He has led battles against overdevelopment on Cape Cod throughout his years there, and was named conservationist of the year by the Massachusetts Wildlife Federation in 1970. Hay taught Nature Writing and Nature and Human Values at Dartmouth College for fifteen years, beginning in 1972; today, he divides his time between Cape Cod and Maine.

"I'm trying to find out where I equate with the world around me," Hay says. "It's not escapism on my part, it's a matter of reality. I like to get inside these things, get my personality somehow acquainted with the rest of life. That's what fascinates me about single subjects like birds or fish or trees.

"I'm saying something on behalf of an absolutely new relationship to nature, to planet Earth. It may not be new to anybody else, but it's new to me."

Homing

With an assurance that came out of unending exchanges with space, the alewives, the bird migrants, even the local trees that swayed and changed with the rhythms of continental weather, would ask me as the seasonal tides went by, if I had learned where I was, and I had to answer: "Not yet." My sense of location was still at a rudimentary stage, and I might have lost all confidence in improving it if I had not decided, after another restless period, that it was better to stay put than to run off. The twentieth-century traveler does not soon come home to roost.

Certainly, if I had paid too much attention to what the economy said about the land I moved to, I would not have stayed there very long in any case. First it was only an abandoned woodlot, full of scrub trees, and worth only twenty-five dollars an acre at best, and years later when the real-estate market and the developers began to take hold it had turned into a potentially profitable investment, which might have kept my family secure somewhere else for the rest of their lives. Either way, it had a cash exchange value that never indicated it was worth living in for its own sake, or even tell me where it was except on a map which ignored the landmarks. Maps, it occurred to me, could not be read by the real pathfinders in this region. That alewives should migrate from some unknown distance at sea to a narrow waterway like Stony Brook was one of its best possible identifications. I wondered if any of us cared about it to the extent that they did. And if a pitch pine found Cape Cod to its liking, or a white pine was such an ancient believer in New Hampshire that its progeny sprang up in the myriads as soon as we had abandoned an old pasture, then we ought to be consulting them about home territory. We had moved to a country that was constantly rediscovered and claimed by beings with universal guidelines. The one thing they could never do was devalue the place they lived in or returned to. That would surely mean the obliteration of their lives and their directional knowledge.

In a world that tells you where is everywhere, it is no simple matter to get your bearings. Although I still had a few human

158

neighbors who were sticking it out and seemed to know the points of the compass, I was not at all sure that they would be able to hang on long enough to supply the rest of us with guidance, and as it turned out, they hardly managed. Yet the fish and songbirds came in from great distances, the mayflower found the appropriate place by old paths and abandoned wagon tracks to spread its leaves and hug the ground, and the sharp pointed beach grass held down the dunes with flexibility and pride, making windblown compasses on the sand.

That same science which could send a camera off to take pictures of Mars, and, later, Saturn, millions of miles away, also told me that a mere beach flea was capable of performing some amazing feats in its own right. There is a little pearly-white creature that lives on the beaches feeding off piles of rotting seaweed, and its name is *Talorchestia megalophthalma,* for its white, protruding eyes. This beach hopper has to live in a humid environment, because though it lives out of water it still has gills. So it feeds at night, to avoid drying out, and burrows in the sand of the upper beach during the day, a hole it abandons to dig another the following night. The eyes are apparently used to ascertain the altitude of the sun, the place of polarization of light in the sky, and the position of the moon. This creature, with its odd, sideways hop, can leap fifty times its own length, which seems to make it a champion among animals. All of which is scarcely an achievement of mind, but it is of such embodied genius that the earth's progress is measured. I could not help being impressed by whatever oriented that little animal, whose house was a temporary hole in the sand, to reference points far beyond its immediate horizon.

You have to suppose, with or without much knowledge, that we don't attach ourselves to a place merely to get away from another one. While human culture seems to be acquiring the role of a substitute for nature, I suspect we also learn from whatever reaches are left inside us to match the earth's, and whatever of its sensual messages we innately receive. Otherwise, why should we roll in the first snow, or rejoice all over again at being visited by the fragile, white blossoms of the shadblow in early spring? Have we not experienced them before?

The first week my wife and I moved in to our newly built house a flock of geese greeted us by flying past the window during a snowstorm, and for some years deer would step out in the late afternoon to browse in a field not fifty yards from the door. In the growing springtime the tender oak leaves hung before us in all their light and shimmering curtains of pink and silvery greens, and I knew scarcely the first thing about their ecology. In fact, when I first started reading about that new discipline, the idea that each life was so aptly fitted to its environment was a little difficult for me. Influenced by a world flung out in all directions, I must have had the notion that anything might be anywhere.

Still, I sensed that we were in the right place when we finally moved in, like those alewife fingerlings that head out to saltwater from the outlet of the ponds they were hatched in. They have never seen the sea before, but its image is in them.

I could now watch the passing weather touch and define the place I lived in. I could reassure myself that we had built at a centrifugal point, which migrants from hundreds or thousands of miles away flew in and identified. I could begin to see that the common forms around me, such as the countless blades of grass, the geometrically distributed leaves on a tree, the barnacles flicking out their feathery feet down by the shore, were displaying the great co-equality of life. They acted as receivers of the sun and of global currents in the waters and the atmosphere, each a reflection of surpassing complexity. Every life that touches on another, or becomes part of another, keeps the earth's fluidity in being.

To locate ourselves, we needed to be located. If the warblers failed to arrive in May, how would we recognize our station between the continents? When we left home we could quickly join the highways and airlines of the world. They had succeeded, along with all forms of modern communication, in making a good deal of local, which is to say connected, travel unnecessary. Despite the telephone, which crossed many voids, there were areas between us and our neighbors, near or far, which we no longer needed to explore, because they were so easily and quickly passed. This also amounted to a loss of local hospitality, in terms of people who could help you on your way and give you directions.

Yet those "compulsive" birds and fish, like the ever moving and changing clouds, kept homing in, roaming and circling, to turn the land and its waters into a center for universal travel. I had to inquire why it was that many of these other forms of life still knew their way across the surface of the globe through such an amazing variety of sensate abilities, while we who were supposed, by all accounts, to share in life's multicellular, neurophysiological connections, and were long-distance flyers into the bargain, seemed to be losing that innate sense of where the earth's great headings were. It would seem that the intellect that probed these matters was in need of more companions.

A phoebe showed up from South America, wagging its tail, pretty much on time by comparison with the year before, and seemed to say, in its dry little voice: "Here am I, ready to nest in your eaves. Let me stay." And I felt that any fool would who knew how much we stood in need of directions from such a reliable partner.

As time went on, I learned to recognize the terns and plovers, the yellowlegs, the sandpipers and songbirds, when they came in off the flyways and sea-lanes of the continent during the spring, while

160

the wintering eiders and brant geese flew away to northern breeding grounds. Some passed through, while others stayed, taking part in a dynamic employment of the range of land and sea. Many shore dwellers knew tidal time as intimately as we know night and day, and they knew sidereal and sun time and could find their way accordingly. The long-distance migrants could take advantage of landmarks, ocean currents, and the wind, and they were able to use information about the earth's magnetic field. They carried earth's directions in them, ancestrally, genetically. Young sooty terns fledged in the Dry Tortugas off the southern tip of Florida migrated in their first autumn all the way across the South Atlantic to the Gulf of Guinea in Africa, which they had never seen before. Nor did they have the benefit of adult leadership, since the adults left their breeding grounds after nesting but appeared to go no farther than the Gulf of Mexico. The young did not fly directly. The route they took was apparently 20 percent longer than that which could be traced on a direct line across the Atlantic, and it was the most favorable one, facing them with less resistance from prevailing winds.

This inherited sense of direction plus the native ability to fly great distances was not a random business but was tied in with the fact that populations had to be homogenous but dispersed at the same time, consistent with the breadth of the ocean and the periodic nature of its food supply. Still, it seemed astonishing and exciting that a young bird, only a few months old, could take off across the Atlantic without any prior knowledge of where to go, a kind of ancestral daring in nature to which we were only half-awake. Was it right to call them limited because their journeys were only unconsciously motivated? To say that human beings needed consciousness to survive while birds did not seemed to put their use of the unconscious on a more accomplished level than ours in some respects, or to make us more wholly conscious than we seemed to be.

The green turtle could find its way from the coast of Brazil to tiny Ascension Island in the middle of the Atlantic, 1,200 miles away, where they nested. Said Archie Carr in his book *So Excellent a Fishe,*: ". . . it really seems impossible that turtles or terns could ever gather at Ascension—and yet they do." Take into account the theory of celestial navigation as it might apply to turtles, or of inertial-sense dead reckoning, or piloting with landmarks unknown to us, or response to the Coriolis force, and then logically knock each of them out, as he did, and you were left with the irreducible fact that either all these factors were involved or that there was some sense in them that we knew nothing about. The great mystery in terns or turtles was their inner synchronization with the changing conditions and ranges of the planet. That may be why some of us, still circling, backtracking, confused by our own directives, might be envious of them.

What did these consistent periods of arrival mean, as I began to note them down in the local phoebes and swallows? Nothing fixed, in a particularly useful sense. (We like it that way, but may be the poorer for it. Besides, the swallows of Capistrano were late last year.) Migrations usually came in waves, and those that were "on time" might be only the vanguard of more to follow. They might be pioneers who were more adept at arriving when expected because they had done it before, which certainly seemed true of the alewives. Many migrants were lost along the way. Many were thrown off course by storms, or moved in the wrong direction by contrary winds, or, if they were fish, by ocean currents. Migration implies searching, or even hundreds of miles of drifting, for some marine species, as much as a fixed and conscious directional movement from one place to another, though some migrants, particularly birds, could be amazingly sure and direct.

Some of this I learned through observation, and more still through natural history and popular science. But it was not only the coordination of these animals with the earth's conditions that lifted me beyond the facts, but also their often mysterious affinity with its elemental reaches, like an arctic tern traveling up to twenty thousand miles a year between the Arctic and the Antarctic, a bird that experienced more daylight than any other creature on earth; it was a bird of light, engaged in an ancestral practice that challenged the planet.

On a whole-earth scale, the timing of these migrations was headed for timelessness. I had been too long confined to dates. (Would I never swing off the undeniable security of the calendar?) Days, on the other hand, days in space, days under the sun, days and nights, shadow and light, forever passing over us, trying out, were the measure an unemployed migrant might be looking for. The need to seek and be centered was as vast as geologic time. And all factors came together on some day in early spring when I heard the high, sharp whistles of a yellowlegs along the shore, announcing its arrival from the southern hemisphere. What could a poor calendar do in the face of that? The bird came in and set me free again.

In part because I was none too practiced in finding my way, easily lost one morning only a few hundred yards offshore when the fog suddenly rolled in, and none too sure of my bearings when I strayed too far from the asphalt, I knew we had much to learn from these explorers, since explore they did. If they were "simple," then so was I. You could doubt, even if we eventually discovered the precise mechanism through which birds found their way, whether we would be able to exploit it, as one account suggested, for our own purposes. By the time we got to the point that we needed some computerized device borrowed from pigeons to tell us how to get to the home loft, it might be too late anyway. Our brains would have lost connection with our legs.

It could be said that we were turning into perpetual migrants with no recognizable place to go to, continually exchanging houses and land, so that Cape Cod could be the same flat, insulated place we were trying to make of Arizona. All life explored its environment, but we seemed to be doing it to a fare-thee-well. I suspected that we needed to return to places we had long since forgotten, such as those the fishes knew.

In response to inner command, not only baby fish to sea turtles but also human beings might still have a sense of the way to head when faced with unknown waters, but we had lost faith in it. Perhaps we were only neglecting directions we felt we no longer needed. Still there was a residue of old seas in us. It stemmed from an ancient part of our natures that we were a little afraid and turned a little wild when watching hordes of fish move inland, or saw the unfolding of a leaf as a new event, off at the edges of experience, or felt the diving of a young seabird from its breeding cliffs for the first time as uncanny in its depth. Nothing was yet found.

I stood on a cliff above the open sea and felt that there was something of me running through its currents and its eddies. I watched continual explosions and disappearances taking place in those rolling waters threaded by sunlight, where unseen legions of fish were roaming, muscularly vibrating under the surface. At times, from a boat offshore I had seen long windrows of foam where bait fish were being chased to the surface, firing the waters with their motion.

It was over the offshore waters too that the white gannets with their six-foot wingspread wheeled from high in the air and then pitched in, sending up jets of spray. Those strikes, those recognitions, went on all around me, and though I had traveled far and wide myself I had hardly begun to recognize them; but they come back, if we are on hand to receive them, repeating their directives for our benefit.

There was, for example, the reappearance of my turtle. As a boy, with the wish in me to capture, I kept a painted turtle in a pen, and it surprised me by laying eggs, thus revealing a secret identity. The eggs were eaten by some marauder, though I had desperately wanted to see them hatch out, and later on the turtle died of unknown causes. I suspect it had something to do with mismanaged captivity. I gave it a formal burial under a big sugar maple that stood over a rock wall. I also carved my initials and the date on the plastron, or undershell, of another one. I can still feel the black, low, shiny-smooth, hemispheric shell, like a water-worn stone, and the little black legs with clawed feet that felt as soft as the pads on a puppy. Red-orange, yellow, and black are first colors for me. Then, many, many years later, with all kinds of human displacements and holocausts in between, I met it again as it ambled slowly across the grass out in front of that same tree where I had buried its compatriot. And I thanked the turtle as an old friend, not only for the pleasure of meeting it again, but for bringing me back into one of those mutual, lasting circles never stopped by the passage of time. It was like another lease, smelling of apple blossoms in the spring, the way they were and would always be.

Though I had a late start in learning my directions from the life around me, the lesson of the turtles was reassuring. They were educators who knew how to take their time. Someone once carved their Cape Cod initials on the shell of a brown and black and yellow box turtle I met while it was peering up and moving slowly ahead through fallen leaves, and so I found it to be twenty years short of the century mark.

That box turtles did not travel more than a few hundred yards or a quarter of a mile from their home territory did not mean that they lacked a deep sense of location. When displaced, which happened when the ignorant and innocent took them away from their native areas to city apartments, they were completely disoriented. So I was grateful that we had box turtles around us, with their scraggy yellow necks and the reassuringly antediluvian look in their little eyes, to help place us, or for that matter, advise us to be satisfied with where we found ourselves.

We needed one kind of turtle to tell us how to cross time in one place, as well as another who was an expert in crossing the seas. This homing business ought not to be wholly entrusted to human beings with nothing more than mechanisms to deal with it. It ought to be left to the professionals, such as turtles, frogs, toads, and salamanders, as well as birds who, in the art of wings, could come and go over thousands of miles. A study of California newts by Victor Twitty—*Of Scientists and Salamanders*—showed that when displaced from their home creek they returned to it, through forests and gullies and over a mountain ridge a thousand feet high. This was their breeding stream. On the other hand, since some of the males captured along the way showed no signs at all of sexual development associated with impending breeding activity, it seems that their ability to home in in such an impressive way had more to do with the right place to live than the right place to breed.

Toads have returned to breed in ponds that have been destroyed by road building. All landmarks were wiped out, and still they possessed their wonderful sense of where home was, or ought to be, in a featureless new landscape.

The homing instincts in these creatures of slow travel was invisible to us, and we did not seem to get close enough with the kind of biological tampering that suppressed their hearing, their sight, or sense of smell. Could we not ask the same kind of investigative question of ourselves? If we were still capable of homing in, after having strayed

164

so far away, it might very well prove to be an attribute which was essential to our future; and we would do well to start practicing the art; though if I were sufficiently relaxed about finding my own way, or not losing it, I might not trouble myself about such things, any more than an experienced fisherman does who steers his way unerringly through fog to his home channel. If toads retained an image in their heads of where a central place was even after it had disappeared, a similar faculty could be useful to us who had erased so many landscapes at will.

We needed a whole earth and a whole sea to find out where we were and dance to its measure, either with the wonderfully swift and graceful style of a tern, or the gait of a turtle, but it should not be too hard to find if everything around us embodied it. Even the common periwinkle down by the shore could claim a kind of global conquest, taking its own good times, like a box turtle. One early spring I found innumerable black specks in the wet sand of the beach. With a hand lens, I could see the round forms of these snails, each not much larger than the wet, pearly quartz grains that surrounded them. The cohesive, rounded surfaces of these grains seemed to give the tiny creatures a temporary medium for growth, before they were large enough to separate and start their slow tracking along the edge of the tidal shore.

The edible periwinkle was not an original native of America but an immigrant like the rest of us. It evidently came over with the Norsemen in their long boats to Newfoundland many centuries ago, but did not cross the Gulf of St. Lawrence to Nova Scotia until 1850. After that it made steady progress, to New Brunswick in 1860, then Portland, Maine, where it was discovered in 1870, and New Haven, Connecticut, in 1879. It had reached the prodigal resort of Atlantic City by 1892, and finally got down to Cape May in 1928. The periwinkle is now found as far south as Chesapeake Bay. So this dark-shelled creature, with its slow roll and two black antennae pointing forward, finally rounded a vast stretch of the North Atlantic for as far as it seemed useful to go.

Symbolically the migrants are also in the stars, sliding across the waters of the sky, traveling past each other, moving on great circles the way the periwinkle did. They made up a heaven of directions for ancient civilizations, extensions of human realities. The Toltecs of Mexico wore a snail shell on their heads. The shell of a snail was an enclosure like a house that held a life inside, and also symbolized birth as the insignia of the moon god. The moon itself was connected with water in the minds of the people, and its hieroglyph was a water vessel. The Mexicans associated the moon's blue light with turquoise. It was also a prototype of growth, and of change in the weather. Even now, if you walk down to the shore and watch the shining moon reflected on the waters, that magic symbolism seems as appropriate as what can be seen through the accurate eye of a telescope. And those who finally landed on the moon to report that it was a pitted wasteland which looked like a "dirty beach" did not alter its regality.

The shell, matter enclosing spirit, was the insignia of the great god Quetzalcoatl, who disappeared in the direction of the East and the land of the dawning sun. It was said that he would one day return like the rising of the morning star in the East, where the moon, with which the star became mythically unified, had died, and would be reborn in the form of a slender crescent tilted in the sky.

The snail identified with the winter solstice, and the slow-moving turtle with the summer solstice, at a time when the sun seems to stand still. The Mayan month of Kayab, when the summer solstice occurs, showed the face of a turtle. The Mayan name for them is *ac,* or *coc ac.* The sea turtles, loggerheads, which are sometimes found stranded and frozen on our shores, and the greens, are also represented on their temple buildings, and when you look up at the constellation of Gemini you can see three stars, which they saw as having the form of a turtle.

Starry nights, moonlit nights, with the sea breathing in the distance, its tremendous gravity poised beyond us, its waves belting the shore. If I know that Cancer the crab scuttles across the tidal flats to its designated part of the sky; if I know Pisces the fish swims off light-years away, I too can follow, from one home to another. To be head-taut with the stars around you, foot secure on soil and stone, to know your direction and return through outer signs, is as new as it is ancient. We are still people of the planet, with all its original directions waiting in our being.

Living with Trees

The seasons turn. Hang on. We are off for another ride. The tide rises higher with the full moon that presides over the flats and imitates daylight between the trees. Spawning fish and sea worms, insects and plants, must be responding to it in any number of magical ways. I myself walk out into the night and feel someone else in me talking to a light beyond him not measured by human control. It is a matter of new dimensions out of old phenomena, a tugging at a half-initiated spirit. However angered and enslaved by circumstances we may feel, the life in us is not allowed to quit. There are changes and reassociations going on in the soil and in the atmosphere, revolutions of a more reliable kind that move us to feelings we never knew we had. To have real rather than sentimental roots is to be in motion.

That will introduce my ignorance of trees, which grow out of life's original sources on terms I do not fully understand. In spite of that, they have been my lifelong companions. I have climbed them and built houses in them. I once built a houseboat of white pine to explore our lake, and it meant a wonderful freedom to me. I have cut them down, and even hugged them, as a boy, in moods of loneliness or affection. They hang on from a past no theory can recover. They will survive us. The air makes their music. Otherwise they live in savage silence, though mites and nematodes and spiders teem at their roots, and though the energy with which they feed on the sun and are able to draw water sometimes hundreds of feet up their trunks and into their twigs and branches calls for a deafening volume of sound. Their major endurance is good to count on, and how would we know a thing about the art of longevity if we cut them all down? If trees have analogies to human families, and I am sure they do, how can we clear-cut all their relatives, young and old, not to mention ancestors and descendants, the stock of generations, and expect them to accompany us as useful resources?

Trees identify their ground in ways that ought to be highly advantageous to people who are beginning to lose the sense of where they are. The fact that much of New England has become heavily forested, since the decline of intimate agriculture and the family farm, has inspired talk of a wasted resource. One would think it was a resource that had come back to tell us what kind of place we live in.

In how many ways, in green and white New Hampshire, did those great hemlocks, white pines, sugar maples, and shining birch not grasp me by the arms and seize my heart so as to give me bearings! How could I have learned to walk without them? Those wonderful trees, a great architecture of living on, their geometric graces in balance with all the extremes of the year! The hemlocks' long, outward-reaching, feathery branches rose and fell in the wind, lithe and limber in their open swinging, sometimes shaken into cranelike dances, while the tall white pines would shudder with their loads of silvery needles glimmering in the sunlight.

Trees are open to the assault of a vast number of insects. All you need to do is examine the leaves of a deciduous tree in midsummer to realize it. Flying insects, crawling insects, chewing and sucking insects, leaf miners, leaf hoppers, borers, mites, myrids, innumerable larvae, never let them alone from their tender emergence until the season when they sense the change in light, stop manufacturing food, and fall off. The respiring, perspiring, food-fixing trees, with their great numbers of leaves all arranged so as to receive as much radiation as possible, have attained a major equilibrium between the uses many other forms of life make of them and their own putting out to feed on the sun. Their structure is also a major statement against adversity, standing in the center of extreme heat and cold, various degrees of freezing and thawing, the chance of lightning, heavy loads of snow, ice storms, and violent winds.

The woods we moved to on this sandy peninsula are primarily composed of oak, including white, black, post, and scrub, with an understory of witch hazel, blueberry, huckleberry, sheep laurel, and viburnum, plus, of course, the pitch pine. Because pitch pines are "intolerant," or greedy for light, they will die when overtopped by other trees. The result is that they are losing out where the oaks have been allowed to grow. Not too many individual pitch pines get a chance to grow to great height, except in sheltered hollows by themselves, and even there the wind sometimes comes roaring in from the open shores at sixty to seventy miles an hour and may crack a big one in two. Sea wind and salt spray tend to keep them down, even for several miles inland, and the oaks, which have been incessantly cut over for centuries, have been weakened by insect borers, fungus, and rot. They do not make the best-quality wood, but as stove wood, or for a fireplace, they burn slowly and well. To split oak logs, especially when the wood fiber snaps in freezing weather, is a pleasure to me; and the cut wood smells good, though "piss oak" is an old term for the black oak.

Their gray trunks are spotted and plated with lichens, some a pale, seashell pink, some gray-blue, others a light green. As you look through them, it is as if they were only a landward extension of the marine world, different forms of the same space. Water, too, is the life of lichens that require a share of mist and dew, of fog and rain. An abundance of fungi and lichens on the trees indicates a wood that has been neither burned nor consistently pruned. In terms of management, it implies lack of care. Since I have not cut down the trees in a very selective way, cleared out the understory, or burned all the dead wood, my neglect has opened up these acres to the greater standards of the wild.

Fungi, which may grow on dead or dying oaks in the form of pink or orange clusters, curling or cuplike, are often hard to tell apart from the lichens. They are plant life which preys on either living or decaying matter, whereas the lichens are agriculturists, cultivating small algae within themselves, protecting them, in return for a share of the food they produce. They are, in fact, made up of an algae and a fungus. These two species join in a single partnership so precise as to produce a lichen that can be scientifically determined as a distinct plant, but both reproduce separately. They have to discover each other, through spores borne widely by the wind and water, so as to produce a lichen of an identifiable species; although friable reindeer moss, which is a lichen, can regenerate from broken fragments if the ground suits them. Lichens take their own slow time about growing, and one you spot on a rock may be centuries old. Apparently, the partnership is not always secure, because, in isolated instances, the alga may be parasitized and killed off by the fungus.

The lichens are delicately beautiful, with colors that complement the bark of the trees. Some are branching, corallike; others have crenellated edges like a flower or a cloud; still others have dotted, wrinkled, or tufted surfaces; and some branch or fork out like flames, or seem to be on the verge of spreading forward like waves. Although they are an original form of life with provisions for an indefinite future, they are unable to tolerate industrial smoke, or air that has been polluted by chemical wastes and gases. They are the signs of a cleaner world.

Lichens belong to an endless suspension in time out of which lives sense their moments of affinity, their opportunities to parasitize, to bond or devour. It is an awesome game, hanging on eternities. The existence of a lichen depends on the centrifugal nature of all contrary forces, on a scale that seems perilous to our short-term sense of things, though exhilarating in the mind. We are just insecure enough to wipe out these primary plants. On the other hand, isn't the human personality itself made up of seemingly irreconcilable components? I wonder how *we* hold together.

So the society of trees—with their companions in a fire of blue light—move on while standing still, tested as they grow by all the combining, convening forces of earth and atmosphere. They are sinewy and tough, erect in their vicissitudes, and their own version of desire, taking their symbionts and predators with them. They do not give in too easily to their destroyers. Cut the oaks down ten times over, and unless you bulldoze out their roots or destroy the ground beyond repair they will sprout up and take up even more space. It is as if they said: "Extermination is not our destiny."

Older religions had no trouble connecting men and trees. A tree was the tree of life, magical bearer of fruit, container of mercurial speech, a trunk with a god inside. At times, when one of them was injured or appeared to wither, a man's life might depend on the outcome. In Europe, certain trees were supposed to have healing properties. Denmark still has oaks that are centuries old. When I was there I was told that a few had reached a thousand years. They had been spared because they grew on the king's land, originally set aside as hunting preserves. An old farmer named Jacobsen, who had some interest in the life of birds, told me that he once rode his horse into an ancient, hollow tree and was able to turn around inside. I was shown a medium-sized oak with a gap in the middle through which children were passed so as to ward off what he called "English sickness," characterized by bad teeth and weak legs—no great compliment to the English.

Old belief may seem weak and childlike to skeptical minds that judge things on the basis of objective verification, but it saw the roots of things. It knew trees, as it sensed itself gone to earth in the process of growth, decline, and death.

In my search for an education in correspondence I owe a great deal to names. *Quercus alba* almost satisfies me in the sight of a white oak, and I could busy myself for years on end with the principles by which a leaf or a needle is adapted to the wind, the leaves of a poplar twisting on flattened stems, the thin needles of a white pine hanging and swaying in silvery compliance with every air that reaches it. Yet the trees are always asking more of me. I know that their staying power needs further exploration, more than one civilization to solve it.

I listen to the wind swishing through the pitch pines and seething or rustling in the oaks, while blue jays bounce cockily overhead, conversing in brassy, insinuating tones. I sense thousands of years between their voices, hanging millennias. Time opens out again, as the trees hold up their pinwheels of needles and their lacy twigs in the carousel of the wind.

Under the full moon, the ground is a network of intricate shadows, meticulously drawn. The trees seem to move across the fluidity of light, extending electric arms and fingers. Their trunks are

171

braced against the wobbling, racing planet. They seem to lift me with them in a sailing of their own. We go between yesterday's wind and rain, today's jubilant, far-reaching light, the revivals of the weather. Another sunset comes to fill the sky with molten colors like tropical birds, crossed by golden, braided strands, an impermanent wall for sight, while next to my bare, encrusted flesh, the bark and trunk with which I meet the air, I am with trees. We wait, and we move out at the same time. It is not only the long-distance migrants who make daring leaps into the unknown.

III

LIVING

"We are wont to imagine rare and delectable places in some remote and more celestial corner of the system, behind the constellation of Cassiopeia's Chair, far from noise and disturbance. I discovered that my house actually had its site in such a withdrawn, but forever new and unprofaned, part of the universe.

. . . My instinct tells me that my head is an organ for burrowing, as some creatures use their snout and fore paws, and with it I would mine and burrow my way through these hills. I think that the richest vein is somewhere hereabouts . . ."

HENRY DAVID THOREAU, *WALDEN*, 1854

ROBERT
FINCH

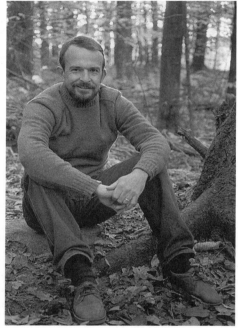

Photograph © Stephen Trimble

Robert Finch is a "New Englander by adoption." He started out as a New Jerseyite, born in 1943 "by the beautiful Passaic River," and moved to West Virginia when he was twelve. He first came to Cape Cod during college summers, working in a sailing camp, and came back again the first year of his marriage. "By then I knew that I wanted to write and that I wanted to write about this place." Finch moved to the Cape in 1971.

He has since published three books of Cape Cod essays: *Common Ground: A Naturalist's Cape Cod* (1981); *The Primal Place* (1983), which includes the two essays that follow; and *Outlands: Journeys to the Outer Edges of Cape Cod* (1986). He feels he has one more book of Cape essays seeded in his journal, and he is editing a Norton anthology of natural history writing. Building a successful collection of essays is not easy, Finch says, "I spent more time actually trying to structure the material than writing it."

Finch comes out of a "strictly humanist background." He taught English at Oregon State University after graduate school in Indiana, and has been on the staff of the Breadloaf Writer's Conference at Middlebury College. He simply found that the material in his journal "seemed to group itself around a natural history focus." Like his Cape neighbor, John Hay, he has been active in both the local Conservation Commission and the Natural History Museum. He jokes that his only academic claim to fame is that he was the first person to teach Cape Cod literature at Cape Cod Community College.

Finch has been working through Thoreau's journals at the same pace as his life, reading the journal passage from a given day that corresponds to his own age that day. He says, "Everybody has to come to terms with Henry David. And I still am. He sort of invented the language, the philosophical language. But I also learned a lot about writing from writers that have nothing to do with nature writing."

Finch feels that writing about the Cape Cod landscape comes both from living there—"what you're used to, what you're fixed on in this place of contrasts"—and from somewhere deeper in the past, "certain shapes of your own imagination." He believes that it is "*personal* essays that make for natural history writing" and that "digression is a very, very positive thing. There is no way you will write like somebody else if you write for yourself."

177

Into the Maze

1

One of the occupational hazards of living in a place like Cape Cod is not always knowing where you are. The sea fog that rolls in regularly over the mud flats and salt marshes is not entirely to blame for such chronic disorientation. Nor are the winter northeasterlies whose heavy surf and storm surges break through barrier beaches, destroy parking lots, silt up harbors, and claim waterfront property all that dislocate us.

Change is the coin of this sandy realm, and as long as we are not too close to it, such change delights us. The seasons flow in their rhythmic variety, a little out of sync with the mainland due to the ocean's moderating influence—which pleases our sense of separateness. With them come in the streaming tides of shorebirds, migrating alewives and striped bass, pack ice in Cape Cod Bay, spring peepers in the bogs, gypsy moths in the oaks, and tourists in the motels and restaurants.

Years flow and bring still broader changes, sometimes surprising, not always welcome. Bald heaths grow up to pine barrens, meadows fill in with juniper, abandoned bogs return to cedar swamps or maple swamps, oaks replace the pines, and a charming water view from the deck or terrace disappears under a rising horizon of leaves.

With these changes some new bird species appear, others grow scarcer. Fish populations fluctuate, ponds slowly silt in. New areas of tidal flats are claimed by spreading salt-marsh grasses, and each year a few more feet of the ocean cliffs topple into the surf, taking a beach cottage or two with them. Major alterations in the shape of the coastline can and do take place within a man's lifetime, adding a feeling of shared mortality in our relationship with this thin spar of glacial leavings. Yet through all this variety of natural change we also sense a continuity, not always to our liking, perhaps, but with a fittingness and perceptible identity of its own, an interplay of great connected forces.

To this natural change, however, we have added our own, in a way that we share with most other parts of the country. In the beginning we may have desired only to "fit in" to this natural scene, to enjoy what it has to offer; and yet in doing so on such a mass scale and on our own terms, we have inevitably introduced forces that have had increasing repercussions of their own.

We move here in winter onto some quiet street and find the following summer that traffic makes it unsafe to cross the road for our mail. A piece of woods where we used to walk our dogs is turned, almost overnight, into roads and building lots. An open stretch of coastal bluffs that once formed a background to our clamming on the mud flats is now clotted with condominiums. Along the Mid-Cape Highway the deer we noticed for years are one day no longer there; in their stead are houses and tennis courts. Back roads and open fields, where fox stalked and woodcock courted, all at once sprout shopping malls, golf courses, new schools, and sewage-treatment plants. And so on.

Countryside is suddenly suburban, suburban areas become densely developed, and in places our highways and urbanized areas begin to take on an aspect that makes us look hard at the exit and street signs in order to reassure ourselves we are not in Boston or New Bedford, yet. Having increased our individual mobility in both the physical and social sense—the speed and ease with which we can travel from place to place as well as the power to choose our hometowns—we find ourselves less and less sure of where it is we have finally arrived.

Sometimes, watching a chickadee or a junco at the window feeder at the end of a winter's day, ruffled and tossed by a wet wind and alone at the coming of darkness, I am tempted to pity its lack of human comforts and security. But the bird at least was born to the condition in which it lives. It is part of an unbroken past of this land and knows where to find itself, despite all human and natural change, during the night and in the morning. Can I say as much for myself?

What is Cape Cod today? Rural? Semirural? Suburban? Seasonally urban? Bits and parts of each, perhaps. For this particular moment, at least, the term *subrural* seems as accurate as any: a patchwork of conflicting claims and uses hanging on to the remnants of a distinct rural culture that now exists almost completely in the past. And once we have named it, what then? What are we to make of it? How are we to know where we are? How are we to get here, once we have arrived?

2

The first step must be to see clearly what is there. This is often more difficult than it might appear, for nature has no guile, which is one of the things that makes it so hard for us to see. The bare, uncompromising face of the land is too much for us to behold, and so we clothe it in myth, sentiment, and imposed expectations.

In West Brewster, for instance, where I live, the scene outside my window looks more like western Connecticut or Minnesota than like what most people think of as a typical Cape Cod landscape. No sandy shores or low dunes, no salt marshes or wide ocean waters stretch out before me. Rather, the house I built here a few years ago sits well inland, tucked into the wooded, hilly terrain of a low, broken line of glacial hills known as the Sandwich Moraine. The moraine begins west at the Cape Cod Canal, rises quickly to a height of just over 300 feet, runs east along the edge of Cape Cod Bay for some forty miles in a descending and increasingly interrupted ridge of loose till and rocks, and eventually peters out into the Atlantic at Orleans, where the arm of the Cape bends north toward Provincetown.

My house is situated near the eastern end of this moraine, on one of the lower crests about eighty feet above mean sea level. It faces south, on the north side of a roughly circular ridge of hills. These hills enclose a steep-sided bowl, or kettle hole, about a quarter of a mile in diameter, known locally as Berry's Hole.

The house is also circular in shape—octagonal, to be exact—with wide overhangs that combine with the higher hills around it to effectively block out most of the sky from inside, even in winter. When it was built, the top of the hill was leveled off and lowered a few feet, so that the yard, which stretches south from the house to the edge of the kettle hole, appears to leap off into nothing, into some great abyss, rather than falling off, as it does, into the rather modest hollow below.

The soil in these hills is sandy by mainland standards, but compared with beach sand it is heavy with clay and studded not only with small stones but also with many large glacial boulders twenty feet or more in length, called glacial erratics. Likewise, there is little of the low seaside vegetation generally associated with a shoreline environment—beach plum, bayberry, pitch pine. Instead, the surrounding slopes are covered with nearly pure stands of oak, that inland tree. For as far as I can see, unbroken stands of black and white oak lift their gray, lichen-spotted trunks and branches crookedly skyward.

Since the soil is relatively poor in nutrients, the oaks tend to be dwarfed in stature and prone to insect infestations. Nevertheless, they combine with the land to create an illusion of size. Stunted and twisted by poor soil, overcutting, and salt-laden winds, these trees possess the crooked look of age. They do not dominate the low hills but fit in proportion to them, so that together they give the impression of a far greater scale than either really has. It is easy to look out at them and see full-sized forests lining the flanks of mountain ranges that stretch for miles across some formidable gorge. The Cape has its own scale, and, to one not used to it, its landscape is full of such tricks of proportion and perspective.

No shorebirds, terns, herons, or crabs inhabit the immediate area. Rather, my wild neighbors are woodland fauna—deer, crows, owls, grouse, and a fox who lives in a burrow halfway down the hollow. Wood thrushes and phoebes nest about the house in summer, whippoorwills sing at night, and red-tailed hawks wheel slowly overhead, sending down sibilant screams from a high October sky. In the spring the slopes of Berry's Hole are covered with star flowers, lady's-slipper orchids, and trailing arbutus, or mayflower, while out of its deep, wet throat comes the tumultuous, insistent, nighttime chorus of wood frogs and spring peepers.

In short, although my home, as the herring gull flies, is less than half a mile inland from the salt marshes and shallow waters of Cape Cod Bay, it often takes an effort of will to remind myself that I live, not deep in the heart of the continent, but on an exposed and vulnerable headland thirty miles out into the open Atlantic, on the thin shores of a narrow land.

3

One May morning, several months after moving into the house, I looked up from the table where I had been typing and saw the stiff, gray, intersecting patterns of the oak trunks and branches outside the window. I saw them, as though for the first time, for what they really were: a maze, a vast, living maze stretching out beyond my lines of sight. And all at once I knew, with a clear and compelling conviction, that what I wanted, what I was seeking here, was entrance, or rather re-entrance, into that maze.

The trees are not an impenetrable thicket. They are, in fact, more like the original Cape Cod forest encountered by the Puritans during the *Mayflower's* initial landfall at Provincetown in November of 1620 and described by Gov. William Bradford as "open and fit to go riding in." I can easily get up from my chair, open the door, cross the yard, and walk down the wooded slope to the bottom of the kettle hole. I have done so many times, before and since, but whenever I do my every noise and movement reveals me as an outsider, an intruder. I jangle my credentials as I go, crashing through the dry leaves, cursing as my jacket or pants leg catches on a barbed strand of catbrier.

What I want is to go silently and smoothly into the maze, without a rustle, as the light fox bounds with inborn agility across the rounded stone walls; as the soft rabbit threads itself surely and painlessly through the brier and viburnum jungles; as the lean, long-legged deer stops to look at me, sideways, with wet black eyes, then steps cleanly and quickly over brush and branch, disappearing up the slope without a word, as though into a fog; as the sharp-shinned hawk flits batlike through the web of branches; as the flicker leaps up and glides out through the layered oak boughs; as the green-and-gold garter snake, warming itself on a stone in the spring sun, suddenly bolts at my presence and, like a sand eel, vanishes in a glistening wriggle down into a crease in the earth and is gone.

These creatures tease me with their unconscious competence, a sureness that implies not so much prowess as belonging, of knowing where and who they are, of being local inhabitants in a way I am not.

Sometimes, frustrated by the unyielding rigidity of these woods in the face of my overtures, I go out with my chainsaw and cut down a few more of the trees around the house, pretending that I am getting firewood. For a few minutes the air is full of a geared thunder that obliterates all perception or participation, until the rashness of my deed catches up with me and I stop, finding myself in a space suddenly cleared and empty, surrounded by a quailed silence, having solved nothing and gotten nowhere.

There is no Gordian knot to cut here. Every part of the maze is a knot tied to every other part. To cut down the trees and scatter the animals, to make broad paths and wide clearings, is not to solve or enter the mystery, but to obliterate it and erect empty designs in its place. Such acts may give me passage and room to move about, but not entrance, and entrance is what I crave.

There is no quick, easy way into this or any other place, no sign pointing out the beginning of the path—no path to point out, for that matter. And yet, as I look out from this house into the round yard bordered by a sea of trees, and beyond them to the unseen shape of the peninsula itself, uncurling like a tendril growing into the sea, I seem to sense that this spot is as good as any from which to begin.

I take the sheet of paper, half-filled with sentences, out of the typewriter and hold it up before my eyes. Turning the sheet sideways, I look over its edge out the window to the trees beyond. When I do, the vertical lines of black ink begin to blur into the dark, rising bars of the trunks. It is a self-conscious gesture, but perhaps that is what it takes—a deliberate change of perspective, a loosening of focus, and a bending of your lines of sight to what it is you would see.

Or perhaps the secret is even simpler, as simple as the insistent, hidden song of the ovenbird, deep in the layered woods around me, that now begins to rise up out of the kettle hole in a ringing and ever more confident crescendo: *teach-er, Teach-er, TEACH-ER,* **TEACH-ER!**

Scratching

1

North of the Mill Ponds, at the lower end of Stony Brook Valley, where the low glacial hills spread apart and slope down toward sea level, the herring river flows through a large tidal gate beneath the state highway, and in so doing changes both its name and its character. From Stony Brook, a shallow, swift-flowing, freshwater stream, it becomes Paine's Creek, a deeper, wider, saltwater tidal river that elbows its way another half a mile north through broad stretches of salt marsh, finally emptying through a break in the shoreline out onto the wide tidal flats of Cape Cod Bay.

One day in early spring, while raking for sea clams at low tide, nearly a mile offshore from the mouth of Paine's Creek, I hit what I first thought was an enormous clam. When I finally managed to work it out of the sand, however, it turned out to be the skull of a blackfish, or pilot whale, a relatively small cetacean that was once commonly hunted in the waters of the Bay. It was a large head, about thirty inches long and nearly two feet wide. Lying there on the dark sands, with its long, sloping upper jaw, it reminded me of one of those hollow-eyed cow skulls that bleakly dot the western deserts. But no desert skull harbors such a diverse community of life as this one did. The whale head was dark and discolored with marine algae. Ribbons of *Fucus,* or rockweed, streamed out of its nostrils. A green crab crawled agilely out of one of the eye sockets to see what had disturbed its home, and the soft, porous, crumbling gray bone was decorated with the mottled, pearly pink hulls of attached slipper shells. There was a flowering of life, if not thought, in this ancient head; it seemed to stare up at me out of the muddy sand with its dark, hollow eyes and speak mutely of another age. As I left it and walked back toward shore with my bucket of clams, I passed lines and broken arcs of post stumps, hairy with fine, dark green seaweed. These were the sheared-off remnants of the many fishing weirs that used to trace their elaborate patterns of poles and nets across these flats. Both were reminders of a time, not too far distant, when blackfish were still a common sight in these waters and men harvested an earnest living, and not merely an occasional bucket of clams, from these shores.

How well-used these long, low flats have been! And how remarkably unscathed they have remained, even today, despite the increasing intensity of human activity. Everywhere over their dark, wet plains are spread the artifacts of past human use: the skeletons of whales, the stumps of fish weirs, crumbling stone breakwaters at the old packet landing, occasional broken bottles of bootlegged whiskey, tossed overboard at night during Prohibition by local rumrunners, the blackened hulks of small wrecks and derelict vessels, alternately buried and exhumed by shifting currents and bars. There even appears, on rare occasions, a human skull.

This, as Thoreau might have put it, is the only genuine Historical Society of Cape Cod, its exhibits artfully arranged in spacious, authentic settings, frequently changed, open to view twice a day, seven days a week, at no charge, their lessons clear and needing no human labels.

2

These expansive tidal flats are peculiar to the Bay side of the Cape. On the south side, or Nantucket Sound shore, meager tides of only three or four feet reveal barely a hundred feet of sea bottom at dead low. In Cape Cod Bay, by contrast, the tides heave and collapse ten feet or more on new moons, sweeping out over the sands for well over a mile in places. I have heard it claimed that Brewster has the largest area of tidal flats of any town on the eastern seaboard.

Every twelve and a half hours the ebbing tide reveals a vast drowned world of extremes, contrasts, specialized needs, appalling barrenness, and, at times, equally appalling abundance. Despite such obvious differences as its marine life forms, the inner flats are in fact a kind of biological wet desert: visually barren, alkaline, lacking in fresh water, and subject to sudden extremes of temperature and moisture. As with the animals of a dry desert during the day, the permanent inhabitants of the flats remain largely underground and out of sight at low tide, emerging only with the twice-daily flood, the aquatic night, to feed and be fed upon. When a dry wind skips over the exposed flats, it can kick up a sandstorm that will sting the face and blind the eyes. And when a stranded fish or bird or marine mammal dies here, the legions of gulls, which have been sitting patiently on the bars amid the exposed whitened shells of sea clams, descend to the feast with the same horrible enthusiasm and grim grace as vultures do.

But it is not just their dramatic display of tidal movement or the harsh extremes to which the life that lives there must adapt that makes the flats a special environment. They also offer us challenges and connections in a wider world. In contrast to the maze of woods around my house, the flats pose a problem of orientation rather than one of entry. Out there one finds few or no landmarks to go by; distance and size become notoriously deceptive, and men have been lost and drowned on the far bars when sudden fogs come in with the tide. Still, these wet plains are of superlative extent and present opportunities for exposure and encounter that are unparalleled in sheer magnitude anywhere else on the Cape's, if not the continent's, shores.

Like most local residents, however, I usually go out on the flats not for adventure but for the pragmatic and rather tame purpose of gathering clams. "Clamming" is a generic term used to refer to the harvesting of a broad spectrum of shellfish, including oysters, soft-shell or steamer clams, mussels, sea clams, razor clams, and even scallops. All of these shellfish can be found to a greater or lesser degree in various parts of the Bay, but in Brewster, owing to local conditions, "clamming" is generally understood to refer to the pedestrian pursuit of a single mollusk—*Venus mercenaria,* the quahog or hard-shell clam.

Some of these conditions are environmental. The lack of proper salinity and substrate, for instance, makes our local flats unsuitable for oysters. But it is more the innate qualities of the quahog that have combined to make it the true shellfish of Brewster's waters. Widely distributed, the quahog tends to occupy a middle ground, neither so far in as the steamer clam (which, easily reached, has been virtually dug out here for years) nor so far out as the sea clam (which can be raked only on the lowest moon tides). It can thus be gathered on most ordinary tides and, in contrast to the swift-burrowing razor clam, requires only moderate skill and effort to catch. Also, unlike the mercurial scallop, the quahog varies little in abundance from year to year.

Plentiful and dependable, these mollusks are also "sociable," in that they lend themselves not only to human enterprise but to encouragement as well. Local quahogs are not, strictly speaking, a wild crop. For years the town has run a seeding program to replenish the supply, enforced quotas, and rotated the sections of flats open to digging on a seasonal basis. In this sense they can be considered "managed," though the quahogs' ultimate success and maintenance still depend far more on the overall health and influence of the Bay than on human manipulation.

Perhaps it is making a virtue of necessity, but the quahog seems not only the most available but also the most rewarding of local shellfish to harvest. Any form of clamming is always much more than the mere taking and eating of a bivalve mollusk, and the quahog's positive qualities go far beyond its reliability and accessibility. When dug up, quahogs have a rich, weathered, blue-and-gray cast to them, as though they had lain beneath the mud for ages, forming slowly like minerals in the earth. Their smooth, rounded, gracefully asymmetrical shape fits satisfyingly in the palm, and their shells are covered with patterns of growth rings that reflect their environment: rippled sand flats, expanding waves, a concentrically spreading world. They take their color and form from where they live and possess that perfection of completeness, of circumference, like the compassed tufts of beach grass whose blade tips draw rings around themselves in the dune sand.

Moreover, we always seem to become part of whatever we pursue. The soft-shell or steamer clam, for instance, is a serious, single-minded creature, deeply entrenched in its hole and industriously siphoning its food through a long, extended "foot." To catch one, the clam digger must squat down or bend far over, keeping his gaze directedly downward to the clam, following it with his thoughts into its dark, mucky hole, and prying it out, not only with his short-handled rake and fingers, but also with the intensity of his concentration. Thus the digging of steamers tends to be a self-absorbed, introspective activity, and when engaged in it I find that I catch a sense of my surroundings only at scattered moments, holding communion for the most part with the clam alone. Whosoever digs that clam loses a portion of the day. He grows a kind of shell over himself that blocks out sunlight.

Quahogging, on the other hand, tends to be a cleaner and more expansive business. Because the quahog is a shallow burrower, the scratcher can remain upright, coaxing rather than wrestling his quarry out of the mud with his long-handled rake. Though his concentration may be in the rake, his consciousness can remain afloat, above the mud, above the water, and he is free to cast his gaze around him and rise to the lures of the day.

3

The true Cape Codder, the local saying goes, spends more time in the water between October and May than from June to September. By this standard, at least, I may claim the status of a genuine resident, for it is during these colder months that the best quahog flats in our town are open to scratching.

Late on an afternoon in early October, when the first set of high-course tides for the month has arrived, I drive down to the Bay shore. It is less than an hour until the end of the tide's long, silent withdrawal, and the flats lie exposed, a glistening expanse of dual nature, half land and half water. I pull on my chest-high waders, hitch my rake and bucket in hand, and, with the sloping sun over my left shoulder, set out after the receding tide.

My footsteps push broad sheens of silver wetness out of the saturated sands as I go. Braided veins of water, draining out from under the slope of the beach like reversed river deltas, weave together and follow me in small, meandering streams. The flats here are strewn with the thin, brittle, empty shells of last year's scallop crop, blown ashore and frozen by winter storms; I crunch them underfoot as I go. Small groups of gulls sit quietly on the higher bars, giving me wary, sidelong glances as I pass. They are content, as always, to wait.

Soon I reach the edge of the withdrawing tide, a few hundred yards from shore. I continue out, sloshing through the shallows, and as the deepening water gradually slows my progress, the abundance of life about my feet increases. Schools of hundreds of minnows nibble at the very hem of the tide, then scatter away in shimmers like wind over water at my approach. Hermit crabs scuttle and paw across the sands in their rocking periwinkle or moon-snail shells. An occasional *Nereis,* or common seaworm, flashes by like an elastic-green ribbon, and everywhere the transparent sand shrimp dart invisibly about, making tiny explosions of sand grains along the bottom.

Farther out I begin to enter the extensive beds of eelgrass, streaming northeastward in the outrunning current. If the inner flats are desertlike, these eelgrass beds are more like plains, or prairies, open eelgrasslands pocked with saltwater holes, where at the lowest tides wide, shallow rivers wind Platte-like through their shifting channels, and tiny burrows of crabs, clams, and plumed worms dot the mud and sand like miniature prairie-dog towns. Now, in the shallow water, early flocks of brant—small, darker versions of Canada geese—graze on the floating tresses of the eelgrass, and two miles out the distant roaring of the surf sounds like unseen herds of bison or antelope.

The eelgrass marks the beginning of the quahog beds. Areas of abundance shift slightly from year to year, but a few rockweed-skirted boulders on the inner flats afford me a rough triangulation to locate myself where past experience has shown the digging to be best. Now about a quarter of a mile out and knee-deep in water, I set down my wire basket, attached by a short rope to a small Styrofoam float, and begin scratching.

"Scratching," like "clamming," can be used in a loose way to denote the gathering of any shellfish with a pronged instrument. Used properly, however, it is part of a very precise terminology and refers only to the harvesting of quahogs (and perhaps sea clams) with a quahog rake. As such, scratching is a highly accurate and descriptive term. Soft-shell and razor clams, which often lie a foot or more below the surface, must be "dug" with a clam rake, a short-handled tool with thick, bent prongs. Scallops, lying on the surface (when they are not swimming around), are more properly "raked" with an instrument very much like a garden rake with a basket attached. But quahogs generally lie just below the surface of the flats and are therefore "scratched" out.

In summer, on the more distant bars, the larger "chowder" quahogs are sometimes found with their "shoulders," or valve hinges, showing white and dry an inch or so above the mud surface. But now, in autumn, when the water in the Bay begins to turn colder, the quahogs dig in deeper, though never more than five or six inches, the depth of the rake tines. In clear, calm water, on a clean bottom, it is relatively easy to tell when you have struck one. Often, though, the water is so choppy and murky that you can hardly distinguish grassy from bare areas. Then you have to be careful not to step inadvertently into one of the "ice holes"—shallow, bare depressions in the eelgrass beds gouged out by the movement of winter ice—which can result in a sudden, chilly bath.

Under these circumstances, and especially when there is an abundance of empty scallop shells partially buried in the sand, it can be quite difficult to tell when a quahog has been struck. This is particularly true of the smaller quahogs, "cherrystones" and "littlenecks," which give less resistance than the large chowders and often feel like small stones or clumps of empty shells. Then you must *feel* them out, using the rake handle and tines as sensory extensions of your arms, as a woodcock feels out earthworms deep in moist ground with his long, prehensile bill, or as an experienced fisherman can distinguish flounder from a skate on his line before pulling it up.

I have become convinced that a well-made quahog rake is one of those perfect tools, superbly fitted to both use and user. A properly made rake stands slightly over four and a half feet tall and has an ash handle tapering slightly in from the end, then thickening again where it enters the socket. The six steel tines are curved and flattened to slip strongly, yet smoothly, through the mud and sand, deep enough to reach the maximum depth to which a quahog will burrow yet spaced widely enough that anything smaller than a legal-size littleneck (two-inch minimum) will fall through.

Not only is a quahog rake perfectly designed for its primary function, but it is also admirably suited for a variety of other manual tasks, such as raking salt hay off the beach for garden mulch, digging potatoes out of their hills, and plucking apples for cider out of roadside trees. It is, moreover, the most effective tool I know for clearing the formidable catbrier thickets on my lot. Its long handle allows me to keep out of harm's way, while its wide, curved tines will hook the thorny vines and pull them out, yard after barbed yard, yet afterward disengage rather easily from the mounded tangle. No other tool I know of is so at home, and of such use at all seasons on land, sea, and in between.

Scratching is in itself a highly satisfying activity. Although it superficially resembles hoeing or cultivating, it is slower, more careful work, something like pulling loose teeth from soft, rotten gums. I love to set myself against an outrunning tide, eelgrass streaming before me, and methodically rake my rows across the sandy bottom. The curved, rusted tines give a soft shush with each sweep through the sand, as I wait for the feel and sound of the hard, rocky scratch that signals a buried quahog. The lines made by the rake remind me of the pronged chalk holders used by my grade-school music teachers to draw wavy staves across the blackboard. My own moving lines strike and reveal the buried quahogs, which sing their sharp squeaks against the metal tines.

Though I usually do not need to go out more than half a mile for quahogs, on very low tides I occasionally go out into the deep channel that runs parallel to the shoreline nearly a mile out, separating the inner flats from the farthest bars where the sea clams are found. Because this channel is only rarely shallow enough to scratch on foot, I often find untouched "honey holes" of cherrystones and littlenecks here, and fill an entire bucket with on-the-half-shell delights in fifteen minutes. There is also something euphoric about raking in such deep channels, standing chest-deep in waders in the cold, clutching water, rocked softly by the gentle swells as the current runs between the bars. As I begin, the long eelgrass streams steadily to the east, then gradually loses tension, weaves indecisively at slack water, swirling about my legs and obscuring the bottom, then slowly unfurls and streams southwesterly as the tide begins to flood.

4

There is so much life other than quahogs in these waters that on some days my rake has trouble attending to business. Its steel tines impale moon snails, green crabs, calico crabs, spider crabs, horseshoe crabs, channeled whelks, and occasionally small winter flounders buried in the mud. Once, in mid-January, I unearthed a burrowed-in sand eel—a thin, silvery fish about five inches long—which flashed and wriggled as though protesting my impertinence. It was a vibration, a bit of tangible panic in the clear, green water that darted and disappeared in an instant into the mud from which I had mis-taken it.

Above the water, I am surrounded in all seasons by birds. The bird life on the flats is not only plentiful but also rather neatly stratified from shore to deep water. In September and October the most common species along the beach are tree swallows in migration. By the hundreds they swoop low over the marsh grass that fringes the shore. Swallows do not fly in the tight formations of most shorebird flocks, but move in loose and shifting cohesions, like raiding parties. At high tide they thread among the tops of the tawny spartina stalks undulating in waves. Frequently they pause and hover for several seconds like hummingbirds, picking off marsh insects that have climbed the grass stems to escape the tide, only to find a sharper death waiting in the swallows' beaks. As they dip and dart, the flock moves in a gradual and staggered, but steady, progression down the beach, like chips of wood carried in a longshore current, traveling slowly on, feeding as they go, toward the declining sun.

By the time these quahog beds open, in October, the peak of the "fall" shorebird migration (which actually begins in early July) has long passed, but over the inner flats at low tide there are still numbers of small sandpipers or "peep," black-bellied plovers, an occasional golden plover, scattered yellowlegs, and small flocks of ruddy turnstones (one of the happier as well as more descriptive of common bird names). Running along the sand, feeding in the shallows and tide pools, these birds stitch together all the varied terrain with their linear footprints and probing bills.

With them, as in all seasons, are the scavenger crows and gulls. Out here on the flats the ubiquitous herring and black-backed gulls give the same general impression of satisfied indolence that they do at the dump or on the pond. Standing about in flocks of varying size, they talk in low, muffled tones and hoarse cackles, like old women— seductive, self-concerned, noncommittal. But these gulls are actually more industrious than they are usually given credit for being. Often, as I go out, I see them standing in the little outrunning tidal streams, paddling backward in the mud like ducks, stirring up small crustaceans and worms, which they quickly snatch up. Better known is their habit of dropping quahogs and other shellfish on rocks, hard mud, parking lots, and even nearby golf-course cart ways to crack them open. Once, while scratching, I became aware of a large flock of gulls swarming and dipping down into the water all around me. They were skillfully plucking long, fringed, *Nereis* seaworms (which emerge from their burrows when the tide begins to flood) from just below the surface. It was a feat at which they were obviously well practiced, though they seemed just as cantankerous and bickering as ever, just as willing to harass and steal from one another as when in the midst of garbage-dump plenty.

From early May to mid-September I see the white sculptured forms of terns fishing for sand eels in short, explosive dives over the shallow Bay waters. By October, however, all but a few of these small seabirds have gathered and left for their wintering quarters in the Caribbean and South America. But an even more dramatic diving display sometimes occurs in late fall when sustained southeasterly winds blow hundreds of migrating gannets into the Bay.

One morning in mid-November, following just such a blow, I spent several hours on the outer bars sea-clamming. About seven o'clock I noticed gannets sailing close in, just beyond the bars: large, white, powerful birds that glided silently on arched, black-tipped wings nearly six feet across. There were forty or fifty in all. Most were white adults, but there were several brown speckled immatures and a few just entering adult plumage that still had some remaining speckles on their wings. They gathered gradually, out of nowhere, like a storm forming over the sea. Circling over an area half a mile north of where I was raking, they shortly began plunging after an invisible school of fish.

Gannets do not drop directly into the water as terns do, but dive diagonally downward from heights of fifty to a hundred feet, keeping their great wings outstretched until the last few feet, when they clap them swiftly shut and disappear in high bursts of white spray. Actually, the splashes were smaller than I might have expected at that height from a bird nearly as large as a swan, but gannets are extremely sleek and tapered in shape. In contrast to a tern flock's dense, combine-like reaping of the waters, theirs was a more circular, freewheeling, dog-fight affair. For half an hour or more I watched their unexcelled display of aerial predation. Finally they flew off or, rather, dispersed, as quietly, separately, and imperceptibly as they had gathered, like a storm that gradually diminishes.

It is also during the fall scratching season that the Cape's winter waterfowl begin to appear in the Bay: arriving flocks of brant and geese, long, dark lines of migrating scoters, and the wintering sea ducks—eiders, mergansers, whistlers, old-squaw, canvasbacks, scaup, the diminutive ruddies and buffleheads—a long and varied parade that is normally not completed until mid-December.

At high tide many of these waterfowl can be seen close inshore, feeding on the eelgrass and tidal organisms. But at low tide, when I am out on the flats, I can commonly see, far out on the horizon to the north, long waves and skeins of brant and scoters wending over the dark waters in massive flocks of a thousand or more, accompanied by a distant cackling din that sounds like a chorus of wood frogs. Nearer in, a line of heavy-bodied eiders, pied drakes and drab females, fly suddenly out of the west and close overhead with rose-tinted breasts, soundless but for the rush of heavy wings. In the middle distance, patches and squadrons of tiny, bobbing buffleheads and sleek mergansers weave, circle, swim, and swing back and forth in endless restlessness across the choppy, blue water. Such casual, distant water ballets suggest an even greater, unseen abundance. How much more uncounted bird life lies within the Bay's circling arm beyond my radius of vision? Yet what I can see fills me with a waterfelt buoyancy that often leaves me standing idle, my rake and bucket forgotten, while patterns of weather wheel overhead.

Finally, though, it is the weather itself that proves to be the most powerful, suggestive, and distracting force out on the flats. In late fall and early winter, the best low tides of the month come just at or after sundown, when the new or full moon is rising. While clamming then, I am frequently taken unawares by remarkable sunset displays. Patches of tinted light, swimming in nearer channels, catch my eye: gradually they spread over the wet, sand-rippled plains, flowing farther out and taking me with them, becoming broad swaths and rafts of glowing colors that congeal and swim slowly westward over the dark waters, meeting at last in a sky of broken sun shafts and rose clouds that move silently above a whitened horizon at the end of a perfectly still day.

At such times even the most prosaic, earthbound clam digger is likely to be caught and held by such atmospheric transformations, compelled into some inarticulate response, if only that of pointing his arm to the obvious and omnipresent in a mute gesture of acknowledgment. No drug, no wine, goes to my head like such moments out on the flats. Engulfed by such splendor, I am plucked out of myself like a hermit crab from his borrowed shell and left stranded, naked and unfinished, on the sands.

Scratching is not a prerequisite for experiencing such moments, but it does seem to make them more likely, or at least more intense. It is as though the rhythm and simplicity of the work has a cathartic, purifying effect on my senses, raking and clearing away the accumulated debris of prolonged indoor activity, making me more receptive to outer beauty. In fact, I sometimes suspect that such states of receptivity are my real reason for going clamming, the hidden quarry I may spend an entire afternoon indirectly stalking, the "honey hole" I may by chance stumble on through such physical and mundane labor.

And yet—perhaps because I am so strongly affected by them and seem to spend so much time placing myself in propitious situations in order to catch them—I often ask myself, even while held transfixed in their grip, What good are they, these moments, however beautiful and rare? What does it matter that light and color play in the sky and on the waters, lifting or driving us into states of open-mouthed wonder and a feeling of sudden evisceration? Do such emotions create action or character? Am I really hitched to any deeper life or wider power because of them? Does experiencing them ennoble my behavior or my responses in the human arena? Do I treat my family better or enjoy my real work more because of them? Do they even make me more ready or more sensitive to other, similar moments? I would like to think they do, but it doesn't seem so.

Moreover, such moments can be downright distracting from the work at hand. Sometimes, late in the day when the net of darkness is tightening over the flats, I find I have the choice of topping off my bucket with a few more quahogs or watching the last moments of an October conflagration on the horizon. This may not matter much when the task is essentially recreational, as it is in my case. But what about when the stakes are mortal, as they largely were in the past? I think of the codfisherman of a century ago, hauling in his handline at the coming of night, alone in his fragile shell of a dory on a wide, landless sea. Was he, I wonder, aware of the impeccable sunset flowering like some unexpected dream in the west, and if he was, did he resent and resist the impulse that drew his tired, salt-reddened eyes from his line and threatened to steal the last precious minutes of light from his catch?

Useless, isolated, and distracting they may be; nevertheless nothing can move us more deeply than such moments of intense and unanticipated apprehension. But what is it that we apprehend? What do we think we see? Flames of eternity? Intimations of immortality? Rhythms of creation? In T. S. Eliot's "The Dry Salvages," the poet speaks of experiencing similar moments off Massachusett's northern cape, Cape Ann, and suggests that they represent "the point of intersection of the timeless with time" or "the moment in and out of time." Is that it? Something genuinely epiphanic, like Emerson's becoming a "transparent eyeball" or Annie Dillard's "tree with lights in it" at Tinker Creek? Do these moments reach beyond the purely physical to something "hidden"? The intensity of our responses may suggest that they do, but for me at least, the actual nature of such experiences, at once overwhelming and ephemeral, will not bear such heavy labels.

Are they, on the other hand, merely physiological? Are we suddenly, simply galvanized by a chance coming together of natural elements that transform us into a kind of bell—a bell rung not by any mystical incarnation but merely by visual and other strictly sensory stimuli, which in turn trigger certain neurochemical and -electric patterns, which in their turn cause us to drop our lower jaws and utter the deity's name? Such a "scientific" explanation may be no more meaningful than any other. In fact, the whole question of beauty may ultimately be no more significant or substantial than that of the number of angels dancing on the head of a pin.

Yet if I had to choose, I would lean toward the latter, more mundane interpretation of these deep and intense moments of natural awareness, not only because it requires no act of faith, but because it strikes me as no less profound or marvelous in its implications than a more transcendental view. It says that our love is not misplaced here on earth, that our sense of wonder and beauty is locked at the very deepest levels into the knotted reality and texture of the physical world from which we wrest a daily living. "The fact," wrote Frost, "is the sweetest dream that labor knows." Who could ask for a more promising vision of life than that? And who would have imagined finding a pearl of such great price in a quahog?

5

Scratching is by and large a solitary business. Nonetheless, it is usually when I am out on the flats that I feel most companionable and sociable. Never do I feel so amiably inclined toward my fellow citizens as when I am raking slightly apart from them.

Usually I go out alone, or at most with one or two companions. Frequently I meet friends out on the flats whom I have not seen for months. We smile and say, "Small world," but it isn't really. Rather it is places like this—tidal creeks, herring runs, tern colonies, crab holes, and clam flats—that are the real thoroughfares and gathering places of our community.

We tend to space ourselves out over a broad area and say little while scratching, content to rake apart in silence, with the wind at our backs, drawing the prongs through the soft, gray muddy sand. What talk there is tends to be of landmarks and seamarks, as though to confirm a sense of where we are out on this wide, unbounded plain. One of us points to a far, dark bar of land to the west, floating like a dead whale on the horizon. "Is that Sandwich?" he shouts. "No, Plymouth," comes the reply. But generally little is said. Wind, light, and space discourage petty conversation, and the tides command our attention.

If we find good digging, we will not move around much but will steadily and methodically reap our own patches of seafloor. The ebb of the tide is our clock, and we ourselves mark its passage as it drops from thigh to knee, knee to calf, until the tops of our wire buckets begin to break above small wavelets and the mounded quahog shells glisten darkly in the nearly horizontal sunlight. As the water falls, the expanse of the Bay drops away from us, leaving each of us alone in a wide, blue field far from land.

This overcrowded peninsula grows increasingly difficult to perceive through its thickening human veneer. Yet out here on these flats I sense there is still space enough for all of us. In the measured rhythm of scratching, in our slow, heavy progress through the water as we move our buckets from place to place, we find an unhurried pace and the sense of a steady, assured, and well-entrenched supply, where every man may go out and find his own limit.

It is getting late. The sky has grown overcast in the course of the afternoon and now grows darker by the minute. I have gotten my limit for the week: ten quarts of chowder quahogs, with perhaps a dozen littlenecks for appetizers. It is time to shoulder my rake and begin the long wade and walk back across the flats to the landing half a mile away.

As I trudge landward, the full, heavy bucket bumps roughly against my waders and grows heavier with each step. This is known as "the quahog's revenge." After a hundred yards or so, I set the bucket down on the hard mud and take a breather. The evening mood is turning all soft and seductive. The gulls laugh softly together on the bars, the low, incoming waters have taken on a silky, saffron color. To the west the strobe-sequined stack of the electric generating plant at the canal spews out a dark, flowing tree of smoke that rises, flattens at the top, and branches out toward the south.

Then, as frequently happens out here at the end of a dull, unremarkable afternoon, the sky explodes with light. The sun drops down out of a shell of dark cloud like a shining, golden yolk and rests for a moment on the horizon, held in a press between earth and sky. Then it falls below the rim of the land and suddenly, as though a switch had been thrown, the whole cowled sky is lit up from beneath, bathing everything in a rosy, diffused glow.

This sudden light gives to the small scattered company of scratchers still out on the flats an unsurpassed intensity of color and outline. Most of them are retired refugees from New Jersey, Long Island, Hartford, or Boston, wearing Swedish rain slickers and wielding rakes made in Parkersburg, West Virginia. But in that moment of searingly pure light I see them in an atavistic posture, an attitude of primitive, intimate contact with the waters surrounding them.

I think of cities I have known, and of the profound loneliness of urban crowds that is the loneliness of a thing unto itself, however large or extensive. True belonging is born of relationships not only to one another but to a place of shared responsibilities and benefits. We love not so much what we have acquired as what we have made and whom we have made it with. There is, at least, the figure for such a love here. And paradoxically, it is in such broad, spacious settings as this—raking the flats, handlining on the banks, working by himself in some common field of endeavor—that a man may feel least alone. The more he allies himself to some varied and interdependent whole, the less he is subject to sudden and wholesale bereavement by chance. His heart rests at the bottom of things; anchored there, he may cast about and never be at sea.

GRETEL
EHRLICH

Photograph © Stephen Trimble

Gretel Ehrlich first came to Wyoming in 1976. She writes, "I had not planned to stay, but I couldn't make myself leave. For the first time I was able to take up residence on earth with no alibis, no self-promoting schemes."

Ehrlich came to Wyoming to make a film about sheepherders with her partner, the man she loved. He died that summer, and she grieved for two years. She stayed on, though, learning herding herself, and then lambing, branding, and calving. She began to write essays about her experiences, and they became the collection *The Solace of Open Spaces* (1985), which won an award from the American Academy of Arts and Letters. Ehrlich describes the book as, "a celebration of everything here. People around here are proud that somebody wrote about them and their state and their way of life."

She says, "I wrote one essay, and then I'd write another, and pretty soon I had fifteen or twenty. Finally it became a book," from which the two reprinted here

198

come. Ehrlich has since published a novel set in Wyoming during World War II (*Heart Mountain*, 1988), and is working on a second volume of essays: "A third or more of the new collection will be about Wyoming, another part of it will be about California—sort of cultural ancient history plus my own personal connections with those places threaded through. The other third will be science, the cosmos.

"This book will be much less personal than *Solace* was—you can only tell your life story once." She laughed, "Thank god that's over with." She has also published two books of poetry and a book of stories (with Edward Hoagland), *City Tales, Wyoming Stories* (1986).

Born in 1946, Ehrlich grew up in Santa Barbara, California. She was educated at Bennington College, the UCLA Film School, and the New School for Social Research. She writes in *Solace* about meeting her husband, Press Stephens, at a John Wayne Film Festival in Cody, Wyoming. They now live "thirty miles from a grocery store, seventy-five from a movie theater," below the raw crags of the Big Horn Mountains. The drafty old rock ranch house faces a treeless field nestled below the upward sweep of the Big Horns. She and Press have lined nearly every wall of their house with bookshelves. Books for insulation. They work the ranch, Press outfits pack trips and Gretel writes, "in a pretty intimate way. In the winter I write a lot sitting in that chair with a notebook, by the fire. It's too cold here for a word processor; the electricity goes on and off all winter."

She says, "There were times I thought I would get a job riding fence in the summers and calving in the winters, and write another book when I have enough material to write about—but now I can't imagine doing anything else. It's like getting used to having a lot of exercise every day—you just can't not do it. It's not easy, but it's a good thing to spend your life doing."

On Water

Frank Hinckley, a neighboring rancher in his seventies, would rather irrigate than ride a horse. He started spreading water on his father's hay and grainfields when he was nine, and his long-term enthusiasm for what's thought of disdainfully by cowboys as "farmers' work" is an example of how a discipline—a daily chore—can grow into a fidelity. When I saw Frank in May he was standing in a dry irrigation ditch looking toward the mountains. The orange tarp dams, hung like curtains from ten-foot-long poles, fluttered in the wind like prayer flags. In Wyoming we are supplicants, waiting all spring for the water to come down, for the snow pack to melt and fill the creeks from which we irrigate. Fall and spring rains amount to less than eight inches a year, while above our ranches, the mountains hold their snows like a secret: no one knows when they will melt or how fast. When the water does come, it floods through the state as if the peaks were silver pitchers tipped forward by mistake. When I looked in, the ditch water had begun dripping over Frank's feet. Then we heard a sound that might have been wind in a steep patch of pines. "Jumpin' Jesus, here it comes," he said, as a head of water, brown and foamy as beer, snaked toward us. He set five dams, digging the bright edges of plastic into silt. Water filled them the way wind fattens a sail, and from three notches cut in the ditch above each dam, water coursed out over a hundred acres of hayfield. When he finished, and the beadwork wetness had spread through the grass, he lowered himself to the ditch and rubbed his face with water.

A season of irrigating here lasts four months. Twenty, thirty, or as many as two hundred dams are changed every twelve hours, ditches are repaired and head gates adjusted to match the inconsistencies of water flow. By September it's over: all but the major Wyoming rivers dry up. Running water is so seasonal it's thought of as a mark on the calendar—a vague wet spot—rather than a geographical site. In May, June, July, and August, water is the sacristy at which we kneel; it equates time going by too fast.

200

Waiting for water is just one of the ways Wyoming ranchers find themselves at the mercy of weather. The hay they irrigate, for example, has to be cut when it's dry but baled with a little dew on it to preserve the leaf. Three days after Frank's water came down, a storm dumped three feet of snow on his alfalfa and the creeks froze up again. His wife "Mike," who grew up in the arid Powder River country, and I rode to the headwaters of our creeks. The elk we startled had been licking ice in a draw. A snow squall rose up from behind a bare ridge and engulfed us. We built a twig fire behind a rock to warm ourselves, then rode home. The creeks didn't thaw completely until June.

Despite the freak snow, April was the second driest in a century; in the lower elevations there had been no precipitation at all. Brisk winds forwarded thunderclouds into the local skies— commuters from other states—but the streamers of rain they let down evaporated before touching us. All month farmers and ranchers burned their irrigation ditches to clear them of obstacles and weeds—optimistic that water would soon come. Shell Valley resembled a battlefield: lines of blue smoke banded every horizon and the cottonwoods that had caught fire by mistake, their outstretched branches blazing, looked human. April, the cruelest month, the month of dry storms.

Six years ago, when I lived on a large sheep ranch, a drought threatened. Every water hole on 100,000 acres of grazing land went dry. We hauled water in clumsy beet-harvest trucks forty miles to spring range, and when we emptied them into a circle of stock tanks, the sheep ran toward us. They pushed to get at the water, trampling lambs in the process, then drank it all in one collective gulp. Other Aprils have brought too much moisture in the form of deadly storms. When a ground blizzard hit one friend's herd in the flatter, eastern part of the state, he knew he had to keep his cattle drifting. If they hit a fence line and had to face the storm, snow would blow into their noses and they'd drown. "We cut wire all the way to Nebraska," he told me. During the same storm another cowboy found his cattle too late: they were buried in a draw under a fifteen-foot drift.

High water comes in June when the runoff peaks, and it's another bugaboo for the ranchers. The otherwise amiable thirty-foot-wide creeks swell and change courses so that when we cross them with livestock, the water is belly-deep or more. Cowboys in the 1800s who rode with the trail herds from Texas often worked in the big rivers on horseback for a week just to cross a thousand head of longhorn steers, losing half of them in the process. On a less-grand scale we have drownings and near drownings here each spring. When we crossed a creek this year the swift current toppled a horse and carried the rider under a log. A cowboy who happened to look back saw her head go under, dove in from horseback, and saved her. At Trapper Creek, where Owen Wister spent several summers in the 1920s and entertained Mr. Hemingway, a cloudburst slapped down on us like a black eye. Scraps of rainbow moved in vertical sweeps of rain that broke apart and disappeared behind a ridge. The creek flooded, taking out a house and a field of corn. We saw one resident walking in a flattened alfalfa field where the river had flowed briefly. "Want to go fishing?" he yelled to us as we rode by. The fish he was throwing into a white bucket were trout that had been "beached" by the flood.

Westerners are ambivalent about water because they've never seen what it can create except havoc and mud. They've never walked through a forest of wild orchids or witnessed the unfurling of five-foot-high ferns. "The only way I like my water is if there's whiskey in it," one rancher told me as we weaned calves in a driving rainstorm. That day we spent twelve hours on horseback in the rain. Despite protective layers of clothing: wool union suits, chaps, ankle-length yellow slickers, neck scarves and hats, we were drenched. Water drips off hat brims into your crotch; boots and gloves soak through. But to stay home out of the storm is deemed by some as a worse fate: "Hell, my wife had me cannin' beans for a week," one cowboy complained. "I'd rather drown like a muskrat out there."

Dryness is the common denominator in Wyoming. We're drenched more often in dust than in water; it is the scalpel and the suit of armor that make westerners what they are. Dry air presses a stockman's insides outward. The secret, inner self is worn not on the sleeve but in the skin. It's an unlubricated condition: there's not enough moisture in the air to keep the whole emotional machinery oiled and working. "What you see is what you get, but you have to learn to look to see all that's there," one young rancher told me. He was physically reckless when coming to see me or leaving. That was his way of saying he had and would miss me, and in the clean, broad sweeps of passion between us, there was no heaviness, no muddy residue. Cowboys have learned not to waste words from not having wasted water, as if verbosity would create a thirst too extreme to bear. If voices are raspy, it's because vocal cords are coated with dust. When I helped ship seven thousand head of steers one fall, the dust in the big, roomy sorting corrals churned as deeply and sensually as water. We wore scarves over our noses and mouths; the rest of our faces blackened with dirt so we looked like raccoons or coal miners. The westerner's face is stiff and dark red as jerky. It gives no clues beyond the discerning look that says, "You've been observed." Perhaps the too-early lines of aging that pull across these ranchers' necks are really cracks in a wall through which we might see the contradictory signs of their character: a complacency, a restlessness, a shy, boyish pride.

I knew a sheepherder who had the words "hard luck" tatooed across his knuckles. "That's for all the times I've been dry," he explained. "And when you've been as thirsty as I've been, you don't forget how something tastes." That's how he mapped out the big ranch he worked for: from thirst to thirst, whiskey to whiskey. To follow the water courses in Wyoming—seven rivers and a network of good sized creeks—is to trace the history of settlement here. After a few bad winters the early ranchers quickly discovered the necessity of raising feed for livestock. Long strips of land on both sides of the creeks and rivers were grabbed up in the 1870s and '80s before Wyoming was a state. Land was cheap and relatively easy to accumulate, but control of water was crucial. The early ranches such as the Swan Land & Cattle Company, the Budd Ranch, the M-L, the Bug Ranch, and the Pitchfork took up land along the Chugwater, Green, Greybull, Big Horn and Shoshone rivers. It was not long before feuds over water began. The old law of "full and undiminished flow" to those who owned land along a creek was changed to one that adjudicated and allocated water by the acre foot to specified pieces of land. By 1890 residents had to file claims for the right to use the water that flowed through their ranches. These rights were, and still are, awarded according to the date a ranch was established regardless of ownership changes. This solved the increasing problem of upstream-downstream disputes, enabling the first ranch established on a creek to maintain the first water right, regardless of how many newer settlements occurred upstream.

Land through which no water flowed posed another problem. Frank's father was one of the Mormon colonists sent by Brigham Young to settle and put under cultivation the arid Big Horn Basin. The twenty thousand acres they claimed were barren and waterless. To remedy this problem they dug a canal thirty-seven miles long, twenty-seven feet across, and sixteen feet deep by hand. The project took four years to complete. Along the way a huge boulder gave the canal diggers trouble: it couldn't be moved. As a last resort the Mormon men held hands around the rock and prayed. The next morning the boulder rolled out of the way.

Piousness was not always the rule. Feuds over water became venomous as the population of the state grew. Ditch riders—so called because they monitored on horseback the flow and use of water—often found themselves on the wrong end of an irrigating shovel. Frank remembers when the ditch rider in his district was hit over the head so hard by the rancher whose water he was turning off that he fell unconscious into the canal, floating on his back until he bumped into the next head gate.

With the completion of the canal, the Mormons built churches, schools, and houses communally, working in unison as if taking their cue from the water that snaked by them. "It was a socialistic sonofabitch from the beginning," Frank recalls, "a beautiful damned thing. These 'western individualists' forget how things got done around here and not so damned many years ago at that."

Frank is the opposite of the strapping, conservative western man. Sturdy, but small-boned, he has an awkward knock-kneed gait that adds to his chronic amiability. Though he's made his life close to home, he has a natural, panoramic vision as if he had upped-periscope through the Basin's dust clouds and had a good look around. Frank's generosity runs like water: it follows the path of least resistance and, tumbling downhill, takes on a fullness so replete and indiscriminate as to sometimes appear absurd. "You can't cheat an honest man," he'll tell you and laugh at the paradox implied. His wide face and forehead indicate the breadth of his unruly fair-mindedness—one that includes not just local affections but the whole human community.

When Frank started irrigating there were no tarp dams. "We plugged up those ditches with any old thing we had—rags, bones, car parts, sod." Though he could afford to hire an irrigator now he prefers to do the work himself, and when I'm away he turns my water as well, then mows my lawn. "Irrigating is a contemptible damned job. I've been fighting water all my life. Mother Nature is a bitter old bitch, isn't she? But we have to have that challenge. We crave it and I'll be goddamned if I know why. I feel sorry for these damned rich ranchers with their pumps and sprinkler systems and gated pipe because they're missing out on something. When I go to change my water at dawn and just before dark, it's peaceful out there, away from everybody. I love the fragrances—grass growing, wild rose on the ditch bank—and hearing the damned old birds twittering away. How can we live without that?"

Two thousand years before the Sidon Canal was built in Wyoming, the Hohokam, a people who lived in what became Arizona, used digging sticks to channel water from the Salt and Gila rivers to dry land. Theirs was the most extensive irrigation system in aboriginal North America. Water was brought thirty miles to spread over fields of corn, beans, and pumpkins—crops inherited from tribes in South and Central America. "It's a primitive damned thing," Frank said about the business of using water. "The change from a digging stick to a shovel isn't much of an evolution. Playing with water is something all kids have done, whether it's in creeks or in front of fire hydrants. Maybe that's how agriculture got started in the first place."

Romans applied their insoluble cement to waterways as if it could arrest the flux and impermanence they knew water to signify. Of the fourteen aqueducts that brought water from mountains and lakes to Rome, several are still in use today. On a Roman latifundium—their equivalent of a ranch—they grew alfalfa, a hot-weather crop introduced by way of Persia and Greece around the fifth century B.C., and fed it to their horses as we do here. Feuds over water were common: Nero was reprimanded for bathing in the canal that carried the city's drinking water, the brothels tapped aqueducts on the sly until once the whole city went dry. The Empire's staying power began to collapse when the waterways fell into disrepair. Crops dried up and the water that had carried life to the great cities stagnated and became breeding grounds for mosquitoes until malaria, not water, flowed into the heart of Rome.

There is nothing in nature that can't be taken as a sign of both mortality and invigoration. Cascading water equates loss followed by loss, a momentum of things falling in the direction of death, then life. In Conrad's *Heart of Darkness,* the river is a redundancy flowing through rain forest, a channel of solitude, a solid thing, a trap. Hemingway's Big Two-Hearted River is the opposite: it's an accepting, restorative place. Water can stand for what is unconscious, instinctive, and sexual in us, for the creative swill in which we fish for ideas. It carries, weightlessly, the imponderable things in our lives: death and creation. We can drown in it or else stay buoyant, quench our thirst, stay alive.

In Navajo mythology, rain is the sun's sperm coming down. A Crow woman I met on a plane told me that. She wore a flowered dress, a man's wool jacket with a package of Vantages stuck in one pocket, and calf-high moccasins held together with two paper clips. "Traditional Crow think water is medicinal," she said as we flew over the Yellowstone River which runs through the tribal land where she lives. "The old tribal crier used to call out every morning for our people to drink all they could, to make water touch their bodies. 'Water is your body,' they used to say." Looking down on the seared landscape below, it wasn't difficult to understand the real and imagined potency of water. "All that would be a big death yard," she said with a sweep of her arm. That's how the drought would come: one sweep and all moisture would be banished. Bluebunch and June grass would wither. Elk and deer would trample sidehills into sand. Draws would fill up with dead horses and cows. Tucked under ledges of shale, dens of rattlesnakes would grow into city-states of snakes. The roots of trees would rise to the surface and flail through dust in search of water.

Everything in nature invites us constantly to be what we are. We are often like rivers: careless and forceful, timid and dangerous, lucid and muddied, eddying, gleaming, still. Lovers, farmers, and artists have one thing in common, at least—a fear of "dry spells," dormant periods in which we do no blooming, internal droughts only the waters of imagination and psychic release can civilize. All such matters are delicate of course. But a good irrigator knows this: too little water brings on the weeds while too much degrades the soil the way too much easy money can trivialize a person's initiative. In his journal Thoreau wrote,

"A man's life should be as fresh as a river. It should be the same channel but a new water every instant."

This morning I walked the length of a narrow, dry wash. Slabs of stone, broken off in great squares, lay propped against the banks like blank mirrors. A sagebrush had drilled a hole through one of these rocks. The roots fanned out and down like hooked noses. Farther up, a quarry of red rock bore the fossilized marks of rippling water. Just yesterday, a cloudburst sent a skinny stream beneath these frozen undulations. Its passage carved the same kind of watery ridges into the sand at my feet. Even in this dry country, where internal and external droughts always threaten, water is self-registering no matter how ancient, recent, or brief.

The Smooth Skull of Winter

inter looks like a fictional place, an elaborate simplicity, a Nabokovian invention of rarefied detail. Winds howl all night and day, pushing litters of storm fronts from the Beartooth to the Big Horn Mountains. When it lets up, the mountains disappear. The hayfield that runs east from my house ends in a curl of clouds that have fallen like sails luffing from sky to ground. Snow returns across the field to me, and the cows, dusted with white, look like snowcapped continents drifting.

The poet Seamus Heaney said that landscape is sacramental, to be read as text. Earth is instinct: perfect, irrational, semiotic. If I read winter right, it is a scroll—the white growing wider and wider like the sweep of an arm—and from it we gain a peripheral vision, a capacity for what Nabokov calls "those asides of spirit, those footnotes in the volume of life by which we know life and find it to be good."

Not unlike emotional transitions—the loss of a friend or the beginning of new work—the passage of seasons is often so belabored and quixotic as to deserve separate names so the year might be divided eight ways instead of four.

This fall ducks flew across the sky in great "V"s as if that one letter were defecting from the alphabet, and when the songbirds climbed to the memorized pathways that route them to winter quarters, they lifted off in a confusion, like paper scraps blown from my writing room.

A Wyoming winter laminates the earth with white, then hardens the lacquer work with wind. Storms come announced by what old-timers call "mare's tails"—long wisps that lash out from a snow cloud's body. Jack Davis, a packer who used to trail his mules all the way from Wyoming to southern Arizona when the first snows came, said, "The first snowball that hits you is God's fault; the second one is yours."

Every three days or so white pastures glide overhead and drop themselves like skeins of hair to earth. The Chinese call snow that has drifted "white jade mountains," but winter looks oceanic to me. Snow swells, drops back, and hits the hulls of our lives with a course-bending sound. Tides of white are overtaken by tides of blue, and the logs in the woodstove, like sister ships, tick toward oblivion.

On the winter solstice it is thirty-four degrees below zero and there is very little in the way of daylight. The deep ache of this audacious Arctic air is also the ache in our lives made physical. Patches of frostbite show up on our noses, toes, and ears. Skin blisters as if cold were a kind of radiation to which we've been exposed. It strips what is ornamental in us. Part of the ache we feel is also a softness growing. Our connections with neighbors—whether strong or tenuous, as lovers or friends—become too urgent to disregard. We rub the frozen toes of a stranger whose pickup has veered off the road; we open water gaps with a tamping bar and an ax; we splice a friend's frozen water pipe; we take mittens and blankets to the men who herd sheep. Twenty or thirty below makes the breath we exchange visible: all of mine for all of yours. It is the tacit way we express the intimacy no one talks about.

One of our recent winters is sure to make the history books because of not the depth of snow but, rather, the depth of cold. For a month the mercury never rose above zero and at night it was fifty below. Cows and sheep froze in place and an oil field worker who tried taking a shortcut home was found next spring two hundred yards from his back door. To say you were snowed in didn't express the problem. You were either "froze in," "froze up," or "froze out," depending on where your pickup or legs stopped working. The day I helped tend sheep camp we drove through a five-mile tunnel of snow. The herder had marked his location for us by deliberately cutting his finger and writing a big "X" on the ice with his blood.

When it's fifty below, the mercury bottoms out and jiggles there as if laughing at those of us still above ground. Once I caught myself on tiptoes, peering down into the thermometer as if there were an extension inside inscribed with higher and higher declarations of physical misery: ninety below to the power of ten and so on.

Winter sets up curious oppositions in us. Where a wall of snow can seem threatening, it also protects our staggering psyches. All this cold has an anesthetizing effect: the pulse lowers and blankets of snow induce sleep. Though the rancher's workload is lightened in winter because of the short days, the work that does need to be done requires an exhausting patience. And while earth's sudden frigidity can seem to dispossess us, the teamwork on cold nights during calving, for instance, creates a profound camaraderie—one that's laced with dark humor, an effervescent lunacy, and unexpected fits of anger and tears. To offset Wyoming's Arctic seascape, a nightly flush of Northern Lights dances above the Big Horns, irradiating winter's pallor and reminding us that even though at this time of year we veer toward our various nests and seclusions, nature expresses itself as a bright fuse, irrepressible and orgasmic.

Winter is smooth-skulled, and all our skids on black ice are cerebral. When we begin to feel cabin-feverish, the brain pistons thump against bone and mind irrupts—literally invading itself—unable to get fresh air. With the songbirds gone only scavengers are left: magpies, crows, eagles. As they pick on road-killed deer we humans are apt to practice the small cruelties on each other.

We suffer from snow blindness, selecting what we see and feel while our pain whites itself out. But where there is suffocation and self-imposed ignorance, there is also refreshment—snow on flushed cheeks and a pristine kind of thinking. All winter we skate the small ponds—places that in summer are water holes for cattle and sheep—and here a reflection of mind appears, sharp, vigilant, precise. Thoughts, bright as frostfall, skate through our brains. In winter, consciousness looks like an etching.

SUE
HUBBELL

Sue Hubbell lives in two places: on a hilltop in the Missouri Ozarks and with her husband in Washington, D.C. She says, "I delight in moving back and forth between two ways of living, one reflecting off the other, clarifying both in the process."

She was born in 1935, and grew up in Kalamazoo, Michigan, where her father was a botanist and park superintendent. (Her brother is Bil Gilbert, a natural history writer who also could have been included in this anthology.) Hubbell "married and had a son and became a librarian at Brown University." When she and her husband Paul quit their jobs in 1972, they wandered the country until they came to the Ozarks. They bought a 105-acre farm, sandwiched between a "swift, showy river" and a small creek: "We had to find something to do to make a living, and Paul said that since we didn't know anything about cows, we might as well become beekeepers.'' Paul left, Sue stayed, and, she says, ''Here I found what I wanted.''

Fifteen years later, she still keeps bees, trying to get ''as close to the bees' lives as they will let me.'' And she writes. Her first book, *A Country Year: Living the Questions* (1986), from which this selection comes, traces ''the natural history of my hilltop from one springtime to the next.'' She wrote it for herself, ''to tease out of the amorphous, chaotic and wordless part of myself'' the reasons for staying in the place after her divorce. Her writing turned out to be ''a record of changes I had thought too subtle for me to have noticed.

''I was astonished when a friend told me I was writing a book. My agent said I was writing a book, too, and in due course my editor at Random House said I was writing a book. After it was published it even looked like a book. And I was still astonished, and remain so to this day.''

Her book led to reacquaintance with an old college friend, her marriage to him, and her present dual life. The Ozark farm has led to a second book, *A Book of Bees . . . and How to Keep Them* (1988). The ''wild things and wild places pull me all the time.'' At the farm, she sleeps under the stars from February to November unless it's a hard winter. She says, ''I want the whole world, and the stars too.''

Hubbell keeps 300 beehives scattered on about twenty farms within thirty-five miles of her own. She rents these ''outyards'' from farmers for a gallon of honey per year. She savors her ''good fortune at actually being able to make a living by associating with a bunch of bugs'' while taking care ''not to count on writing for a living—it's like playing the slot machines in Las Vegas.''

Hubbell admits, ''It is a very complex life, but, after all, I've never wanted a simple one.'' She would like to decrease her numbers of beehives, to reduce the need for frequent sales trips. She says, ''the bees keep me humble.'' Her Ozark neighbors still call her ''The Bee Lady.''

Spring

The river to the north of my place is claimed by the U.S. Park Service, and the creek to the south is under the protection of the Missouri State Conservation Department, so I am surrounded by government land. The deed to the property says my farm is a hundred and five acres, but it is probably something more like ninety. The land hasn't been surveyed since the mid-1800s and it is hard to know where the boundaries are; a park ranger told me he suspected that the nineteenth-century surveyor had run his lines from a tavern, because the corners seem to have been established by someone in his cups.

The place is so beautiful that it nearly brought tears to my eyes the first time I saw it twelve years ago; I feel the same way today, so I have never much cared about the number of acres, or where the boundary lines run or who, exactly, owns what. But the things that make it so beautiful and desirable to me have also convinced others that this is prime land, too, and belongs to them as well. At the moment, for instance, I am feeling a bit of an outsider, having discovered that I live in the middle of an indigo bunting ghetto. As ghettos go, it is a cheerful one in which to live, but it has forced me to think about property rights.

Indigo buntings are small but emphatic birds. They believe that they own the place, and it is hard to ignore their claim. The male birds—brilliant, shimmering blue—perch on the garden posts or on top of the cedar trees that have taken over the pasture. From there they survey their holdings and belt out their songs, complicated tangles of couplets that waken me first thing in the morning; they keep it up all day, even at noon, after the other birds have quieted. The indigo buntings have several important facts to tell us, especially about who's in charge around here. The dull brown, sparrowlike females and juveniles are more interested in eating; they stay nearer the ground and search the low-growing shrubs and grasses for seeds and an occasional caterpillar, but even they know what's what. One day, walking back along the edge of the field, I came upon a young indigo bunting preoccupied with song practice. He had not yet dared take as visible a perch as his father would have

214

chosen, but there he was, clinging to a bare twig and softly running through his couplets, getting them all wrong and then going back over them so quietly that had I not been within a few feet of him I would not have heard.

Another time I discovered that the back door of the honey house had blown open and the room was filled with a variety of winged creatures. Most were insects, but among them I found a half-grown indigo bunting who had blundered in and was trying to find his way out, beating his small wings against the screened window. Holding him carefully, I stroked the back of his neck to try to soothe him, but discovered that his heart was not beating in terror. Perhaps he was so young that he had not learned fear, but I prefer to think that like the rest of his breed he was simply too pert and too sure of his rights to be afraid. He eyed me crossly and tweaked my giant thumb with his beak to tell me that I was to let him go right this minute. I did so, of course, and watched him fly off to the tall grasses behind the honey house, where I knew that one family of indigo buntings had been nesting.

Well, they think they own the place, and their assurance is only countered by a scrap of paper in my files. But there are other contenders, and perhaps I ought to try to take a census and judge claims before I grant them title. There are other birds who call this place theirs—buzzards, who work the updrafts over the river and creek, goldfinches, wild turkey, phoebes and whippoorwills. But it is a pair of cardinals who have ended up with the prize piece of real estate—the spot with the bird feeder. I have tapes of birdsongs, and when I play them I try to skip the one of the cardinal, because the current resident goes into a frenzy of territorial song when he hears his rival. His otherwise lovely day is ruined.

And what about the coyote? For a while she was confident that this was her farm, especially the chicken part of it. She was so sure of herself that once she sauntered by in daylight and picked up the tough old rooster to take back to her pups. However, the dogs grew wise to her, and the next few times she returned to exercise her rights they chased her off, explaining that this farm belonged to them and that the chicken flock was their responsibility.

When I start thinking about it that way—that those who inhabit the land and use it have a real claim to it in a nonlegal sort of way—the whole question gets complicated.

A long time ago, before I came to live in the Ozarks, I spent a springtime working on a plot of university research land. I was young and in love, and most tasks seemed happy ones, but the project would have captured my fancy anyway. There were three contrasting habitats being studied: upland forest, bottomland and sandy waste. My job was to dig up a cube of earth from each place every week, sift it, count and rough-classify the inhabitants visible to the naked eye, and then plot the population growth. The resulting curve, a joyous, vibrant freshening of life, matched the weather and my own pulse beat.

That particular love has quieted, and I have not excavated cubes of earth on this place, but I know what is going on down there: Millions of little bodies are fiercely metabolizing and using the land. I dare not even think what numbers I would come up with if I added a pocket lens or microscope to my census-taking tools. But there are other residents I can count who do have arguable title here. There are twenty hives of bees back by the woodlot in my home beeyard, each hive containing some 60,000 bees. That makes 1,200,000 bee souls flitting about making claim to all the flowers within two miles.

On the other hand, there are the copperheads, who make walking the fields a boot affair, and all their snakish kin. How am I to count them and judge their claims? There are the turtles who eat the strawberries in the garden, the peepers who own the pond. What about raccoon and skunk and deer rights? What about the bobcat who denned in the cliff by the river and considers my place to be the merest sliver of her own?

It begins to make me dizzy even trying to think of taking a census of everybody who lives here; and all of them seem to have certain claims to the place that are every bit as good as and perhaps better than mine.

Up the road there is a human squabble going on over some land less happily situated. Rather than lying between two environmentally benign government stake-outs, that land and all that surrounds it is in private hands. One owner wants to bulldoze and develop, and so the boundary question is becoming a sticky one. There is talk about having an expensive survey made to establish who owns what. As a spinoff, I suppose that corners will be set and lines run, and then I may know whether this farm is a hundred and five acres or ninety or some other definity.

The indigo buntings probably won't care.

I met Paul, the boy who was to become my husband, when he was sixteen and I was fifteen. We were married some years later, and the legal arrangement that is called marriage worked well enough while we were children and while we had a child. But we grew older, and the son went off to school, and marriage did not serve as a structure for our lives as well as it once had. Still, he was the man in my life for all those years. There was no other. So when the legal arrangement was ended, I had a difficult time sifting through the emotional debris that was left after the framework of an intimate, thirty-year association had broken.

I went through all the usual things: I couldn't sleep or eat, talked feverishly to friends, plunged recklessly into a destructive affair with a man who had more problems than I did but who was convenient, made a series of stupid decisions about my honey business and pretty generally botched up my life for several years running. And for a long, long time, my mind didn't work. I could not listen to the news on the radio with understanding. My attention came unglued when I tried to read anything but the lightest froth. My brain spun in endless, painful loops, and I could neither concentrate nor think with any semblance of order. I have always rather enjoyed having a mind, and I missed mine extravagantly. I was out to lunch for three years.

I mused about structure, framework, schemata, system, classification and order. I discovered a classification Jorge Luis Borges devised, claiming that

> *A certain Chinese encyclopedia divides animals into:*
>
> *a. Belonging to the Emperor*
> *b. Embalmed*
> *c. Tame*
> *d. Sucking pigs*
> *e. Sirens*
> *f. Fabulous*
> *g. Stray dogs*
> *h. Included in the present classification*
> *i. Frenzied*
> *j. Innumerable*
> *k. Drawn with a very fine camel-hair brush*
> *l. Et cetera*
> *m. Having just broken the water pitcher*
> *n. That from a long way off look like flies.*

Friends and I laughed over the list, and we decided that the fact that we did so tells more about us and our European, Western way of thinking that it does about a supposed Oriental world view. We believe we have a more proper concept of how the natural world should be classified, and when Borges rumples that concept it amuses us. That I could join in the laughter made me realize I must have retained some sense of that order, no matter how disorderly my mind seemed to have become.

My father was a botanist. When I was a child he reserved Saturday afternoons for me, and we spent many of them walking in woods and rough places. He would name the plants we came upon by their Latin binomials and tell me how they grew. The names were too hard for me, but I did understand that plants had names that described their relationships one to another and found this elegant and interesting even when I was six years old.

So after reading the Borges list, I turned to Linnaeus. Whatever faults the man may have had as a scientist, he gave us a beautiful tool for thinking about diversity in the world. The first word in his scheme of Latin binomials tells the genus, grouping diverse plants which nevertheless share a commonality; the second word names the species, plants alike enough to regularly interbreed and produce offspring like themselves. It is a framework for understanding, a way to show how pieces of the world fit together.

I have no Latin, but as I began to botanize, to learn to call the plants around me up here on my hill by their Latin names, I was diverted from my lack of wits by the wit of the system.

Commelina virginica, the common dayflower, is a rangy weed bearing blue flowers with unequal sepals, two of them showy and rounded, the third hardly noticeable. After I identified it as that particular *Commelina,* named from a sample taken in Virginia, I read in one of my handbooks, written before it was considered necessary to be dull to be taken seriously:

> *Delightful Linnaeus, who dearly loved his little joke, himself confesses to have named the day-flowers after three brothers Commelyn, Dutch botanists, because two of them—commemorated in the showy blue petals of the blossom—published their works; the third, lacking application and ambition, amounted to nothing, like the third inconspicuous whitish third petal.*

There is a tree growing in the woodland with shiny, oval leaves that turn brilliant red early in the fall, sometimes even at summer's end. It has small clusters of white flowers in June that bees like, and later blue fruits that are eaten by bluebirds and robins. It is one of the tupelos, and people in this part of the country call it black-gum or sour-gum. When I was growing up in Michigan I knew it as pepperidge. Its botanic name is *Nyssa sylvatica. Nyssa* groups the tupelos, and is derived from the Nyseides—the Greek nymphs of Mount Nysa who cared for the infant Dionysus. *Sylvatica* means "of the woodlands." *Nyssa sylvatica,* a wild, untamed name. The trees, which are often hollow when old, served as beehives for the first American settlers, who cut sections of them, capped them and dumped in the swarms that they found. To this day some people still call beehives "gums," unknowingly acknowledging the common name of the tree. The hollow logs were also used for making pipes that carried salt water to the salt works in Syracuse in colonial days. The ends of the wooden pipes could be fitted together without using iron bands, which would rust.

This gives me a lot to think about when I come across *Nyssa sylvatica* in the woods.

I botanized obsessively during that difficult time. Every day I learned new plants by their Latin names. I wandered about the woods that winter, good for little else, examining the bark of leafless trees. As wildflowers began to bloom in the spring, I carried my guidebooks with me, and filled a fat notebook as I identified the plants, their habitats, habits and dates of blooming. I had to write them down, for my brain, unaccustomed to exercise, was now on overload.

One spring afternoon, I was walking back down my lane after getting the mail. I had two fine new flowers to look up when I got back to the cabin. Warblers were migrating, and I had been watching them with binoculars; I had identified one I had never before seen. The sun was slanting through new leaves, and the air was fragrant with wild cherry (*Prunus serotina: Prunus*—plum, *serotina*—late blooming) blossoms, which my bees were working eagerly. I stopped to watch them, standing in the sunbeam. The world appeared to have been running along quite nicely without my even noticing it. Quietly, gratefully, I discovered that a part of me that had been off somewhere nursing grief and pain had returned. I had come back from lunch.

Once back, I set about doing all the things that one does when one returns from lunch. I cleared the desk and tended to the messages that others had left. I had been gone for a long time, so there was quite a pile to clear away before I could settle down to the work of the afternoon of my life, the work of building a new kind of order, a structure on which a fifty-year-old woman can live her life alone, at peace with herself and the world around her.

One spring evening a couple of years ago, I was sitting in the brown leather chair in the living room reading the newspaper and minding my own business when I became aware that I was no longer alone.

Looking up, I discovered that the three big windows that run from floor to ceiling were covered with frogs.

There were hundreds of them, inch-long frogs with delicate webbed feet whose fingerlike toes ended in round pads that enabled them to cling to the smooth surface of the glass. From their toe structure, size and light-colored bellies, I supposed them to be spring peepers, *Hyla crucifer,* and went outside for a closer look. I had to be careful where I put my feet, for the grass in front of the windows was thick with frogs, waiting in patient ranks to move up to the lighted surface of the glass. Sure enough, each pinkish-brownish frog had a back criss-crossed with the dark markings that give the species its scientific name. I had not known before that they were attracted to light.

I let my newspaper go and spent the evening watching them. They did not move much beyond the top of the windows, but clung to the glass or the moldings, seemingly unable to decide what to do next. The following morning they were gone, and I have never seen them at the windows since. It struck me as curious behavior.

These window climbers were silent; we usually are only aware of spring peepers at winter's end—I first hear their shrill bell-like mating calls in February from the pond up in the field. The males produce the calls by closing their mouths and nasal openings and forcing air from their lungs over the vocal cords into their mouths, and then back over the vocal cords into the lungs again. This sound attracts the females to the pond, and when they enter the water the males embrace them, positioning their vents directly above those of the females. The females then lay their eggs, which the males fertilize with their milt.

It is a clubby thing, this frog mating, and the frogs are so many and their calls so shrill and intense that I like to walk up to the pond in the evenings and listen to the chorus, which, to a human, is both exhilarating and oddly disturbing at close range. One evening I walked there with a friend, and we sat by the edge of the pond for a long time. Conversation was inappropriate, but even if it had not it would have been impossible. The bell-like chorus completely surrounded us, filled us. It seemed to reverberate with the shrill insistence of hysteria, driving focused thought from our heads, forcing us not only to hear sound but to feel it.

Comparing notes as we walked back to my cabin, we were startled to discover that we had both wondered, independently, whether that was what it was like to go mad.

A slightly larger cousin of the spring peeper that belongs to the same genus, the gray tree frog, commonly lives in my beehives during the summer months. These frogs cling under the protective overhang of the hive cover, and as I pry up the lid, they hop calmly to the white inner cover and sit there placidly eying me.

They are a pleasing soft grayish-green, marked with darker moss-colored patches, and look like a bit of lichen-covered bark when they are on a tree. Having evolved this wonderfully successful protective coloration, the safest behavior for a gray tree frog in a tight spot is to stay still and pretend to be a piece of bark. Sitting on the white inner cover of the beehive, the frog's protective coloration serves him not at all, but of course he doesn't know that, and not having learned any value in conspicuously hopping away, he continues to sit there looking at me with what appears to be smug self-satisfaction and righteous spunk.

Last evening I was reading in bed and felt rather than heard a soft plop on the bed next to me. Peering over the top of my glasses, I saw a plump, proud gray tree frog inspecting me. We studied each other for quite a time, the gray tree frog seemingly at ease, until I picked him up, carried him out the back door and put him on the hickory tree there. But even in my cupped hands he moved very little, and after I put him on the tree he sat quietly, blending in beautifully with the bark. A serene frog.

The sills in my bedroom are rotten, so I supposed that he had found a hole to come through and wondered if he'd had friends. I looked under the bed and discovered three more gray tree frogs, possibly each one a frog prince. Nevertheless, I transferred them to the hickory.

There was something in the back of my mind from childhood Sunday-school classes about a plague of frogs, so I took down my Bible and settled back in bed to search for it. I found the story in Exodus. It was one of those plagues that God sent to convince the Pharaoh to let the Jews leave Egypt.

And the Lord spake unto Moses, Go unto Pharaoh, and say unto him, Thus saith the Lord, Let my people go, that they may serve me.

And if thou refuse to let them go, behold, I will smite all thy borders with frogs:

And the river shall bring forth frogs abundantly, which shall go up and come into thine house, and into thy bedchamber, and upon thy bed . . .

This was exciting stuff; my evening had taken on a positively biblical quality. I was having a plague of frogs, and had obviously had another the evening that the spring peepers had crawled up the living-room windows. Actually, I enjoyed both plagues, but Pharaoh didn't. The writer of Exodus tells us that Pharaoh was so distressed by frogs in his bed that he called Moses and said,

Intreat the Lord, that he may take away the frogs from me, and from my people; and I will let the people go, that they may do sacrifice unto the Lord.

A fussy man, that Pharaoh, and one easily unnerved.

I once knew a pickerel frog, *Rana palustris,* frog of the marshes, who might have changed Pharaoh's mind. The pickerel frog was an appealing creature who lived in my barn one whole summer. He was handsome, grayish with dark, square blotches highlighted with yellow on his legs. I found him in the barn one morning trying to escape the attentions of the cat and the dogs. At some point he had lost the foot from his right front leg, and although the stump was well healed, his hop was awkward and lopsided. I decided that he would be better off taking his chances with wild things, so I carried him out to the pond and left him under the protective bramble thicket that grows there. But the next day he was back in the barn, having hopped the length of a football field to get there. So I let him stay, giving him a dish of water and few dead flies.

All summer long I kept his water fresh, killed flies for him and kept an eye out for his safety. Pickerel frogs sometimes live in caves, and I wondered if the dim light of the barn and the cool concrete floor made him think he had discovered a cave where the service was particularly good. That part of the barn serves as a passageway to my honey house, and I grew accustomed to seeing him as I went in and out of it. I came to regard him as a tutelary sprite, the guardian of the honey house, the Penate Melissus.

Then one day the health inspector came for his annual tour. Like Pharaoh, the health inspector is a fussy man. Once he gave me a hard time because there were a few stray honeybees in the honey house. Bees, he explained patiently, were insects, and regulations forbade insects in a food-processing plant. I pointed out, perhaps not so patiently, that these insects had made the food, and that until I took it from them, they were in continuous, complete and intimate contact with it. He gave up, but I know he didn't like it. So I wasn't sure how he'd react to the pickerel frog squatting outside the honey-house door with his bowl of water and mason-jar lid full of dead flies. But the health inspector is a brisk man, and he walked briskly by the frog and never saw him. I was thankful.

Years ago, in an introductory biology class, I cut up a frog, carefully laying aside the muscles, tracing the nerves and identifying the organs. I remember that as I discarded the carcass I was quite pleased with myself, for now I knew all about frogs and could go on to learn the remaining one or two things about which I still had some small ignorance. I was just about as smug as a gray tree frog on a white beehive.

In the years after that, and before I moved to the Ozarks, I also lived a brisk life, and although I never had much reason to doubt that I still knew all about frogs, I don't think I ever thought about them, for, like the health inspector, I never saw any.

Today my life has frogs aplenty and this delights me, but I am not so pleased with myself. My life hasn't turned out as I expected it would, for one thing. For another, I no longer know all about anything. I don't even know the first thing about frogs, for instance. There's nothing like having frogs fill up my windows or share my bed or require my protection to convince me of that.

I don't cut up frogs anymore, and I read more poetry than I did when I was twenty. I just read a couplet about the natural world by an anonymous Japanese poet. I copied it out and put it up on the wall above my desk today:

> *Unknown to me what resideth here*
> *Tears flow from a sense of unworthiness and*
> *gratitude.*

WENDELL
BERRY

Photograph © Debra Cook

Wendell Berry was born in 1934 in Henry County, Kentucky, and lives there today by choice. After returning with his family in 1964, he bought a small farm on the Kentucky River, farming organically, working his fields with draft horses after 1973. His grandparents and great-grandparents all lived within four miles of this farm. He writes, "The place and [its] history have been inseparable, and there is a sense in which my own life is inseparable from the history and the place."

Berry was educated at the University of Kentucky and at Stanford, where he wrote his first novel, *Nathan Coulter* (1960). He traveled in Europe and taught at New York University before his return to Kentucky. His last post in academia ended in 1977, when he resigned from the University of Kentucky to farm and write full time. Berry writes novels, poetry, criticism, and essays. The land and his concern for "the life and health of the world" run throughout his writing.

What motivates his work is "a desire to make myself responsibly at home in this world and in my native and chosen place. . . . This is a long-term desire, proposing not the work of a lifetime but of generations." He believes that, "it is only in the processes of the natural world and in analogous and related processes of human culture that the new may grow usefully old and the old be made new."

Berry's 1977 book of essays, *The Unsettling of America*, poses eloquent and powerful connections between culture, character, and the ecological and agricultural crises. His philosophy is elemental and nourishing: "If we do not live where we work, and when we work, we are wasting our lives, and our work, too." He continues to explore these themes in *The Gift of Good Land* (1981). *Recollected Essays* (1981) collected pieces from five of his earlier books.

In part to argue against building a dam in the Red River Gorge, in 1971 Berry published *The Unforeseen Wilderness: An Essay on Kentucky's Red River Gorge*, with photographs by Gene Meatyard. One of his least-known works, the following selection is the book's final chapter.

The Gorge is a place farther from Berry's farm than one might think of calling home. But Berry writes in "the long-legged house," a riverbank cabin originally built by his great-uncle. The water that flows past his cabin comes from a big chunk of Kentucky; it includes water from the Red River, on the far side of the Bluegrass from Henry County and, as he makes clear here, it is all connected. The Gorge and its river are both part of the "familiar mysteries."

The Journey's End

Five years have passed since I first looked out over the Red River country from the fire tower on Pine Ridge. During that time I have come to know it a little. I think of the growth of that knowledge, small as it is, as one of the landmarks of my life—a happening both large and altogether good. I have spent not a single moment there that I look back on with regret. I have not a single feeling about it that is vague or uncertain. My times there answer to memory with a purity and clarity that is like the water of its streams. For all my remaining ignorance of it, for all in it that is dark, I think of it as a *clear* place.

At first I experienced mostly its strangeness. I remember the curious uneasiness I would feel, then, when I went any distance off the trails. It was not that I was lost. But the lay of the land was strange to me and when I cast myself loose in it, certain as I might be of where I was, I would begin to *feel* lost. Sometimes my return to the trail would partake somewhat of the nature of a retreat, the Unknown having surrounded me on three sides and begun to close in.

I thought of it then as a strange place, a place strange *to me*. The presumptuousness of that, it now occurs to me, is probably a key to the destructiveness that has characterized the whole history of the white man's relation to the American wilderness. For it is presumptuous, entirely so, to enter a place for the first time and pronounce it strange. Strange to whom? Certainly to its own creatures—to the birds and animals and insects and fish and snakes, to the human family I know that lives knowingly and lovingly there—it is not strange. To them as it was to the Indians who once lived in its caves and in the bottomlands near its creek mouths it is daily reality, regular stuff.

To call a place strange in the presence of its natives is bad manners at best. At worst, it partakes of the fateful arrogance of those explorers who familiarize the "strange" places they come to by planting in them the alien flag of the place they have left, and who have been followed, always, by the machinery of conquest and exploitation and destruction.

The strangeness, as I recognized after a while, for I went in flying no flag and riding no machine, was all in me. It was my own strangeness that I felt, for I was a man out of place. And I believe that only in that realization lay the possibility that I would come to know the Red River Gorge even a little. If I had continued to look upon the place as strange I would clearly have had only two choices: stay out of it altogether, or change it, destroy it as I found it and make it into something else. But once I learned to look upon myself as a stranger there, it became possible for me to return again and again without preaching to the natives, or making treaties with them, or swindling them out of their property, or cutting down any timber, or buying a lot on which to build a drive-in restaurant. It became possible for me to leave the place as it is, to want it to be as it is, to be quiet in it, to learn about it and from it. Lacking any such disciplining and humbling sense of being strangers, wanderers away from home, the European conquerors entered America like so many English sparrows or Japanese beetles, free of controls, cultural or natural, that would have brought their lives into harmony with this land. And they and their descendants have lived here for the most part as strangers, and for the most part out of control, ever since. And now, by "generously" gospelizing the technology of exploitation and waste, they are teaching other peoples how to be strangers, even in their own homes.

Slowly, almost imperceptibly, the experience of strangeness was transformed into the experience of familiarity. The place did not become predictable; the more I learned of it, the less predictable it seemed. But my visits began to define themselves in terms of recurrences and recognitions that were pleasant in themselves, and that set me free in the place. I began to depart from the trails with a comfortable notion of where the contour of a slope or the fall of a stream would be likely to take me. Better than that, I soon began to think that there could not be many better places to get lost in. And finally—to my regret, as it turned out—I realized that, as much crossed as that country is by roads and trails, it is very likely impossible to get lost in it, at least not for long at a time.

Its mysteries remained—for though we pretend otherwise, the unknown increases with the known. But mystery is not the same as strangeness. A mystery can be familiar. In this scientific age, when our "practical intelligence" gropes so destructively toward the "use" of everything, we should remember that it is possible to be comfortably ignorant. It is possible, and I am sure it will prove to be necessary, to make peace with what we cannot understand.

As my knowledge of the place grew I began to have a sense of the meaning—or the anti-meaning—of its planned destruction, which carried me far beyond the mere principles of conservation and preservation. I began to feel in the presence and substance of its life the complexity and the magnitude of its death. And I realized that in the story of the Gorge I had forsaken the role, and the immunity, of an observer. Or rather that role had forsaken me, for I had become personally involved. The death of the Gorge, for some of my fellow Kentuckians, would be merely an act of "progress"—a cause that they may themselves be dying for. But to me, because I *knew* something of what would die, it would be a great personal loss. It was too late to be objective. I had spent some of my best days there, not as an observer or writer but as a creature bemused by the creation. The Gorge had become part of my life. I knew that whether I continued to go there or not it would remain meaningful and important to me. I knew that there would be a certain irreplaceable comfort that I would draw from the knowledge that it was preserved and cherished and enjoyed by members of my species.

And the more I saw and understood of the condition of the watershed and of the river itself, the more clear it seemed to me that the damming of the Gorge would be not only destructive and meaningless, but useless upon the very terms of the argument for its destruction: the conditions that are responsible for flooding will cause the rapid siltation and destruction of the proposed reservoir; the conditions that bring water shortages into prospect for the downstream cities preclude the possibility that these shortages will be forestalled for very long by the building of reservoirs.

It is a fact that the entire Kentucky River system, which the central part of the state complacently depends upon for its future water supply, is deteriorating rapidly because of strip mining, because of bad farming, because of industrial and agricultural pollutants, because of urban sewage. It is deteriorating, that is to say, because almost nobody cares, or cares to know, where water comes from, so long as it keeps coming. The going assumption is that people so ignorant and thoughtless and silly and greedy may simply call upon the Army Corps of Engineers in order to receive a clean and abundant supply of water from reservoirs in the mountains. A much likelier outcome is that they will be drinking an ever stronger mixture of sewage and mine acid and mud and cropspray and various other defecations of the industrial paradise.

The proposed dam in the Red River Gorge is not the definitive solution to any problem, upstream or down. Like many another project that has been offered to the people as a lasting monument of human progress, it is an illusion, expedient and temporary, that will only delay, perhaps catastrophically, the achievement of a real solution. It is a cheap shortcut. And in the art of earth-keeping, as in any other art, shortcuts always leave out essential steps. If the destruction of the steep land of the watershed were stopped; if that land were adequately forested and grassed, as it could be in a comparatively short time; if sane methods were made to prevail in mining and forestry and agriculture—that work alone would go far toward assuring an adequate water supply from the Kentucky River. And flood control on the scale now contemplated would certainly become unnecessary. Seriously damaging floods of the Kentucky River are a modern phenomenon. They are manmade, caused by the abuse of the watershed. As evidence, I need cite only the fact that well into this century, in the lower part of the valley, houses were confidently built upon sites that now are flooded about every ten years. Flood control as we now know it is no more than a subsidization of the crimes and abuses of the exploiters, the burden as usual falling upon innocent taxpayers.

Upon their disreputable argument of "use" the Engineers and the pushers of "development" have erected the even more disreputable argument of "recreation." For in addition to putting nature to man's work, a dam, they say, will put it to his pleasure. Modern Americans, as we all know, are crowded and stifled in the cities, and are therefore most excruciatingly in need of recreation. And what is to be the form of this recreation? Why, it is to be crowded and stifled in the country. Relief from the suburbs of brick and Bedford stone is to be found in suburbs of canvas and aluminum. Relief from traffic in the streets is to be sought amid traffic on a lake. The harried city dweller, who has for fifty weeks coveted his neighbor's house and his neighbor's wife, may now soothe his nerves for two weeks in coveting his neighbor's trailer and his neighbor's boat—also in putting up with his neighbor's children, listening to his neighbor's radio, breathing his neighbor's smoke, walking on his neighbor's broken bottles. While he is doing all these things he is surrounded by "the beauty of nature," which is a big item in recreation. And a lake, the developers agree, provides a lot more beauty of nature than a river because it has room for more and bigger and faster boats, making it possible for more people to see more beauty of nature in less time. It really takes a lot of beauty of nature to make up for the smoke and smog of our stifling cities, and so if it only takes ten minutes to go in a boat where it once took all day to go on foot that is all to the good. Why be satisfied with mere yards of natural beauty when you can have *miles* of it?

If dam building is an illusion deeply rooted in our history and culture, recreation is no more than a blatant commercial fraud, rooted in nothing deeper than the anxieties of a people chronically unsettled and upset by the experience of belonging nowhere. It is a gimmick for selling tents, trailers, stoves, lanterns, sleeping bags, cooking pots, boats, motors, campsites, fishing tackle, gasoline, suntan lotion, bug dope, bad food, hay fever pills, water skis, etc., etc. to people who have no needs that they understand, and who therefore want everything.

The net product of all this wheeling and dealing, this going to and fro in the earth, is only a well-advertised, countrified, glamorized version of life in a suburb.

And to make the quiet lonely dells and ravines and glades of the Red River country accessible to motorized crowds is quite simply to remove the reasons for going there. People who want to see the beauty of nature from motorboats and automobiles would obviously be just as well pleased, and as fully recreated, at a drive-in movie.

The Gorge, dammed, would be like *Hamlet* rewritten for the feeble-minded. And as illusory—for what is lost is the experience of the living thing itself. Motor vehicles translate the landscape, alter the experience of it, in precisely this way, making possible oversimplifications that are dangerous. A man who has always looked from a motorized perspective will have too comfortable a view of the world; it will seem to him more answerable to his convenience than it is in fact. An artificial lake is a river's oversimplification. A motorboat is an over-simplification of the lake.

I would be among the last, I hope, to discourage anybody from going to the woods. In the name of sanity, let us *all* go, the oftener the better. But let us go without motors. Let us go by rowboat or sailboat or canoe, or on horseback or on skis or on foot. Let us admit that the simple quiet we seek cannot be found with a motor. Motorized, we can only arrive at the uproar we meant to escape.

There are endless ways to amuse oneself and be idle, and most of them lie outside the woods. I assume that when a man goes to the woods he goes because he needs to. I think he is drawn to the wilderness much as he is drawn to a woman: it is, in its way, his opposite. It is as far as possible unlike his home or his work or anything he will ever manufacture. For that reason he can take from it a solace—an understanding of himself, of what he needs and what he can do without— such as he can find nowhere else. Though one would surely never want to deny the possibility of being amused, amusement is far from all there is to it. A man drawn to the woods does not go there for what has come to be known as recreation. Why should anything once created need to be re-created? Why should a man, who did not and could not create himself, assume that he can re-create himself? By recreation we mean no such thing, but only distraction from what we ought to be paying attention to: the probable effects of our behavior.

Going to the woods and the wild places has little to do with recreation, and much to do with creation. For the wilderness is the creation in its pure state, its processes unqualified by the doings of people. A man in the woods comes face to face with the creation, of which he must begin to see himself a part—a much less imposing part than he thought. And seeing that the creation survives all wishful preconceptions about it, that it includes him only upon its own sovereign terms, that he is not free except in his proper place in it, then he may begin, perhaps, to take a hand in the creation of himself.

But the Red River Gorge is not now a place of unqualified wilderness. It is not a virgin forest. It has been mined a bit for iron, cut over by loggers, farmed in the few places where the land is not too steep to plow, cut through by roads, and it is surrounded by the mud and garbage machine known as the Affluent Society. To some that makes an argument for destroying the place as it is now, and as it will become if let alone. These people say that if the Gorge is not a *virgin* wilderness then it is not a wilderness, and there is therefore no reason why it should not be flooded and "developed."

One obvious answer is that, because of the depredations of such arguers, Kentucky now has almost no wilderness that is, strictly speaking, virgin. If we want to have a mature forest in which the ecology is unimpaired, then we must realize that we are no longer privileged to have it merely by preserving it. Now, if we want it, we will have to grow it. If we want it we must bow to its conditions, get out of its way, invite it to return.

As long as we insist on relating to it strictly on our own terms—as strange to us or subject to us—the wilderness is alien, threatening, fearful. We have no choice then but to become its exploiters, and to lose, by consequence, our place in it. It is only when, by humility, openness, generosity, courage, we make ourselves able to relate to it on *its* terms that it ceases to be alien. Then it begins to be familiar to us. We begin to see that it is at least partly beneficent. We see that we belong to it, and have our place in it. We see that its terms are the only terms, that in the final sense we have no terms, that our terms are a fiction of our pride.

But it if has become familiar, if we have begun to feel at home in it, that is not because it has become comfortable or predictable or in any way prejudiced in our favor. (It is prejudiced in favor of *life,* leaving it up to us to qualify if we can.) It has not even become less fearful. But the nature of our fear has changed. We no longer fear it as we fear an enemy or as we fear malevolence. Now we fear it as we fear the unknown. Our fear has ceased to be the sort that accompanies hate and contempt and the ignorance that preserves pride; it has begun to be the fear that accompanies awe, that comes with the understanding of our smallness in the presence of wonder, that teaches us to be respectful and careful. And it is a fear that is accompanied by love. We have lost our lives as in our pride we wanted them to be, and have found them as they are—

much smaller than we hoped, much shorter, much less important, much less certain, but also more abundant and joyful. We have ceased to think of the world as a piece of merchandise, and have begun to know it as an endless adventure and a blessing.

A man is standing at night in his lighted living room, feeling that in that light he knows himself, that his ends and aims are clear, that his past is coherent and his future certain. And suddenly the roof and the walls are swept away, and he sees that he inhabits a darkness reaching out to the remote lights of the sky. And then he sees that the outer stars also circumscribe a sort of living room, and that beyond them there is a darkness even darker and more immense. All the assumptions based on the premise of his own importance break loose from him. He is like a man stripped of his armor and his arms. He is left naked and humble, brought to earth.

But slowly it dawns on him that his life is a fact, whole and firm, risen up from the ground among other lives. And he feels a blessedness and a joyousness in that. He begins to trust in his own existence. He is a creature among the other creatures, and if the world promises him pain and grief and death, it also promises him health and joy and life. And if he lives generously enough even his suffering and his death may pass back into the world, its increment, to sustain the lives that will follow his. Knowing nothing but what is circumstantial, he begins to live by a sort of faith in circumstance—a faith that his life and the lives of the other creatures all belong together and sustain each other within the life of the creation which is their order and their blessing.

Courage and generosity are the moral conditions of his faith. He must trust himself to the world, freely, openly, without preconditions. He must look upon whatever happens as an opportunity. And he must never exploit his circumstances, for that is to exploit the world and his own life. Then the unity of his life with other lives is broken, the blessedness of the creation is withdrawn, and he is left alone.

A man who loves the world only insofar as it conforms to his expectations (insofar, that is, as he can understand it or use it, as engineers use it) is like an adolescent lover who loves a girl because (he thinks) she loves him. He is encapsuled in himself, and he misses the whole adventure.

A man who loves the world beyond his understanding, welcoming its unexpected blessings and depending on them, in spite of its unexpected trials and dangers, has the wisdom of a man long married to a beloved woman.

I do not believe that the Red River Gorge can

be preserved simply by making a law to preserve it. It cannot be preserved simply by defeating the dam builders for the issue is not, finally, that of the dam. To advocate the preservation of the Gorge is to advocate a profound change in the American mind. It is to go directly against the mentality that has so far been dominant in our national experience, and that has made us unable, as a people, to value any object or any act for itself, but only for its economic worth. We value the Red River Gorge not in terms of what it is, but in terms of what it can be marketed for. By this logic everything becomes expendable.

But what appears, on a ledger, to be economic sanity is ecological madness. A man who would value a piece of land strictly according to its economic worth is precisely as crazy, or as evil, as a man who would make a whore of his wife. If we were a civilized people we would not dam the Red River Gorge or overgraze a pasture or strip mine a mountain or pollute a river any more than we would sell our wives and children, because we would understand that the real values of a wilderness or a pasture or a mountain or a river, like the real values of wives and children, are not transferable or transformable. They have no monetary equivalents. The Red River Gorge cannot be transformed into a lake and it cannot be replaced by a lake any more than a wife or child can be transformed into or replaced by an insurance payment. In order for the lake to come to be, what is there now must be destroyed. The processes of economics can be reversed: what has been sold can be bought back. But when the laws of economics are applied to the creation, as they have been relentlessly throughout our history, they do not work economically but are irreversibly destructive: all transactions are final. The Red River Gorge can be destroyed within a length of time and for a price that are calculable and relatively small. The price and the time of its restoration surpass human reckoning. Should our generation destroy the Gorge, and should our grandchildren decide that we were in error, they cannot restore it; at most they can only begin a process of restoration that may take thousands of years.

If we are to preserve anything worth preserving —including, perhaps, our own lives—the economic mentality will have to give way to a mentality that will be ecological. Whereas the economic mentality holds that you give in order to get something commensurate with what you gave, the ecological mentality would center on the awareness that you get—and *far* more than you can ever earn or deserve or understand—in order to give. The nature of the economic mentality is exploitive; its motive is greed. The nature of the ecological mentality would be preserving; its motive would be generosity. That we receive so abundantly as we do from the creation does not merely imply a moral obligation to give, but the giving is the condition of the getting: you can only have, in the fullest sense, what you are prepared to give up; you can only preserve what you have become willing and glad for others to have when you are dead. "Whosoever shall seek to save his life shall lose it; and whosoever shall lose his life shall preserve it." The economic mentality assumes—the message is tirelessly repeated in advertisements and in political praises of affluence—that the blessings of our lives are luxuries and superficialities, ostentations and fashions. The ecological mentality would recognize that in reality the blessings are air and water and food and love and warmth and light and darkness and sleep and the lives of other people and other creatures. There is a story of a Zen master who said, "My miracle is that when I feel hungry I eat, and when I feel thirsty I drink." And in the thirty-third psalm it is said that "the earth is full of the goodness of the Lord." We must recover that sense of holiness in the world, and learn to respect and forbear accordingly. Failing that, we have literally everything to lose.

Early in 1968 the state's newspapers were taking note of the discovery, in one of the rock houses in the Gorge, of a crude hut built of short split planks overlaying a framework of poles. The hut was hardly bigger than a pup tent, barely large enough, I would say, to accommodate one man and a small stone fireplace. One of its planks bore the carved name: "D. boon." There was some controversy over whether or not it really was built by Daniel Boone. Perhaps it does not matter. But the news of the discovery and of the controversy over it had given the place a certain fame.

The find interested me, for I never cease to regret the scarcity of knowledge of the first explorations of the continent. Some hint, such as the "Boone hut" might provide, of the experience of the Long Hunters would be invaluable. And so one of my earliest visits to the Gorge included a trip to see the hut.

The head of the trail was not yet marked, but once I found the path leading down through the woods it was clear to me that I had already had numerous predecessors. And I had not gone far before I knew their species: scattered more and more thickly along the trail the nearer I got to the site of the hut was the trash that has come to be more characteristic than shoe-prints of the race that produced (as I am a little encouraged to remember) such a man as D. boon. And when I came to the rock house itself I found the mouth of it entirely closed, from the ground to the overhanging rock some twenty-five feet above, by a chain-link fence. Outside the fence the ground was littered with polaroid negatives, film spools, film boxes, food wrappers, cigarette butts, a paper plate, a Coke bottle.

And inside the fence, which I peered through like a prisoner, was the hut, a forlorn relic overpowered by what had been done to protect it from collectors of mementos, who would perhaps not even know what it was supposed to remind them of. There it was, perhaps a vital clue to our history and our inheritance, turned into a curio. Whether because of the ignorant enthusiasm of souvenir hunters, or because of the strenuous measures necessary to protect it from them, Boone's hut had become a doodad—as had Boone's name, which now stood for a mendacious TV show and a brand of fried chicken.

I did not go back to that place again, not wanting to be associated with the crowd whose vandalism had been so accurately foreseen and so overwhelmingly thwarted. But I did not forget it either, and the memory of it seems to me to bear, both for the Gorge and for ourselves, a heavy premonition of ruin. For are those who propose damming the Gorge, arguing *convenience,* not the same as these who can go no place, not even a few hundred steps to see the hut of D. boon, without the trash of convenience? Are they not the same who will use the proposed lake as a means of transporting the same trash into every isolated cranny that shoreline will penetrate? I have a vision (I don't know if it is nightmare or foresight) of a time when our children will go to the Gorge and find there a web-work of paved, heavily littered trails passing through tunnels of steel mesh. When people are so ignorant and destructive that they must be divided by a fence from what is vital to them, whether it is their history or their world, they are imprisoned.

On a cold drizzly day in the middle of October I walk down the side of a badly overgrazed ridge into a deep, steep hollow where there remains the only tiny grove of virgin timber still standing in all the Red River country. It is a journey backward through time, from the freeway droning both directions through 1969, across the old ridge denuded by the agricultural policies and practices of the white man's era, and down into such woods as the Shawnees knew before they knew white men.

Going down, the sense that it is a virgin place comes over you slowly. First you notice what would be the great difficulty of getting in and out, were it not for such improvements as bridges and stairways in the trail. It is this difficulty that preserved the trees, and that even now gives the hollow a feeling of austerity and remoteness. And then you realize that you are passing among poplars and hemlocks of a startling girth and height, the bark of their trunks deeply grooved and moss-grown. And finally it comes to you where you are; the virginity, the uninterrupted wildness, of the place comes to you in a clear strong dose like the first breath of a wind. Here the world is in its pure state, and such men as have been here have all been here in their pure state, for they have destroyed nothing. It has lived whole into our lifetime out of the ages. Its life is a vivid link between us and Boone and the Long Hunters and their predecessors, the Indians. It stands, brooding upon its continuance, in a strangely moving perfection, from the tops of the immense trees down to the leaves of the partridge berries on the ground. Standing and looking, moving on and looking again, I suddenly realize what is missing from nearly all the Kentucky woodlands I have known: the summit, the grandeur of these old trunks that lead the eyes up through the foliage of the lesser trees toward the sky.

At the foot of the climb, over the stone floor of the hollow, the stream is mottled with the gold leaves of the beeches. The water has taken on a vegetable taste from the leaves steeping in it. It has become a kind of weak tea, infused with the essence of the crown of the forest. By spring the fallen leaves on the stream bed will all have been swept away, and the water, filtered once again through the air and the ground, will take back the clear taste of the rock. I drink the cool brew of the autumn.

And then I wander some more among the trees. There is a thought repeating itself in my mind: This is a great Work, this is a great Work. It occurs to me that my head has gone to talking religion, that it is going ahead more or less on its own, assenting to the Creation, finding it good, in the spirit of the first chapters of Genesis. For no matter the age or the hour, I am celebrating the morning of the seventh day. I assent to my mind's assent. It *is* a great Work. It is a *great* Work—begun in the beginning, carried on until now, to be carried on, not by such processes as men make or understand, but by "the kind of intelligence that enables grass seed to grow grass; the cherry stone to make cherries."

Here is the place to remember D. boon's hut. Lay aside all questions of its age and ownership—whether or not he built it, he undoubtedly built others like it in similar places. Imagine it in a cave in a cliff overlooking such a place as this. Imagine it separated by several hundred miles from the nearest white men and by two hundred years from the drone, audible even here, of the parkway traffic. Imagine that the great trees surrounding it are part of a virgin wilderness still nearly as large as the continent, vast rich unspoiled distances quietly peopled by scattered Indian tribes, its ways still followed by buffalo and bear and panther and wolf. Imagine a cold gray winter evening, the wind loud in the branches above the protected hollows. Imagine a man dressed in skins coming silently down off the ridge and along the cliff face into the shelter of the rock house. Imagine his silence that is unbroken as he enters, crawling, a small hut that is only a negligible detail among the stone rubble of the cave floor, as unobtrusive there as the nest of an animal or bird, and as he livens the banked embers of a fire on the stone hearth, adding wood, and holds out his chilled hands before the blaze. Imagine him roasting his supper meat on a stick over the fire while the night falls and the darkness and the wind enclose the hollow. Imagine him sitting on there, miles and months from words, staring into the fire, letting its warmth deepen in him until finally he sleeps. Imagine his sleep.

When I return again it is the middle of December, getting on toward the final shortening, the first lengthening of the days. The year is ending, and my trip too has a conclusive feeling about it. The ends are gathering. The things I have learned about the Gorge, my thoughts and feelings about it, have begun to have a sequence, a pattern. From the start of the morning, because of this sense of the imminence of connections and conclusions, the day has both an excitement and a comfort about it.

As I drive in I see small lots staked off and a road newly graveled in one of the creek bottoms. And I can hear chain saws running in the vicinity of another development on Tunnel Ridge. This work is being done in anticipation of the lake, but I know that it has been hastened by the publicity surrounding the effort to keep the Gorge unspoiled. I consider the ironic possibility that what I will write for love of it may also contribute to its destruction, enlarging the hearsay of it, bringing in more people to drive the roads and crowd the "points of interest" until they become exactly as interesting as a busy street. And yet I might as well leave the place anonymous, for what I have learned here could be learned from any woods and any free-running river.

I pull off the road near the mouth of a hollow I have not yet been in. The day is warm and overcast, but it seems unlikely to rain. Taking only a notebook and a map, I turn away from the road and start out. The woods closes me in. Within a few minutes I have put the road, and where it came from and is going, out of mind. There comes to be a wonderful friendliness, a sort of sweetness I have not known here before, about this day and this solitary walk—as if, having finally understood this country well enough to accept it on its terms, I am in turn accepted. It is as though, in this year of men's arrival on the moon, I have completed my own journey at last, and have arrived, an exultant traveler, here on the earth.

I come around a big rock in the stream and two grouse flush in the open not ten steps away. I walk on more quietly, full of the sense of ending and beginning. At any moment, I think, the forest may reveal itself to you in a new way. Some intimate insight, that all you have known has been secretly adding up to, may suddenly open into the clear—like a grouse, that one moment seemed only a part of the forest floor, the next moment rising in flight. Also it may not.

Where I am going I have never been before. And since I have no destination that I know, where I am going is always where I am. When I come to good resting places, I rest. I rest whether I am tired or not because the places are good. Each one is an arrival. I am where

I have been going. At a narrow place in the stream I sit on one side and prop my feet on the other. For a while I content myself to be a bridge. The water of heaven and earth is flowing beneath me. While I rest a piece of the world's work is continuing here without my help.

Since I was here last the leaves have fallen. The forest has been at work, dying to renew itself, covering the tracks of those of us who were here, burying the paths and the old campsites and the refuse. It is showing us what to hope for. And that we can hope. And *how* to hope. It will always be a new world, if we will let it be.

The place as it was is gone, and we are gone as we were. We will never be in that place again. Rejoice that it is dead, for having received that death, the place of next year, a new place, is lying potent in the ground like a deep dream.

Somewhere, somewhere behind me that I will not go back to, I have lost my map. At first I am sorry, for on these trips I have always kept it with me. I brood over the thought of it, the map of this place rotting into it along with its leaves and its fallen wood. The image takes hold of me, and I suddenly realize that it is the culmination, the final insight, that I have felt impending all through the day. It is the symbol of what I have learned here, and of the process: the gradual relinquishment of maps, the yielding of knowledge before the new facts and the mysteries of growth and renewal and change. What men know and presume about the earth is part of it, passing always back into it, carried on by it into what they do not know. Even their abuses of it, their diminishments and dooms, belong to it. The tragedy is only ours, who have little time to be here, not the world's whose creation bears triumphantly on and on from the fulfillment of catastrophe to the fulfillment of hepatica blossoms. The thought of the lost map, the map fallen and decaying like a leaf among the leaves, grows in my mind to the force of a cleansing vision. As though freed of a heavy weight, I am light and exultant here in the end and the beginning.

IV

WIDENING THE CIRCLES

"The story of my people
and the story of this place are one single story.
No man can think of us without thinking of this
place. We are always joined together."

TAOS PUEBLO INDIAN MAN

PETER
MATTHIESSEN

Photograph © Barbara Hall

Peter Matthiessen was born in New York City in 1927. He graduated from Yale and spent the early 1950s in Paris, writing fiction, co-founding the *Paris Review*, and living the life of an expatriate novelist. In the mid-fifties, he moved to Long Island and worked for several years as a commercial fisherman and charter boat captain, after which he began his travels in the wilderness, initially under the sponsorship of the *New Yorker*. These trips have led to a dozen books of non-fiction, with themes circling around native peoples, wildlife, and environmental ruin, in locations from New Guinea to East Africa, from the Bering Sea to Peru.

The Snow Leopard, Matthiessen's journey through Nepal to the

242

Tibetan Plateau—studying the blue sheep, searching the mountains with zoologist George Schaller for the rare cat, and searching in himself for some understanding of the death of his wife, Deborah, the year before—won a National Book Award in 1978. *Sand Rivers*, a trek into the Selous Game Reserve in Tanzania, won the John Burroughs Medal in 1982. Matthiessen's other books of travel and natural history include *Wildlife In America* (1959); *The Cloud Forest: A Chronicle of the South American Wilderness* (1961); and *The Tree Where Man Was Born* (1972). His two best-known novels are *At Play in the Fields of the Lord* (1965) and *Far Tortuga* (1975). The Philadelphia Academy of Science recognized his natural history writing with its 1984 gold medal.

Matthiessen published two books in 1986, *Men's Lives*, about the difficulties of Long Island's fishermen today, and *Nine-Headed Dragon River*, about his practice of Zen Buddhism. He now is ready to return to fiction, and says, "I hope to stay there.

"I prefer writing fiction; I find it exhilarating. I've always thought of nonfiction as a livelihood, my way of making a living so I could write fiction. Nonfiction may be extremely skillful, it may be cabinetwork rather than carpentry, but it's still assembled from facts, from research, from observation; it comes from outside, not from within. It may be well made or badly made, but it's still an assemblage. If you're an honest journalist, you're inevitably confined by the facts; you can't use your imagination beyond a certain point.

"*The Snow Leopard* in a sense was an exception, because that was a strange kind of mythic trek across time as well as distance. 'The writer' seemed to disappear into that landscape, and the book seemed to generate itself, just as a novel does."

The following selection comes from *The Snow Leopard* beginning with the arrival of Matthiessen and Schaller (GS) at Crystal Mountain Monastery, Shey Gompa, after a 250-mile trek across the Himalaya.

243

At Crystal Mountain

November 1

This Black Pond Camp, though well below the Kang Pass, lies at an altitude of 17,000 feet, and an hour after the sun sinks behind the peaks, my wet boots have turned to blocks of ice. GS's thermometer registers −20° Centigrade (4° below zero Fahrenheit) and though I wear everything I have, I quake with cold all night. Dawn comes at last, but making hot water from a pot of ice is difficult at this altitude, and it is past nine before boots are thawed and we are under way.

The snow bowl is the head of an ice river that descends a deep canyon to Shey. In the canyon we meet Jang-bu and Phu-Tsering, on their way up to fetch some food and pots: Dawa, they say, is down again with acute snow blindness.

Sherpa tracks in the frozen shadows follow the glassy boulders of the stream edge, and somewhere along the way I slip, losing the hoopoe feather that adorned my cap. The river falls steeply, for Shey lies three thousand feet below Kang La, and in the deep snow, the going is so treacherous that the sherpas have made no path; each man flounders through the drifts as best he can. Eventually, from a high corner of the canyon, rough red-brown lumps of human habitation come in view. The monastery stands like a small fort on a bluff where another river flows in from the east; a mile below, the rivers vanish into a deep and dark ravine. Excepting the lower slopes of the mountainside behind the monastery, which is open to the south, most of this treeless waste lies under snow, broken here and there by calligraphic patterns of bare rock, in an atmosphere so wild and desolate as to overwhelm the small huddle of dwellings.

High to the west, a white pyramid sails on the sky—the Crystal Mountain. In summer, this monument of rock is a shrine for pilgrims from all over Dolpo and beyond, who come here to make a prescribed circle around the Crystal Mountain and attend a holy festival at Shey. What is stirring about this peak, in snow time, is its powerful shape, which even today, with no clouds passing, makes it appear to be forging through the blue. "The power of such a mountain is so great and yet so

244

subtle that, without compulsion, people are drawn to it from near and far, as if by the force of some invisible magnet; and they will undergo untold hardships and privations in their inexplicable urge to approach and to worship the centre of this sacred power. . . . This worshipful or religious attitude is not impressed by scientific facts, like figures of altitude, which are foremost in the mind of modern man. Nor is it motivated by the urge to 'conquer' the mountain. . . ."

A gravel island under Shey is reached by crossing ice and stones of a shallow channel. At the island's lower end are prayer walls and a stone stockade for animals; farther on, small conduits divert a flow of river water to a group of prayer mills in the form of waterwheels, each one housed separately in its own stone shrine. The conduits are frozen and the wheels are still. On top of the small stupas are offerings of white quartz crystals, presumably taken from the Crystal Mountain in the summer, when the five wheels spin five ancient prayer drums, sending OM MANI PADME HUM down the cold canyon.

On the far side of a plank bridge, a path climbs the bank to two big red-and-white entrance stupas on the bluff: I go up slowly. Prayer flags snap thinly on the wind, and a wind-bell has a wooden wing in the shape of a half-moon that moves the clapper; over the glacial rumble on the river stones, the wistful ring on the light wind is the first sound that is heard here at Shey Gompa.

The cluster of a half-dozen stone houses is stained red, in sign that Shey is a monastery, not a village. Another group of five small houses sits higher up the mountain; above this hamlet, a band of blue sheep may be seen with the naked eye. Across the river to the north, stuck on a cliff face at the portals of the canyon, is a red hermitage. Otherwise, except for prayer walls and the stone corrals, there are only the mighty rock formations and dry treeless mountainside where snow has melted, and the snow and sky.

I move on slowly, dull in mind and body. Gazing back up the Black River toward the rampart of icy cornices, I understand that we have come over the Kanjirobas to the mountain deserts of the Tibetan Plateau: we have crossed the Himalaya from south to north. But not until I had to climb this short steep path from the wintry river to the bluffs did I realize how tired I was after thirty-five days of hard trekking. And here I am, on this first day of November, standing before the Crystal Monastery, with its strange stones and flags and bells under the snows.

The monastery temple with its attached houses forms a sort of open court facing the south. Two women and two infants, sitting in the sun, make no sign of welcome. Fearing Kham-pa brigands, the women had locked themselves into their houses a few days ago, when Jang-bu and GS first appeared, and plainly they are still suspicious of our seemingly inexplicable mission. The younger woman is weaving a rough cloth on an ancient loom. When I say, *"Namaste!"* she repeats it, as if trying the word out. Three scraggy *dzos* and an old black nanny goat excepted, these are the only sentient beings left at Shey, which its inhabitants call Somdo, or "Confluence," because of the meeting of rivers beneath its bluff—the Kangju, "Snow Waters" (the one I think of as Black River, because of the black pond at its head, and the black eagle, and the black patterns of its stones and ice in the dark canyon), and the Yeju, "Low Waters" (which I shall call White River, because it comes down from the eastern snows).

For cooking hut and storeroom, Jang-bu has appropriated the only unlocked dwelling. Like all the rest, it has a flat roof of clay and saplings piled on top with brushwood, a small wooden door into the single room, and a tiny window in the western wall to catch afternoon light. The solitary ray of light, as in a medieval painting, illumines the smoke-blackened posts that support the roof, which is so low that GS and I must bend half-over. The earth floor is bare, except for a clay oven built up in three points to hold a pot, with a hole near the floor to blow life into the smoky fire of dung and brushwood. Jang-bu and Phu-Tsering's tent is just outside the door, while Dawa will sleep inside with the supplies. GS pitches his blue tent just uphill from the hut, while I place mine some distance away, facing east up the White River toward the sunrise.

The cooking hut is the sometime dwelling of the brother of the younger woman, Tasi Chanjun, whom the sherpas call Namu, meaning hostess. (Among Tibetans as among Native Americans, it is often rude to address people by their formal name.) Her little boy, aged about four, is Karma Chambel, and her daughter, perhaps two, is Nyima Poti. Nyima means "sun" or "sunny"—Sunny Poti! The old woman's name is Sonam: her husband, Chang Rapke, and her daughter Karima Poti have gone away to winter in Saldang, and Sonam lives alone in the abandoned hamlet up the mountain. Namu says that before the snows there were forty people here, including twenty-odd monks and two lamas: all are gone across the mountains to Saldang, from where—is this a warning to outlandish men who come here without women?—her husband will return in a few days. Namu's husband has the key to the Crystal Monastery, or so she says, and will doubtless bring it with him when he comes to visit, in four or five days, or in twenty. Namu is perhaps thirty years old, and pretty in a sturdy way, and self-dependent. She speaks familiarly of B'od but not Nepal; even Ring-mo is a foreign land, far away across Kang La.

That the Lama is gone is very disappointing. Nevertheless, we are extremely happy to be here, all the more so since it often seemed that we would never arrive at all. Now we can wake up in the morning without having to put on wet boots, break camp, get people moving; and there is home to return to in the evening. There are no porters harassing our days, and we are sheltered more or less, from evil weather. The high pass between Shey and the outside world lies in the snow peaks, ghostly now in the light of the cold stars. "God, I'm glad I'm not up there tonight," GS exclaims, as we emerge from the smoky hut, our bellies warm with lentil soup. We know how fortunate it was that the Kang Pass was crossed in this fine, windless weather, and wonder how long fair skies will hold, and if Tukten and Gyaltsen will appear. It is November now, and everything depends upon the snows.

November 2

. . . This morning I bathe inside my sunny tent, and sort out gear. Dawa is still groaning with snow blindness, but Jang-bu and Phu-Tsering have crossed Black River to hunt scraps of low shrub juniper for firewood, and GS is up on this Somdo mountainside viewing his sheep; he returns half-frozen toward midmorning. After a quick meal of chapatis, we set off on a survey of other sheep populations in the region, heading eastward up the Saldang trail, which follows the north bank of the White River. Like the Saure and other east-west rivers in this season, this one is snowbound on the side that faces north, and across the water, we can see snow tracks of marmot, wandering outward in weird patterns from a burrow; perhaps the animals, sent underground too early by those blizzards of the late monsoon, had gone out foraging. But they are hibernated now, there is no fresh marmot sign, the land seems empty.

Snow clouds come up over the mountains, and the shining river turns to black, over black rocks. A lone black *dzo* nuzzles the stony earth. GS has picked up scat of a large carnivore and turns it in his hand, wondering aloud why fox sign, so abundant at Black Pond, is uncommon here at lower altitudes. "Too big for fox, I think. . . ."

As GS speaks, I scan the mountain slopes for bharal: on these rolling hills to the east of Somdo, we have not seen even one. Abruptly, he says, "Hold it! Freeze! Two snow leopard!" I see a pale shape slip behind a low rise patched with snow, as GS, agitated, mutters, "Tail's too short! Must have been foxes—!"

"No!" I say. "Much too big—!"

"Wolves!" he cries out. "Wolves!"

And there they are.

Moving away without haste up an open slope beyond the rise, the wolves bring the barren hills to life. Two on the slope to northward frisk and play, but soon they pause to look us over; their tameness is astonishing. Then they cut across the hill to join three others that are climbing a stone gully. The pack stops each little while to gaze at us, and through the telescope we rejoice in every shining hair: two silver wolves, and two of faded gold, and one that is the no-color of frost:

this frost-colored wolf, a big male, seems to be leader. All have black tail tips and a delicate black fretting on the back. "That's why there's no sign of fox or leopard!" GS says, "and that's why the blue sheep stay near the river cliffs, away from this open country!" I ask if the wolves would hunt and kill the fox and leopard, and he says they would. For some reason, the wolves' appearance here has taken us by surprise; it is in Tibet that such mythic creatures belong. This is an Asian race of *Canis lupus,* the timber wolf, which both of us have seen in Alaska, and it is always an exciting animal: the empty hills where the pack has gone have come to life. In a snow patch are five sets of wolf tracks, and old wolf scats along the path contain brittle gray stuff and soft yellow hair—blue sheep and marmot.

Down the path comes an old woman who has walked alone from Saldang, over the Shey Pass to the east; we are as surprised by her appearance as she is by ours. The old woman has seen the five *jangu,* and two more, but seems less wary of the wolves than of big strangers.

We wonder about the solitary *dzo,* not more than a half mile from the place where we had turned the wolves back toward the east. Later Namu says that wolves kill two or three *dzos* every year, and five or six sheep at a time in the corrals. She sets out upriver to fetch her *dzo,* and is back with the lone beast just before sundown.

November 3

There is so much that enchants me in this spare, silent place that I move softly so as not to break a spell. Because the taking of life has been forbidden by the Lama of Shey, bharal and wolves alike draw near the monastery. On the hills and in the stone beds of the river are fossils from blue ancient days when all this soaring rock lay beneath the sea. And all about are the prayer stones, prayer flags, prayer wheels, and prayer mills in the torrent, calling on all the elements in nature to join in celebration of the One. What I hear from my tent is a delicate wind-bell and the river from the east, in this easterly wind that may bring a change in the weather. At daybreak, two great ravens come, their long toes scratching on the prayer walls.

The sun refracts from the white glaze of the mountains, chills the air. Old Sonam, who lives alone in the hamlet up the hill, was on the mountain before day, gathering the summer's dung to dry and store as cooking fuel; what I took for lumpish matter straightens on the sky as the sun rises, setting her gaunt silhouette afire.

Eleven sheep are visible on the Somdo slope above the monastery, six rams together and a group of ewes and young; though the bands begin to draw near to one another and sniff urine traces, there is no real sign of rut. From our lookout above Sonam's house, three more groups—six, fourteen, and twenty-six—can be seen on the westward slopes, across Black River.

Unable to hold the scope on the restless animals, GS calls out to me to shift the binoculars from the band of fourteen to the group of six sheep directly across the river from our lookout. "Why are those sheep running?" he demands and a moment later hollers, "Wolves!" All six sheep are springing for the cliffs, but a pair of wolves coming straight downhill are cutting off the rearmost animal as it bounds across a stretch of snow toward the ledges. In the hard light, the blue-gray creature seems far too swift to catch, yet the streaming wolves gain ground on the hard snow. Then they are whisking through the matted juniper and down over steepening rocks, and it appears that the bharal will be cut off and bowled over, down the mountain, but at the last moment it scoots free and gains a narrow ledge where no wolf can follow.

In the frozen air, the whole mountain is taut; the silence rings. The sheep's flanks quake, and the wolves are panting; otherwise, all is still, as if the arrangement of pale shapes held the world together. Then I breathe, and the mountain breathes, setting the world in motion once again.

Briefly, the wolves gaze about, then make their way up the mountainside in the unhurried gait that may carry them fifty miles in a single day. Two pack mates join them, and in high yak pasture the four pause to romp and roll in dung. Two of these were not among the five seen yesterday, and we recall that the old woman had seen seven. Then they trot onward, disappearing behind a ridge of snow. The band of fourteen sheep high on this ridge gives a brief run of alarm, then forms a line on a high point to stare down at the wolves and watch them go. Before long, all are browsing once again, including the six that were chased onto the precipice.

Turning to speak, we just shake our heads and grin. "It was worth walking five weeks just to see that," GS sighs at last. "That was the most exciting wolf hunt I ever saw." And a little later, exhilarated still, he wonders aloud if I remember "that rainy afternoon in the Serengeti when we watched wild dogs make a zebra kill in that strange storm light on the plain, and all those thousands of animals running?" I nod. I am still excited by the wolves seen so close yesterday, and to see them again, to watch them hunt blue sheep in such fashion, flying down across the cliffs within sight of our tents at Shey Gompa—what happiness!

After years of studying the carnivores, GS has become fascinated by the Caprini—the sheep and the goats—which have the attraction of inhabiting the remote high mountains that he loves. And among the Caprini, this "blue sheep" is a most peculiar species, which is one reason we have come so far to see it. It is presumed that the sheeps and goats branched off from a common ancestor among the Rupicaprini, the so-called goat-antelopes, which are thought to have evolved somewhere south of the Himalaya; this generalized ancestor may have resembled the small goat-antelope called the goral, which we saw last month in the dry canyon of the Bheri. Besides the six species of true goat (*Capra*) and the six of true sheep (*Ovis*), the Caprini include three species of tahr (*Hemitragus*), the aoudad or Barbary sheep (*Ammotragus*), and the bharal or Himalayan blue sheep (*Pseudois*), all of which exhibit characters of both sheep and goat. The tahrs, which in their morphology and behavior appear to be intermediate between goat-antelopes and true goats, are classified as goats, and the sheeplike aoudad is mostly goat as well. *Pseudois,* too, looks very like a sheep; it recalls the Rocky Mountain sheep, not only in its general aspect but in type of habitat—rolling upland in the vicinity of cliffs. Certain specimens, GS says, possess the interdigital glands on all four feet which were thought to be a diagnostic character of *Ovis,* and the species lacks the strong smell, beard, and knee callouses that are found in *Capra.* Nevertheless, GS considers it more goat than sheep, and hopes to establish this beyond all doubt by observation of its behavior in the rut.

Hunters' reports account for most of what is known of the wild goats and sheep of Asia, which may be why the classification of *Pseudois* is still disputed. Since the blue sheep is now rare in world collections, the one way to resolve the question is to observe the animal in its own inhospitable habitat—above timberline, as high as 18,000 feet, in the vicinity of cliffs—in one of the most remote ranges of any animal on earth: from Ladakh and Kashmir east across Tibet into northwest China, south to the Himalayan crest, and north to the Kuenlun and Altyn mountains. In Nepal, a few bharal are found on the western and southern flanks of Dhaulagiri (this is the population that we saw near the Jang Pass), as well as in the upper Arun Valley, in the east, but most are found here in the northwest, near the Tibetan border.

This morning, through the telescope, I study blue sheep carefully for the first time. Like the Rocky Mountain sheep, they·are short-legged, strong, broad-backed animals, quick and neat-footed, with gold demonic eyes. The thick-horned male is a handsome slaty blue, the white of his rump and belly set off by bold black face marks, chest, and flank stripe, and black anteriors on all four legs; the black flank stripes like the horns, become heavier with age. The female is much smaller, with dull pelage and less contrast in the black, and her horns are spindly, as in female sheep. Those of the males, on the other hand, are heavy, curving upward, out, and back. Also, the basio-occipital bone at the base of the skull is goatlike, and so are the large dew claws and the prominent markings on the fore sides of the legs. In this confused situation, the rutting behavior will be a deciding factor, yet from the limited reports available, even the rutting is ambiguous. For example, the courting sheep rarely raises its tail above the horizontal, whereas the goat may arch it back onto its rump: perhaps for lack of the odorous tail-gland secretions of true goats that the arched tail may help disseminate, but tahr and bharal compromise by erecting the tail straight up into the air.

Although the male herds are still intact—this sociability of rams is a trait of Caprini—the males are mounting one another, as much to establish dominance as in sexuality; among many sheep and goats, the juvenile males and the females are quite similar in appearance, and tend to imitate the behavior of the other, so that rams may fail to differentiate between them, treating all of these subordinates alike. A few are displaying the bizarre behavior (heretofore unreported) that GS calls "rump-rubbing," in which one male may rub his face against the hind end of another. In the vicinity of females, the male "kicks"—a loose twitch of the leg in her direction that appears to be a mounting preliminary and may also serve to display his handsome markings. Also, he thrusts his muzzle into her urine stream, as if to learn whether or not she is in estrus, and licks in agitation at his penis. But the blue sheep stops short of certain practices developed by the markhor of Pakistan and the wild goat (the ancestor of the domestic goat, ranging from Pakistan to Greece), both of which take their penises into their mouths, urinate copiously, then spit on their own coats; the beard of the male goat is an adaptive character, a sort of urine sponge that perpetuates the fine funky smell for which the goats are known.

The itch of the rutting season has begun, and even the young animals play at butting and sparring, as if anxious not to miss the only lively time in the blue sheep's year. GS wonders at the scarcity of the young, concluding that a 50 percent mortality must occur in the first year, due as much to weakness or disease caused by poor range condition as to predation by wolves and leopard. Perhaps one juvenile in three attains maturity, and this may suffice to sustain the herds, which must adjust numbers to the limited amount of habitat that remains snow-free all the year. This region of the Tibetan Plateau is a near desert of rock and barren slopes dominated by two thorn shrubs, *Caragana* and a bush honeysuckle, *Lonicera;* the blue sheep will eat small amounts of almost any growth including the dry everlasting and the oily juniper, and the adaptations of the Caprini for hard, abrasive forage permit limited browsing of this thorny scrub as well. But excepting a few tufts among the thorns, almost all the native grasses that are its preferred food have been eradicated by the herds of yak and sheep and goats that are brought here from distant villages in summer, and the overgrazing has led already to erosion.

. . . November 6

The nights at Shey are rigid, under rigid stars; the fall of a wolf pad on the frozen path might be heard up and down the canyon. But a hard wind comes before the dawn to rattle the tent

251

canvas, and this morning it is clear again, and colder. At daybreak, the White River, just below, is sheathed in ice, with scarcely a murmur from the stream beneath.

The two ravens come to tritons on the gompa roof. *Gorawk, gorawk,* they croak, and this is the name given to them by the sherpas. Amidst the prayer flags and great horns of Tibetan argali, the gorawks greet first light with an odd musical doublenote—*a-ho*—that emerges as if by miracle from those ragged throats. Before sunrise every day, the great blackbirds are gone, like the last tatters of departing night.

The sun rising at the head of the White River brings a suffused glow to the tent canvas, and the robin accentor flits away across the frozen yard. At seven, there is breakfast in the cook hut—tea and porridge—and after breakfast on most days I watch sheep with GS, parting company with him after a while, when the sheep lie down, to go off on some expedition of my own. Often I scan the caves and ledges on the far side of Black River in the hope of leopard; I am alert for fossils, wolves, and birds. Sometimes I observe the sky and mountains, and sometimes I sit in meditation, doing my best to empty out my mind, to attain that state in which everything is "at rest, free, and immortal. . . . All things abided eternally as they were in their proper places . . . something infinite behind everything appeared." (No Buddhist said this, but a seventeenth-century Briton.) And soon all sounds, and all one sees and feels, take on imminence, an immanence, as if the Universe were coming to attention, a Universe of which one is the center, a Universe that is not the same and yet not different from oneself, even from a scientific point of view: within man as within mountains there are many parts of hydrogen and oxygen, of calcium, phosphorus, potassium, and other elements. "You never enjoy the world aright, till the Sea itself flows in your veins, till you are clothed with the heavens, and crowned with the stars: and perceive yourself to be the sole heir of the whole world, and more than so, because men are in it who are every one sole heirs as well as you."

I have a meditation place on Somdo mountain, a broken rock outcrop like an altar set into the hillside, protected from all but the south wind by shards of granite and dense thorn. In the full sun it is warm, and its rock crannies give shelter to small stunted plants that cling to this desert mountainside—dead red-brown stalks of a wild buckwheat (*Polygonum*), some shrubby cinquefoil, pale edelweiss, and everlasting, and even a few poor wisps of *Cannabis.* I arrange a rude rock seat as a lookout on the world, set out binoculars in case wild creatures should happen into view, then cross my legs and regulate my breath, until I scarcely breathe at all.

Now the mountains all around me take on life; the Crystal Mountain moves. Soon there comes the murmur of the torrent, from far away below under the ice: it seems impossible that I can hear this sound. Even in windlessness, the sound of rivers comes and goes and falls and rises, like the wind itself. An instinct comes to open outward by letting all life in, just as a flower fills with sun. To burst forth from this old husk and cast one's energy abroad, to fly. . . .

Although I am not conscious of emotion, the mind-opening brings a soft mist to my eyes. Then the mist passes, the cold wind clears my head, and body-mind comes and goes on the light air. A sun-filled Buddha. One day I shall meditate in falling snow.

I lower my gaze from the snow peaks to the glistening thorns, the snow patches, the lichens. Though I am blind to it, the Truth is near, in the reality of what I sit on—rocks. These hard rocks instruct my bones in what my brain could never grasp in the Heart Sutra, that "form is emptiness, and emptiness is form"—the Void, the emptiness of blue-black space, contained in everything. Sometimes when I meditate, the big rocks dance.

The secret of the mountains is that the mountains simply exist, as I do myself: the mountains exist simply, which I do not. The mountains have no "meaning," they *are* meaning; the mountains *are*. The sun is round. I ring with life, and the mountains ring, and when I can hear it, there is a ringing that we share. I understand all this, not in my mind but in my heart, knowing how meaningless it is to try to capture what cannot be expressed, knowing that mere words will remain when I read it all again, another day.

Toward four, the sun sets fires on the Crystal Mountain. I turn my collar up and put on gloves and go down to Somdo, where my tent has stored the last sun of the day. In the tent entrance, out of the wind, I drink hot tea and watch the darkness rise out of the earth. The sunset fills the deepening blues with holy rays and turns a twilight raven into the silver bird of night as it passes into the shadow of the mountain. Then the great hush falls and cold descends. The temperature has already dropped well below freezing, and will drop twenty degrees more before the dawn.

At dark, I walk past lifeless houses to the cooking hut where Phu-Tsering will be baking a green loaf; the sherpas have erected two stone tables, and in the evenings the hut is almost cozy, warmed by the dung and smoking juniper in the clay oven.

As usual GS is there ahead of me, recording data. Eyes watering, we read and write by kerosene lamp. We are glad to see each other, but we rarely speak more than a few words during a simple supper, usually rice of a poor bitter kind, with tomato or soy sauce, salt and pepper, sometimes accompanied by thin lentil soup. After supper I watch the fire for a time until smoke from the sparking juniper closes my eyes. Bidding goodnight, I bend through the low doorway and go out under the stars and pick my way around the frozen walls to my cold tent, there to remain for twelve hours or more until first light. I read until near asphyxiated by my small wick candle in its flask of kerosene, then lie still for a long time in the very heart of the earth silence, exhilarated and excited as a child. I have yet to use the large packet of *Cannabis* that I gathered at Yamarkhar and dried along the way, to see me through long lightless evenings on this journey: I am high enough.

"Regard as one, this life, the next life, and the life between," wrote Milarepa. And sometimes I wonder into which life I have wandered, so still are the long nights here, and so cold.

November 7

High on the mountain, I come upon a herd of twenty-seven blue sheep that includes males and females of all ages; until today the Somdo rams formed their own herd.

At the sight of man, the bharal drift over a snow ridge toward the north. I trail this promising mixed party, hoping to make observations for GS, who is working near Tsakang. Eventually, the sheep lie down on a steep grassy slope that plunges toward the mouth of Black River Canyon, and I withdraw to a point where they cannot see me, letting them calm themselves before attempting to go closer.

For a long time I sit very still. To a nearby rock comes a black redstart, bobbing in spry agitation and flaring its rufous tail. Then choughs come sqealing on the wind, lifting and dancing in a flock of fifty or more: the small black crows, in escadrilles, plummet from view, filling the silence with a rush of air.

In my parka I find a few wild walnuts from Rohagaon, and crack them open with a stone. From this point of mountain, I can see in four directions. Eastward, the White River comes down out of the snow—this is the direction of Saldang. To the south, the Black River Canyon climbs into the Kanjirobas. To the west is the great pyramidal butte of Crystal Mountain, parting the wind that bears uneasy clouds down the blue sky. Northward, beyond Somdo mountain, on a hidden plateau above the canyons, lies the old B'on stronghold at Samling.

The Somdo herd has moved uphill, above 15,000 feet. Since the wind is from the south, bearing my scent, I traverse a half mile to the east before starting to climb; by the time the climb is finished, the wind has shifted to the north, and I can wriggle to a point not one hundred yards away from the nearest animal.

To be right among the sheep like this is stirring. I lie belly down, out of the wind, and the whole warm mountain, breathing as I breathe, seems to take me in. All the sheep but two are lying down, and four big rams a little uphill from the rest face me without alarm. The sun glows in the coarse hairs of their blue coats as they chew their cud, carved faces sweeping back to the huge cracked horns. These males are big and heavy, broad across the back, strong, handsome animals: although I am downwind of the herd, there is no smell at all.

One of the males senses me, for there is an elegant arch to his neck, and his eyes and ears are wide in that relaxed readiness that reminds me unaccountably of Tukten: what can our evil monk be doing now? The other sheep are dozing. Most of the young animals lie with their rumps downhill, in my direction (the reverse is true of the adults, which expect threats from below), and two sprawl out in an adolescent manner, heads laid back along their flanks. This is the morning lull, observed each day; they will not browse again for at least an hour. I back down behind the rise, to wait. An hour later, when I stalk them once again, they are just getting to their feet. A female squats to urinate, and a male thrusts his muzzle into the fluid and then into her vulva, upon which he extends his neck in seeming ecstasy and curls his upper lip, eyes closed, the better to savor his findings. Another ram follows a different female, and he, too, pokes his nose against her rump; a third turns his head along his flank as if to seek out his own penis, in the way of goats, then loses interest.

Now the animals begin to graze, twisting their necks to search out grass tufts under the bush honeysuckle; a few browse the small yellow-green leaves of the shrub itself. Led by a female—and in this mixed herd a female usually leads—they move downhill a little as they feed, until they disappear below the rise. When they reappear, they come directly toward the hummock where I lie. Suddenly the creatures are so close that I must lower the binoculars inch by inch so as not to flare them, drawing my chin deep into the thin growth of the mountainside, hoping my brown hair may be seen as marmot. On they come, browsing a little, males sniffing ignominiously after the females, the two calves of the year bringing up the rear.

The lead female comes out of the hollow not ten yards up the hill, moving a little way eastward. Suddenly, she gets my scent and turns quickly to stare at my still form in the dust below. She does not move but simply stands, eyes round. In her tension, the black marks on her legs are fairly shivering; she is superb. Then the first ram comes to her, and he, too, scents me. In a jump, he whorls in my direction, and his tail shoots straight up in the air, and he stamps his right forefoot, venting a weird harsh high-pitched whinny—*chir-r-rit*—more like a squirrel than any ungulate. (Later I described this carefully to GS—so far as we know, the first datum on the voice of the blue sheep.) Boldly this ram steps forward to investigate, and the rest follow, until the mountain blue is full of horned heads and sheep faces, sheep vibrations—I hold my breath as best I can. In nervousness, a few pretend to browse, and one male nips edgily at a yearling's rump, coming away with a silver tuft that shimmers in the sun. Unhurriedly, they move away, rounding the slope toward the east. Soon the heads of two females reappear, as if to make sure nothing is following. Then all are gone.

On the way down the mountain, I stop outside Old Sonam's yard in the upper village. In sooty rags and rough-spun boots, wearing the coral-colored beads of her lost girlhood, Sonam is sitting legs straight out in the dry dung, weaving a blanket on a crazy handloom rigged to rocks and sticks, bracing the whole with old twine soles pushed stiff against a stone. Her wool has a handsome and delicate pattern, for there is design in the eye of this old wild one. I admire her sudden grin, strong back, and grimy hide indifferent to the cold.

Once Sonam was an infant with red cheeks, like Sunny Poti. Now she works close in the last light, as cold descends under a faint half-moon. Soon night will come, and she will creep through her narrow door and eat a little barley; what does she dream of until daybreak, when she goes out on her endless quest for dung? Perhaps she knows better than to think at all, but goes simply about the business of survival, like the wolf; survival is her way of meditation. When I ask Jang-bu why Sonam lives alone all winter in the upper village when she might use an empty house near Namu, he seems astonished. "She has the habit of that place," he says.

. . . November 9

From the path that leads beyond Tsakang, along the precipices of the Black River Canyon, there is a stirring prospect of the great cliffs and escarpments, marching northward toward the point where this Yeju-Kangju flows into the great Karnali River. The path is no more than a ledge in many places and, on the northward face of each ravine, is covered by glare ice and crusted snow. Even on the southward face, the path is narrow, and concentrating hard on every step, I come upon what looks like a big pug mark. Because it is faint, and because GS is too far ahead to summon back, and because until now we have found no trace of leopard, I keep quiet; the mark will be there still when we return. And just at this moment, looking up, I see that GS has paused on the path ahead. When I come up, he points at a distinct cat scrape and print. The print is faded, but at least we know that the snow leopard is here.

Mostly we spend the day apart, meeting over the clay oven for breakfast and supper, but whenever we act like social animals, the impulse has brought luck. A little farther on there is another scrape, and then another, and GS, looking ahead to where the path turns the cliff corner into the next ravine, says, "There ought to be a leopard scat out on that next point—it's just the sort of place they choose." And there it is, all but glowing in the path, right beneath the prayer stones of the stupa—the Jewel in the Heart of the Lotus, I think, unaccountably, and nod at my friend, impressed. "Isn't that something?" GS says, "To be so delighted with a pile of crap?" He gathers the dropping into one of the plastic bags that he keeps with him for the purpose and tucks it away into his rucksack with our lunch. Though the sign is probably a week old, we are already scanning the sunny ledges and open caves on both sides of the river that we have studied for so many days in vain.

On the ledge path we find two more scats and a half dozen scrapes, as well as melted cat prints in the snow on the north face of the ravines. Perhaps this creature is not resident but comes through on a hunting circuit, as the wolves do: the wolves have been missing now for near a week. On the other hand, this labyrinth of caves and ledges is fine haunt for leopard, out of the way of its enemy, the wolf, and handy to a herd of bharal that is resident on the ridge above and often wanders down close to these cliffs. Perhaps, in the days left to us, we shall never see the snow leopard but it seems certain that the leopard will see us.

Across the next ravine is the second hermitage, of earth red decorated in blue-gray and white. It lacks stacked brush or other sign of life, and its white prayer flags are worn to wisps by wind. In the cliffs nearby are smoke-roofed caves and the ruins of cells that must have sheltered anchorites of former times; perhaps their food was brought them from Tsakang. This small gompa, half-covering a walled-up cave, is tucked into an outer corner of a cliff that falls into Black Canyon, and like Tsakang it faces south, up the Black River. Because the points of the Shey stupas are just visible, its situation is less hallucinatory than the pure blue-and-white prospect at Tsakang, but the sheer drop of a thousand feet into the gorge, the torrent's roar, the wind, and the high walls darkening the sky all around make its situation more disturbing. The hermitage lies on the last part of a pilgrim's path that climbs from Black River and circles round the Crystal Mountain, striking Black Canyon once again on the north side of this point and returning to Shey by way of Tsakang; but most of the path is lost beneath the snows.

Taking shelter on the sunny step, leaning back into the warmth of the wooden door, I eat a green disc of Phu-Tsering's buckwheat bread that looks and tastes like a lichened stone mandala from the prayer walls. Blue sheep have littered this small dooryard with their dung, a human hand has painted a sun and moon above the lintel, yet in this forlorn place, here at the edge of things, the stony bread, the dung and painted moon, the lonely tattering of flags worn to transparence by the wind seem as illusory as sanity itself. The deep muttering of boulders in Black River—why am I uneasy? To swallow the torrent sun, and wind, to fill one's breath with the plenitude of being . . . and yet . . . I draw back from that sound, which seems to echo the dread rumble of the universe.

Today GS is stumbling on the ledges. He specu-
lates about atmospheric ions that affect depression, as in the mistral winds
of southern France (there are recent speculations that negative ions, which
seem to be positive in their effect on animals and plants, may be somehow related to *prana,* the "life energy), and we agree that one is clumsy
when depressed, but he feels that his own stumbling is a sign of incipient
sickness, a cold coming on or the like. Perhaps he is right, perhaps I imagine
things, but earlier on this same ledge, as if impelled, my boots sought out
the loose stones and snow-hidden ice, and I felt dull and heavy and afraid;
there was a power in the air, a random menace. On the return, an oppression has lifted, I am light and quick. Things go better when my left foot
is on the outside edge, as it is now, but this cannot account for the sudden limberness, the pleasure in skirting the same abyss that two hours ago
filled me with dread. Not that I cease to pay attention; on the contrary,
it is the precise bite and feel and sound of every step that fills me with
life. Sun rays glance from snow pinnacles above and the black choughs
dance in their escadrilles over the void, and dark and light interpenetrate
the path, in the all-pervading presence of the Present.

. . . November 11

In the east, at dark, bright Mars appears, and
soon the full moon follows the sun's path, east to west across a blue-black
sky. I am always restless in the time of the full moon, a common lunatic,
and move about the frozen monastery, moon-watching. Rising over the
White River, the moon illuminates the ghostly prayer flag blowing so softly
on the roof of the still hut, and seems to kindle the stacked brushwood;
on its altar stone my small clay Buddha stirs. The snow across the river
glows, and the rocks and peaks, the serpentine black stream, the snows,
sky, stars, the firmament—all ring like the bell of Dorje-Chang. *Now!* Here
is the secret! *Now!*

At daybreak, when the blue-black turns to sil-
ver in the east, the moon sets with the darkness in the west. On frozen
sun rays, fourteen pigeons come to pick about the yard, pale blue-gray
birds with a broad band across the tail that fills with light as they flutter
down upon the rigid walls. Like all wild things at Crystal Mountain, the
hill pigeons are tame, and do not fly as I draw near, but cock their gentle
dovelike heads to see me better.

I climb the mountain with the sun, and find the
mixed herd high up on the slope; I try angling toward them, then away
again, zig-zagging as I climb. For some reason, this seems to reassure them,
for after watching me awhile, and perhaps concluding that I am not to
be taken seriously, they go on about their business, which this morning
is unusually dull. I keep on climbing. Far below, the torrent, freed from
daybreak ice, carries gray scree down out of the mountains.

In hope of seeing the snow leopard, I have made a wind shelter and lookout on this mountain, just at the snow line, that faces north over the Black Canyon all the way to the pale terraces below Samling. From here, the Tsakang mountainsides across Black River are in view, and the cliff caves, too, and the slopes between ravines, so that most of the blue sheep in this region may be seen should they be set upon by wolf or leopard. (GS estimates a population of 175 to 200 animals on the mountainsides in the near vicinity of Shey.) Unlike the wolves, the leopard cannot eat everything at once, and may remain in the vicinity of its kill for several days. Therefore our best hope is to see the griffons gather, and the choughs and ravens, and the lammergeier.

The Himalayan griffon, buff and brown, is almost the size of the great lammergeier; its graceful turns against the peaks inspire the Tibetans, who, like the vanished Aryans of the Vedas, revere the wind and sky. Blue and white are the celestial colors of the B'on sky god, who is seen as an embodiment of space and light, and creatures of the upper air become B'on symbols—the griffon, the mythical garuda, and the dragon. For Buddhist Tibetans, prayer flags and windbells confide spiritual longings to the winds, and the red kites that dance on holidays over the old brown city of Kathmandu are of Tibetan origin as well. There is also a custom called "air burial," in which the body of the deceased is set out on a wild crag such as this one, to be rended and devoured by the wild beasts; when only the bones are left, these are broken and ground down to powder, then mixed into lumps of dough, to be set out again for passing birds. Thus all is returned into the elements, death into life.

Against the faces of the canyon, shadows of griffons turn. Perhaps the Somdo raptors think that this queer lump on the landscape—the motionless form of a man in meditation—is the defunct celebrant in an air burial, for a young eagle, plumage burnished a heraldic bronzy-black, draws near with its high peeping, and a lammergeier, approaching from behind, descends with a sudden rush of feathers, sweeping so close past my head that I feel the break of air. This whisper of the shroud gives me a start, and my sudden jump flares the dark bird, causing it to take four deep slow strokes—the only movement of the wings that I was ever to observe in this great sailer that sweeps up and down the Himalayan canyons, the cold air ringing in its golden head.

Dark, light, dark: a raptor, scimitar-winged, under the sun peak—I know, I know. In such a light, one might hope to see the shadow of that bird upon the sky.

The ground whirls with its own energy, not in an alarming way but in slow spiral, and at these altitudes, in this vast space and silence, that energy pours through me, joining my body with the sun until small silver breaths of cold, clear air, no longer mine, are lost in the mineral breathing of the mountain. A white down feather, sun-filled, dances before me on the wind: alighting nowhere, it balances on a shin-ing thorn, goes spinning on. Between this white feather, sheep dung, light, and the fleeting aggregate of atoms that is "I," there is no particle of differ-ence. There is a mountain opposite, but this "I" is opposite nothing, opposed to nothing.

I grow into these mountains like a moss. I am bewitched. The blinding snow peaks and the clarion air, the sound of earth and heaven in the silence, the requiem birds, the mythic beasts, the flags, great horns, and old carved stones, the rough-hewn Tartars in their braids and homespun boots, the silver ice in the black river, the Kang, the Crys-tal Mountain. Also, I love the common miracles—the murmur of my friends at evening, the clay fires of smudgy juniper, the coarse dull food, the hard-ship and simplicity, the contentment of doing one thing at a time: when I take my blue tin cup into my hand, that is all I do. We have had no news of modern times since late September, and will have none until Decem-ber, and gradually my mind has cleared itself, and wind and sun pour through my head, as through a bell. Though we talk little here, I am never lonely; I am returned into myself.

Having got here at last, I do not wish to leave the Crystal Mountain. I am in pain about it, truly, so much so that I have to smile, or I might weep. I think of D and how she would smile, too. In another life—this isn't what I know, but how I feel—these mountains were my home; there is a rising of forgotten knowledge, like a spring from hidden aquifers under the earth. To glimpse one's own true nature is a kind of homegoing, to a place East of the Sun, West of the Moon—the homegoing that needs no home, like that waterfall on the upper Suli Gad that turns to mist before touching the earth and rises once again into the sky.

. . . November 14

Crossing Black River, I climb the west slope trail, out of the night canyon, into the sun. In the matted juniper is a small busy bird, the Tibetan tit-warbler, blue-gray with a rufous cap, and an insistent call note, *t-sip*: what can it be insisting on, so near the winter?

On this bright morning, under the old moon, leopard prints are fresh as petals on the trail. But perhaps two hundred yards short of the trip line to GS's camera, the tracks appear to end, as if the cat had jumped aside into the juniper; the two prints closer to the trip line had been made the day before. Beyond the next cairn, where the path rounds the ridge high above the river and enters the steep snow-covered ravine below Tsakang, more fresh tracks are visible in the snow, as if the snow leopard had cut across the ridge to avoid the trip line, and resumed the path higher up, in this next ravine. Close by one print is an imprint of lost ages, a fernlike fossil brachiopod in a broken stone.

From Tsakang comes the weird thump of a *damaru,* or prayer drum, sometimes constructed of two human skulls; this instrument and the *kangling* trumpet, carved from the human thigh bone, are used in Tantrism to deepen meditation not through the encouragement of morbid thoughts but as reminders that our time on earth is fleeting. Or perhaps this is the hollow echo of the cavern water, dripping down into the copper canister; I cannot be sure. But the extraordinary sound brings the wild landscape to attention: somewhere on this mountainside the leopard listens.

High on the ridge above Tsakang, I see a blue spot where GS is tracking; I come up with him in the next hour. "It fooled me," he calls by way of greeting. "Turned up the valley just below the trip line, then over the ridge, not one hundred yards from where I was lying, and down onto the path again—typical." He shifts his binoculars to the Tsakang herd, which has now been joined by the smaller bands of the west slope. "I've lost the trail now, but that leopard is right here right this minute, watching us." His words are borne out by the sheep, which break into short skittish runs as the wind makes its midmorning shift, then flee the rock and thorn of this bare ridge, plunging across deep crusted snow with hollow booming blows, in flight to a point high up on the Crystal Mountain. Blue sheep do not run from man like that even when driven.

The snow leopard is a strong presence; its vertical pupils and small stilled breaths are no more than a snow cock's glide away. GS murmurs, "Unless it moves, we are not going to see it, not even on the snow—those creatures are really something." With our binoculars, we study the barren ridge face, foot by foot. Then he says, "You know something? We've seen so much, maybe it's better if there are some things that we *don't* see." He seems startled by his own remark, and I wonder if he means this as I take it—that we have been spared the desolation of success, the doubt: is this *really* what we came so far to see?

When I say, "That was the haiku-writer speaking," he knows just what I mean, and we both laugh. GS strikes me as much less dogmatic, more open to the unexplained than he was two months ago. In Kathmandu, he might have been suspicious of this haiku, written on our journey by himself:

Cloud-men beneath loads.
A dark line of tracks in snow.
Suddenly nothing.

Because his sheep, spooked by the leopard, have fled to the high snows, GS accompanies me on my last visit to Tsakang. There we are met by Jang-bu, who has come as an interpreter, and by Tukten, who alone among the sherpas has curiosity enough to cross the river and climb up to Tsakang of his own accord. Even that "gay and lovable fellow," as GS once said of Phu-Tsering, "hasn't the slightest curiosity about what I am doing; he'll stand behind me for two hours while I'm looking and taking notes and not ask a single question."

Once again, the Lama of Shey lets us wait on the stone terrace, but this time—for we are here by invitation—the aspirant monk Takla has prepared sun-dried green yak cheese in a coarse powder, with *tsampa* and buttered tea, called *so-cha,* served in blue china cups in the mountain sun. The sharp green cheese and bitter tea, flavored with salt and rancid yak butter, give character to the *tsampa,* and in the cool air, this hermit's meal is very very good.

Takla lays out red-striped carpeting for us to sit upon, and eventually the Lama comes, wrapped in his wolf skin. Jang-bu seems wary in the Lama's presence, whereas Tukten is calm and easy and at the same time deferential; for the first time since I have known him, indoors or out, he doffs his raffish cap, revealing a monk's tonsure of close-cropped hair. Tukten does most of the translation as we show the Lama pictures from our books and talk animatedly for several hours. Lama Tup-juk asks about Tibetan lamas in America, and I tell him about Chögyam Trungpa, Rinpoche (*rinpoche,* or "precious one," signifies a high lama), of his own Karma-pa sect, who left Tibet at the age of thirteen and now teaches in Vermont and Colorado. For GS, he repeats what he had told me about the snow leopard and the argali, pointing across Black River at the slopes of Somdo.

Horns high, flanks taut, the blue sheep have begun a slow descent off the Crystal Mountain, in a beautiful curved line etched on the snow. The leopard is gone—perhaps they saw it go. Through binoculars, now and again, a ram can be seen to rear up wildly as if dancing on the snow, then run forward on hind legs and descend again, to crash its horns against those of a rival.

In the high sun, snows shift and flow, bathing the mind in diamond light. Tupjuk Rinpoche speaks now of the snow leopard, which he has seen often from his ledge, and has watched carefully; to judge from the accuracy of all his observations: he knows that it cries most frequently in mating time, in spring, and which caves and ledges it inhabits, and how it makes its scrape and defecation.

Before we leave, I show him the plum pit inscribed with the sutra to Chen-resigs that was given me by Soen Roshi, and promise to send him my wicker camp stool from the tea stall on the Yamdi River. The Lama gives me a white prayer flag—*lung-p'ar,* he calls it, "wind pictures"—printed with both script and images from the old wood blocks at Shey; among the Buddhist symbols is an image of Nurpu Khonday Pung-jun, the great god of mountains and rivers, who was here, says the Lama, long before the B'on-pos and the Buddhists: presumably this was the god who was vanquished by Drutob Senge Yeshe and his hundred and eight snow leopards. Nurpu is now a Protector of the Dharma, and his image on flags such as this one is often placed on bridges and the cairns in the high passes, as an aid to travelers. The Lama folds it with greatest concentration, and presents it with the blessing of his smile.

The Lama of the Crystal Monastery appears to be a very happy man, and yet I wonder how he feels about his isolation in the silences of Tsakang, which he has not left in eight years now and, because of his legs, may never leave again. Since Jang-bu seems uncomfortable with the Lama or with himself or perhaps with us, I tell him not to inquire on this point if it seems to him impertinent, but after a moment Jang-bu does so. And this holy man of great directness and simplicity, big white teeth shining, laughs out loud in an infectious way at Jang-bu's question. Indicating his twisted legs without a trace of self-pity or bitterness, as if they belonged to all of us, he casts his arms wide to the sky and the snow mountains, the high sun and dancing sheep, and cries, "Of course I am happy here! It's wonderful! *Especially* when I have no choice!"

In its wholehearted acceptance of *what is,* this is just what Soen Roshi might have said: I feel as if he had struck me in the chest. I thank him, bow, go softly down the mountain: under my parka, the folded prayer flag glows. Butter tea and wind pictures, the Crystal Mountain, and blue sheep dancing on the snow—it's quite enough!

Have you seen the snow leopard?
No! Isn't that wonderful?

GARY
NABHAN

Gary Nabhan claims to be named after his hometown of Gary, Indiana, where he was born in 1952. He also sometimes claims that he never writes about a plant unless he has eaten it first, so you never can tell just how serious his claims are. He was educated at Cornell and Prescott colleges; as a desert ethnobotanist, he received a Ph.D. from the University of Arizona.

Nabhan did not set out to be a desert person. His first fascination was with the plains and prairies. "I really feel a lot more baffled and intrinsically mystified by the grasslands." But when Nabhan moved to Arizona to study the desert grasslands, the botanist whose interests most closely matched his was a Sonoran Desert specialist. "The desert won me over without any predilection on my part," says Nabhan.

He has always been a writer—for years mostly of journals and poetry, taking workshops from Gary Snyder and William Stafford. His research interests in desert-adapted plants, native agriculture, and Sonoran

Desert ecology, led to his first book of essays, *The Desert Smells Like Rain: A Naturalist in Papago Indian Country* (1982). *Gathering the Desert* (1985), sketches of twelve Sonoran Desert plants seen as "calories, cures, and characters," won the John Burroughs Medal. One essay from each book appears here.

Nabhan and his wife, Karen, also a plant ecologist, helped found Native Seeds/SEARCH, an organization working to conserve native crops and their wild relatives. They live in Tempe, where he is assistant director for research at the Desert Botanical Garden. As Nabhan says, "I think it would be a shame if there was a specialized natural history profession. Why do we have to be only one thing?" He describes himself as a chameleon. "I take on different accents and syntax when I'm talking with Piman (Indian) speakers, evolutionary biologists, desert rats, Hispanic farmers—to the point where I don't even know that I'm doing it. I don't think I'm attempting to imitate someone, but to really understand what they are saying, I just throw myself into it."

Nabhan would like to see natural history writers take more risks, even if it might mean a temporary failure or two. For models, he points to Native American writers: "The real bright spot, now, is that so many people are working with their own cultural traditions. Essentially they are bilingual writers and bilingual thinkers sharing their traditions with the outside world."

Nabhan's interest in "how the conservation of wild plants and wild lands is crucial to the long-term sustainability of North American agriculture" is the theme of his most recent book, *Enduring Seeds: Native American Agriculture and Wild Plant Conservation* (1989).

Where Has All the
Panic Gone?

Few men knew how to use the badger claws to make the river banks cave in. He had already buried the claws once, and the river had begun to rise toward them, digging, imitating the badger. Here, on the main channel of the Río Colorado, the banks crumbled. He had found the three claws and was moving them back, away from where the floods had reached yesterday, hoping that the waters would rise even more, following the claws. As the river continued to fatten this June, leveling and saturating the ground, he would keep it digging toward the buried badger claws. Then, when the floods subsided in July, he would begin to plant the *shimcha* grain there, on the newly watered, fertilized mudflats.

He knew not what the other Yuman-speaking peoples did, but among his band, he was one of a handful of men that went to the effort to obtain the claws, to plant this grain. His motivations were simple. His children loved the taste of the sun-dried cakes made from the ground *shimcha* grain. He himself preferred to have them now and then as a change of pace when their diet grew monotonous.

So he took the time to purify himself to be worthy of killing a badger. For four days in early June, he had not eaten nor coupled with his wife. Waking before daylight, he would go to the river channel, swim and wash his hair, to return before the others rose. He stayed away from where the women gossiped over their grindstones, and sat quietly at the margins of the camp throughout the day.

The fourth evening, an elder gestured to him to go. He left camp, to return to a badger hole he had spotted not too long before. He kept a club poised in his right hand and probed the den with a slender, curved stick held in his left. Dirt fell from the roof of the tunnel into the den. Confused, the badger lumbered out. In a moment it was dead, skull cracked at the neck. Three claws were cut off the front feet.

Now, a good place for planting *shimcha* could be assured. He looked out over the inundated flats, imagining what it might look like in another three months: thousands of long seedheads, golden, nodding in the sun.

What the migrating birds did not bother, he would harvest. He would work through the stand, rubbing dried-out seed-heads over pottery pans. His wife would parch the harvest, loosening the chaff and roasting the grain, then winnow away the refuse. She would grind some of the grain fine, like pinole. Part they would give to relatives, who would soon come to camp and gamble with them. Part they would keep for periodic use, storing it in huge bird's nest granaries. These enormous baskets would be lifted to the tops of storage platforms, secure from floods and wild camp robbers. And part of the harvest he would save as seed for planting. He would weave willow splints into a *sawa* basket, shaped like a bulging bag, fill it with seeds, then stitch it shut. This he would hide in a place safe and dry, for future sowing.

Centuries passed.

It held the odor of wild animals. The young Cocopa Indian men were uneasy. It was a cave high in the Trigo Mountains, above the east bank of the Río Colorado. Haskell Yowell had found the cave while prospecting, and thought that the bat guano which filled it, or something else hidden inside it, might have economic value. He and his brother Dudley had driven north from Yuma twenty-eight kilometers, then up into the range as far as the road would go. From the road, up to the cave, they rigged a little trolley to carry down the slope whatever was worth pulling out of the cave. Then, with gunnysacks and spades in hand, they all climbed to the cave mouth and smelled it. They stood outside the cave for a minute, sharing a cigarette. The Indians spoke quietly to one another in a mixture of Spanish and Cocopa.

"Well, go ahead, boys," one of the Yowells said. "That bat dung, there on the floor, let's bag it up. If we don't find anything else here, at least we can use some of that as fertilizer back on the farm."

A half meter of guano disappeared into the gunny sacks. They scraped the floor of the front of the rock shelter clean, then moved back into more cramped space. An inner chamber, perhaps five meters deep and ten wide, was nearly filled with guano to its ceiling. They shoveled out layer after layer, filling one sack after another. As their spades reached toward the inner chamber floor, one of them nudged something spongy. They stopped.

"What's that?" asked Dudley. "Here, give me that shovel."

As they scraped the guano loose, he could see that it was straw wrapped around some object. Peeling back this covering of slender grass stems and leaves, they found a plump, twined-woven bag. It was stitched shut.

267

Bringing it out into full light, Dudley loosened the bag's stitching. He tipped it, and a stream of golden grain spilled onto the ground.

"Why, look at all those seeds! It's jammed full of them! Must be at least a pound packed in there!"

"You boys know what they are?" Dudley queried.

They shook their heads faintly.

"Well, it don't look like anything that grows up here."

"C'mon, let's tighten these stitches and close it," Dudley deliberated. "I want to get this to the state archaeologists, to see if they've ever seen anything like it before. Let's start digging again, see if there's pots or arrows or anything else under there. If you hit anything, stop, and we'll take a good slow look at it."

No other artifacts were found in the cave or in any other caves nearby. Haskell and Dudley eventually mailed the bag and its contents to the Arizona State Museum in Tucson to learn of its antiquity and identity. Museum curators found a similar, modern bag made by Yuman Indians in their collections and sifted through the seeds. In addition to the golden grain, they sorted out seventy grams of seeds of green-striped cushaw squash, and about the same amount of three different kinds of tepary beans. The grain itself weighed three quarters of a kilogram, and a sample of it was ample for radiocarbon dating. Though still appearing to be in a perfect state of preservation, the seeds were estimated to be 603 years old, plus or minus 140 years.

So they tried to figure out what this grain was that filled the bulk of the woven bag.

Most of the 2-millimeter-long seeds, golden to cream-colored, roundish and shiny, looked like a grain that the Indians were once said to have sown and eaten. It was called *Panicum sonorum,* the Sonoran panicgrass or millet. Just before the turn of the century, several delta explorers noted that the Cocopa and nearby tribes grew and ground it, making cakes, gruel or mush. Another panicgrass, smaller, darker, more slender in seed shape, was found in the bag among these domesticated grains. Perhaps it was a wild form of Sonoran panicgrass that spontaneously came up as a weed in their plantings, or a closely related species. The Indians had once spoken of a spontaneous panicgrass, more bitter in flavor, harder to thresh than the form they usually had sown. Yet most of the seeds in the bag were cleanly threshed, with hardly any chaff left in the whole cache.

Great care had been taken to clean and store this grain. By whom? Why?

The grain of the Trigo Mountains cave cache revealed little else. The tepary beans, however, spoke to Dr. Lawrence Kaplan, who compared them to many other collections. The only historic Indian community known to grow both kinds of dark-colored teparies found in the cave was the Cocopa band that now lives around Somerton, Arizona.

When the Cocopa men went back to the ramshackle houses where their families lived near the border, they still had the bagful of seeds on their minds. They mentioned them to one elderly man, an invalid with an amputated leg.

"Seeds? We used to harvest many kinds of grasses off the delta, some of which we would plant there. But the river is different now. They have dammed it upstream. It has gone into their ditches. The floods don't flow the same way anymore, so those grasses can't grow."

But these grains. They must have been special. The guano miners tried to describe them to the tired old man.

"Who knows what they were? We'd have to hold them, taste them to know. It's been years since our people harvested those things. They're gone. It's too bad Sam Spa is not living among us anymore. He would have known. He's gone too. . . ."

Sam Spa died in 1951 at the age of seventy-seven. Although of mixed ancestry—Pai Pai, Diegueño, and Mountain Cocopa—he grew up among the Kwakwarsh Cocopa band below the Río Hardy-Río Colorado confluence. For his first dozen years, he was immersed in the life of the delta. In his last dozen years, he recalled those days to a young student out of Harvard, William Kelly. His recollections had a tragic undertow to them. He remembered a river, a culture, and a riverine agriculture that had already been broken, dammed off, and invaded.

Sam Spa became a man amidst mudflat fields of *shimcha* that extended for more than eight kilometers along the channels of the delta. A patchwork of five plots, each one sown by a different man, edged the channels. The plots were up to 500 meters wide, and might take each man six days to plant, for the work was tedious and tricky. Young boys could hardly help with it.

The difficulty was that the panicgrass came up thickest if sown by hand. The planter needed his limbs to balance himself, to hold walking sticks to help pull himself forward through the mud, to steady his footing or to lift himself up if he fell. The seeds were therefore blown from his mouth, and scattered by the wind or draining water. These seeds for planting were carried in a gourd hung around the neck.

After planting, the Cocopa men would remain nearby, fishing or weeding amaranth greens out of tepary and corn fields, while the women gathered the all-important mesquite beans and other

269

wild plants. The Cocopa hardly weeded the panicgrass fields compared to those of other crops. Floodwashed panicgrass fields, Spa recalled, seldom had as many weeds as the more permanent plots of ditch-irrigated beans, grains, or cucurbits. Some wild panic, sour to the taste, may have come up as a crop mimic among the domesticated *shimcha,* but this they could have accidentally sown themselves. Or the droppings from last year's harvest could have volunteered. The Cocopa had little time and no reason to bother with them. They soon had to leave for the mountains to gather upland plants, and would return just before the crop would be ripe.

In October, as they trailed back to the delta, the river volume was only a sixth of what it was when the badger claws were left in the banks during June. Side channels had dried up. The mudflats were deeply cracked in some places. In the sandier loams where the panicgrass had been sown, the top few inches were dry to the touch, and shallow-rooted wild plants had begun to wilt.

The panicgrass itself had several tillers as well as a meter-high main stalk to support. All its energy was being shifted to the filling of the grain. As many as 2500 grains per plant were fleshing out and drying in the still-hot autumn sun.

Sam Spa remembered that one man's *shimcha* harvest might fill a small bird's nest basket, roughly sixty centimeters tall and the same in diameter. Other Cocopa recalled yields at least six times that per plot. Alice Olds said she knew of a *shimcha* field that supplied thirty-five to fifty-five kilos of grain for her family one year. It was not their mainstay, but with grains containing around twelve and a half percent protein, *shimcha* was probably a fine addition to their seasonally variable diet.

The Cocopa was not the only Colorado River tribe that planted panicgrass as part of its flood-recession agriculture. When Hernando Alarcón made the first European contact with several Yuman-speaking peoples in 1540, he observed that "these people have, besides their maize, certain cucurbits and some grain like millet."

The comparison of Sonoran panicgrass to millet is fruitful, for this Old World grain is also a *Panicum* with similarly shaped grain. In certain of the Yuman tribes, it was said to have been scattered along the lower Colorado by the god Kumastamxo. Sonoran panicgrass cultivation did extend north beyond the Yuman speakers, into the Río Colorado floodplain territory of the Chemehuevi, north of Parker, Arizona. In 1904, desert scientist Robert Forbes boated down the Colorado, and found that the Chemehuevi had an agriculture at least as diverse as that of the tribes downstream:

> The Indians, especially the Chemehuevis, who
> are at present the most successful farmers on the

river, grow beans, cowpeas, watermelons, Turk-
ish winter muskmelons, martynias [devil's claw],
a soft maize easily ground in their metates and
maturing in about seven weeks, a black sweet
corn, winter squashes, pumpkins, a little wheat,
barnyard millet and a seedy grass, *Panicum
sonorum,* useful for both grains and horse
forage. . . . Grass and millet seed is sown in the
soft mud as soon as the river subsides, a method
strikingly similar to that employed by the Egyp-
tians in sowing their great forage crop, Berseem,
along the Nile.

This crop diversity was soon to disappear along
with the natural diversity of the lower Río Colorado. In retrospect, it is
hard to underestimate the extent to which the lower Colorado was a vortex of life more powerful and productive than anything else between the
Continental Divide and the Pacific Coast. Yet we know from naturalist
Aldo Leopold that its richness was once so stunning that he could not emotionally bear to return there after the river had been starved by dams
upstream. His account of a 1922 canoe trip through delta lagoons, already
after the river had partially healed from the first few modern engineering
operations to hit it, describes a calm before the storm of more pervasive
assault:

'He leadeth me by still waters' was to us only
a phrase in a book until we had nosed our canoe
through the green lagoons. . . . At each bend, we
saw egrets standing in the pools ahead, each
white statue matched by its white reflection.
Fleets of cormorants drove their black prows in
quest of skyward in alarm. . . . All this wealth
of fowl and fish was not for our delectation
alone. Often we came upon bobcats . . . families
of raccoons . . .[and] coyotes. All game was of
incredible fatness. Every deer laid down so
much tallow that the dimple along his backbone
would have held a small pail of water.

The origin of all this opulence was not far to
seek. Every mesquite and every tornillo was
loaded with pods. There were great patches of
a legume resembling coffeeweed [*Sesbania?*]; if
you walked through these, your pockets filled
up with shelled beans. . . . The dried-up

mudflats bore an annual grass, the grain-like seeds of which could be scooped up by the cupful. . . .

If the grains of the annual grass which Aldo Leopold scooped up were by happenstance Sonoran panicgrass, he may have been the last scientist to see it growing in the Colorado River watershed. The river had already undergone dramatic changes in the two decades prior to Leopold's visit, but those which were to occur in the next two decades bordered on being irrevocable. For the Cocopa farmers and for the panicgrass seeds dependent upon the rich floods of the delta, their world was drying up.

They had always responded to the uncertainty of delta life, but this time, the scale of change was devastating. For the four centuries prior to 1890, the Río Colorado would meander from one side of the floodplain to the other. Yet it generally followed the same course between Yuma and the Sea of Cortez. Yields of crops fluctuated with the amount of runoff issued from the watersheds, but the delta's subsistence resources for the Cocopa and their neighbors remained rich. In addition to their range of crops, the Cocopa could draw upon fish, fowl, and wild stands of plants such as mesquite and Palmer's saltgrass. They had even gone to work for Fort Yuma riverboats between 1852 and 1877, only to return to delta farming and foraging when the railroads economically displaced the river transport.

By the 1890s, there remained at least 1200 Cocopa on or near the delta, but around them, a population of recent arrivals was becoming more evident. These Anglo-Americans were confident that they could organize themselves to engineer the great river into a series of ditches that would turn the Salton Sink into the Imperial Valley Irrigation District. The Cocopa and others had never tapped the river's "potentialities of production" that were "beyond any land in our country which has ever known the plow." The Imperial Valley boosters were sure that in the West, water is power. In a 1900 *Sunset,* they prophesied that "whoever shall control the right to divert these turbid waters will be the master of this empire."

Cocksure that they could control the Río Colorado, they began a flow of water into the Imperial Valley in June 1901, which ignited a land boom. Yet their canal headgate at the river was poorly placed for remaining free of silt and diverting sufficient water toward the below sea level areas of the old Salton Sink. In 1902, 1904, and 1905 new canals were dug. The more successful placement of the 1905 canal, combined with a spring of heavy rains in watersheds upstream, enabled the Imperial Valley developers to become the greatest overachievers of river diversion in the history of the West. By August 1905, the entire flow of the Colorado was washing a kilometer wide across the Imperial Valley, draining down to where prehistoric Lake Cahuilla once stood. As the Valley gained the Salton Sea, the Cocopa were left high and dry on the old delta.

Cocopa problems didn't stop when the riverbank was finally patched in February 1907, allowing some flow to return to the delta. The Wi Ahwir Cocopa had already begun to move to Mexicali when their habitat had been starved by the diversion. With the building of the Laguna diversion dam twenty kilometers above Fort Yuma in 1909, floods on the delta were decisively lessened, and the river never returned to what had been its main channel in previous decades. This forced the Hwanyak Cocopa who had remained in the midst of the delta to abandon their homes and resettle in Somerton, Arizona, and in San Luis del Río Colorado, Sonora.

Thus when Aldo Leopold canoed the delta lagoons in 1922, it was depopulated and degraded compared to what it had been in decades and centuries before. The evening-out of the river volume had allowed the establishment of introduced plants that may have been no more than scattered, inconspicuous seedlings from his canoe. These plants, known as salt cedars or tamarisks, were introduced from the Old World as an erosion-control plant. They rapidly began to colonize areas of reduced flow on the lower Colorado, and within decades became established hundreds of kilometers upstream on the Gila as well. Their rapid spread during the late 1920s probably reduced habitat available for the germination of panicgrass.

As the Great Depression hit America, a permanent depression hit the Cocopa, and with them, cultivated panicgrass. The Vacanora Canal, completed in 1929, established a channel for delta drainage that has persisted ever since. It was kilometers away from the old panicgrass fields. By the end of the 1931 drought, the old fields were bone-dry, since no summer rise in flow occurred that year at all. Then, with the completion of Hoover Dam in 1935 and Parker Dam in 1938, the river became so tamed that both Mexicans and Arizonans felt confident enough to begin to level and develop the delta for their own purposes. Scientists Edward Castetter and Willis Bell were soon to realize that these dams would in most years eliminate the remaining river overflow and make the practice of ancient flood-recession agriculture on the delta virtually impossible. Though they asked to see panicgrass fields during their fieldwork on the lower Colorado in the 40s, seeing a handful of seed in Yuma was the closest they came.

At the same time, the Cocopa were being swallowed up by those who had come into their homeland to practice "modern agriculture." An ejido farming cooperative was set up for the Baja California Cocopa by the Mexican government in 1936, but mestizos had completely taken it over by 1943, and many of the Cocopa who had at first participated in it moved to Sonora. There, they watched as San Luis del Río Colorado grew from its 1921 population of 175 people, to 4079 in 1950, to more than 50,000 by 1970. The binational Cocopa population dwindled to half of what it was prior to the disruption of the delta.

In the abstract terms of human ecology, the Cocopa had lost their traditional ecological niche. By 1970, nine out of ten Cocopa men in the U.S. were unemployed. Their meager reservation lands near Somerton were often leased out, so that few grew any food for direct consumption by their families. The foodstuffs that they received as government aid, conceded food scientists Doris Calloway and June Gibbs, were of "dilute nutritional quality." On their modern diets of snack foods and groceries purchased with food stamps from minimarkets, contemporary Cocopa lack vitamins A, B-6, C, pantothenic acid, calcium, and iron. There are high incidences of liver, kidney, gall bladder, and respiratory diseases among them, as well as unusual frequencies of colitis, gastroenteritis, rabies, tuberculosis, and trachoma for a human population in a "modern country." The inactivity resulting from these ailments reinforces obesity, which in turn aggravates the extremely high expression of diabetes among adult Cocopa. As if ill physical health weren't enough, alcoholism and suicides frequently take their toll in the American Cocopa community.

The Mexican Cocopa communities are perhaps as poor, but they retain some traditional activities such as fishing to a greater extent. Nevertheless, they too live between a rock and a hard place in the most marginal areas of the valleys of Mexicali and San Luis. When we visited the El Mayor Cocopa in 1983, almost the entire floor of their village was under water, with only hillside houses and those upon stilts remaining intact. They had no land that could be cultivated, since most everything below the cliffs had been flooded for months. The dams built upstream half a century before had accumulated so much silt in their reservoirs that they had become worthless for containing floods of any size. Thus water was let out over the dams, inundating what was left of Cocopa country.

We showed middle-aged Cocopa men and women specimens of panicgrass. For most of them, it was an historic oddity. One older woman, though, asked for a few seeds to examine. She ground a few grains between her teeth.

"*That* would be good grinding, good flour, good food. . . ." She tucked the other seeds into her apron pocket. Ancient seeds had returned to her community, but there was literally no place to plant them. The exotic salt cedars would again infest the mud flats as soon as

274

these once-in-a-generation floods subsided.

William Kelly called them "the scattered descendants of a once powerful people who, from earliest known times . . . occupied the rich delta country of the Colorado River." For decades, there has hardly been an opportunity for this scattering to reseed itself on the delta.

What happens to the human spirit when that which once gave it productivity and meaning is cut off, disrupted upstream? A *Look* reporter recorded the uninterested Cocopa responses to the promise of new opportunities and adequate housing from a BIA bureaucrat. A fourteen-year-old Cocopa boy reiterated to his grandmother that she might get a new house.

"When?" she asked.

"Maybe two years," he replied.

"Oh, I die before then," she said. "Sometimes we feel we're already dead."

As Sonoran panicgrass was dying on the delta, the Depression was forcing out of business a number of the southern California vegetable growers who had responded to the irrigated land boom during the previous decades. The hard times which followed in the United States forced a change of plans for one southern Californian. And it was he who was soon to find a refugium of panicgrass left 800 kilometers south of the delta, in the barrancas of Chihuahua.

A family by the name of Gentry lost most of their California cantaloupe business as the Depression unraveled the financial supports for farmers. The two young Gentrys, after seeing their farming future crumble, considered college. With the family's financial position weakened, but both of them hard workers, the Gentry brothers resolved themselves to taking turns, one working packing fruit and vegetables while the other pored over the textbooks.

One of these brothers, Howard Scott Gentry, extended his interests beyond agriculture to biology and to the natural and cultural history of northwest Mexico. He came under the influence of the great geographer Carl Sauer at the University of California at Berkeley. Sauer and anthropologist A. L. Kroeber had each traveled down to southern Sonora in 1930, "rediscovering" a Uto-Aztecan-speaking tribe which they called the Varohio. These scholars must have immediately seen in Gentry a perceptive student tough enough to endure fieldwork in the back country with a minimum of institutional support. They encouraged him to visit this tribe in the canyon country above Sonora's coastal plains. With his savvy for biology and experience in farming, perhaps he could learn more of the Varohio plant lore and subsistence skills which their own brief visits had not been able to decipher.

Gentry found something quite different from the winter rainfall-dominated open desert he had known in southern California. He journeyed upstream from the southern edge of the Sonoran Desert, through a Sinaloan thornscrub transition, into the Short Tree Forest of the barrancas of the Sierra Madre. These subtropical deciduous forests covered an "endless series of ridges flanking their mother sierras," in a topography so complex "that a traveler scarcely knows just where he is hidden or from which direction the tortuous trail has brought him."

There, in the barrancas where cardon and old man's cactus abound, where thick-trunked pochote trees dangle cotton-like kapok from their branches, where lianas, orchids, and epiphytic bromeliads tangle, Gentry went to camp and hunted for plants and animals with the least-known tribes in northwestern Mexico.

Other than Gentry's ethnographic notes on these Indians, now known as the Warihio, there has been little published on their culture or agriculture. Numbering around 2000, they speak two dialects of a language closely related to that of the Tarahumara, the mountain tribe that is the second largest in the Uto-Aztecan family. They have maintained certain traditions which the more vulnerable Mayo and Piman tribes to the south, west, and north have long abandoned. While many of them have visited or worked in the coastal plain cities of Navajoa and Obregón, their life in the foothills and barrancas is one in which ancient subsistence skills have still retained their usefulness in the twentieth century.

Occasionally, Gentry would go with a guide and perhaps a pack mule or two into the remote headwaters of the Río Mayo watershed. On one such trip in September 1935, Gentry came upon a handful of houses stuck out on a steep projection in a deep barranca. This was the ranchería named Sahuacoa. There, elderly Warihios let him collect a rare grain plant that others had told him of—*sagui*. He soon realized that this was the same cereal that Edward Palmer had collected from Indian fields on the Colorado delta decades earlier—*Panicum sonorum*. His notes, jotted down while the Hoover Dam was placing the headstone on the grave of panicgrass cultivation in the U.S., give the sense that here there was something very valuable indeed:

> It is planted in the milpas or in the small gardens and like weywi [grain amaranth] is valued as a pinole or prepared and eaten in the same way. While generally known to both the Warihio and the barrancan Mexicans, it appears to be quite scarce and its culture is being lost. I found it only upon one occasion, tended in the small milpa of an old couple in Sahuacoa, near Guaseremos.

They had only a few dozen plants, but sold me a few entire plants for samples. The plants were about 1 meter tall with large panicles of seeds just beginning to mature in late September. Like the corn, they had germinated in June with the first summer rains and would therefore require some 90 days to mature. . . . With weywi and conivari [cultivated *Hyptis suaveolens*] it may have preceded maize in the Warihio culture. . . .

When I first spoke with Gentry about Sonoran panicgrass, it was more than forty years after he had pressed twelve robust plants from that milpa to make dried herbarium specimens. In the intervening period, he had worked for twenty years as the USDA's principal plant explorer collecting viable seeds from around the world for government genetic-conservation programs. He had thought about *sagui* many times since, he said, with a sad look in his eyes. He had collected herbarium specimens of a rare plant, but unlike his habitual activity later in life, he had not worried about obtaining any viable seed. The seeds on the herbarium sheets were now dead.

"I've always wondered if those Indians back in the barrancas still have seed of that domesticated *sagui.* Young man, if you ever travel down into the Río Mayo, be sure to tell me you're going beforehand. I want to give you directions on how to get to that Rancho Sahuacoa."

If young Howard Scott Gentry had felt challenged by the Sauer and Kroeber suggestion to visit Warihio country, a few of us in Tucson could hardly stand still long enough to make plans to visit Sahuacoa. For a combination four-wheel-drive and mule trip, I had no trouble enlisting Barney Burns, who already knew the region from his studies of tree-rings and traditional crafts, and Tom Sheridan, an ethnologist who doesn't feel comfortable unless he's in backcountry Sonora. Barney and I had already spoken with Warihio farmers at an all-night pascola dance near Buropaco. Yes, they knew *sagui,* though none of them personally grew it. Yes, perhaps someone still did, "mas allá," back in the sierras. That was all the extra encouragement we needed.

In early autumn of 1978, Barney's Scout brought us into Alamos, Sonora, in the middle of the night. Leaving around noon the following day, we hightailed it for the sierras, across the Río Mayo, up the switchbacks of the Cuesta de Algodones, to reach the pine forest on the Sonora-Chihuahua border—just in time for a nightful of rain. With the arroyos running, we were unsure how much farther we could drive.

It turned out that we were just a short distance from the Byerly Ranch—the part-time home of an English-speaking European immigrant whose father had started an experimental orchard back in the sierras. His men directed us to nearby El Limón, a mestizo village where we rented mules and gained the help of a young guide who wanted to visit his aunt in Guaseremos.

When our group left early the next morning, knapsack stuffed with seedbags and a plantpress strapped over the mule's ass, I quickly came to realize why panicgrass may have had a chance to persist here. For every horizontal kilometer of distance, we had to ride five. Sometimes ten. The switchbacks wound us down a seemingly endless slope, we splashed across a meter-deep stream bubbling with rapids, and then started the climb up to a ridge that always seemed beyond our reach. In land like this, how could cash crops replace ancient ones? Who would ever have the endurance to carry their harvest out through such terrain? The Colorado delta, though once considered inaccessible, was at least flat. Once it had been drained, modern agriculture quickly made its inroads. Back here, it would be hard to find more than ten contiguous flat hectares within a fifty-kilometer stretch. What fields we saw were patchily placed, slash-and-burn milpas on 30- to 45-degree slopes. Each was an ideal island within which to preserve crop genetic diversity for each was situated differently in terms of elevation, slope, aspect, and distance to others. Throw in the various house gardens where Warihio families still raised chia-like conivari, native tobacco, and amaranth, and we could see that the barrancas still harbored a diversity of habitats in which native crops could thrive.

Yet what we didn't see was panicgrass. It had remained as a minor crop, but farmers simply did not sow it every year. We learned that it grew best in the ashes of freshly torched, first year slash-and-burn milpas. We learned that Warihios attempted to rogue out most of the hairy, darker-seeded weedy plants, preferring the sweet taste of the domesticated *sagui*. They would harvest the mature plants by cutting the seedheads free with knives or sickles. These are placed on woven mats to dry in the sun, and then are rolled over with a smooth-sided stone to loosen the grain. The wind-winnowed cereal grain is later ground for tortillas, tamales, and atole, as well as the pinole which Gentry had reported. Yes, they still ate *sagui* foods, but didn't have any seed on hand for planting. No, they had no mature living plants to show us.

At Sahuacoa, the old couple which Gentry had met had long ago passed from the scene. Houses and garden fences were in a state of deterioration, sliding down the slope.

Yet not far away, at Guaseremos, both panicgrass and grain amaranth were still occasionally grown by a sizable, healthy

Warihio farming community. We visited the family of our quiet teen-age guide. He whispered something to his aunt. She went into the house, and returned with a bag holding half a kilo of seed.

The seeds within that bag have prospered. In a 1980 growout by Pat Williams of the U.S. Soil Conservation Service, more than thirty-five kilos were produced in a small plot—a yield equivalent to 440 kilograms per hectare. Looking out over the dense stand of nodding seedheads, I had to smile: they didn't look as though they minded being north of the border again.

But much of that harvest didn't stay north of the border. Anthropologist Eric Powell has distributed several kilos to Warihio villages and ranchos that had lost their supply of seeds. I have taken seed to their neighboring tribes, which historic documents suggest once had a milletlike grain called *sabi* as well. The seed supply has made it into U.S. and Sonoran state seed banks. And a cupful has been given to the Cocopa in Baja California.

In his herbarium in Papago Park, an anomalous chunk of desert stuck between downtown Phoenix, Tempe, and Scottsdale, Howard Gentry held a vial of golden viable seed up to the light. He went to one of his herbarium cabinets, and pulled a half-century-old specimen filed neatly in a manila folder.

"Ho! Old *Panicum sonorum*—there it is." He compared the seeds. "God, those were robust plants in that little milpa at Sahuacoa. Each filled up a whole sheet. Some were more than a meter tall, I'd say. You said you got out from Byerly's ranch clear to Tucson in one day! Pshaw, young man! It must have taken me five! Well, you brought me some seeds! Well, look at that!"

Where the Birds Are Our Friends: The Tale of Two Oases

I n this lake there lived a monster, much larger than a man, who hated people, and killed them when they came for water."

THE MONSTER OF QUITOVAC

A cloudless sky, a bone-dry road. After miles of eating dust on a drive parallel to the border, we had arrived at what seemed an apparition—a little pocket of greenery in an otherwise harsh grey habitat.

Soon my old Papago friend Remedio had found his way down the trail to the pond. The next thing I knew, he was crawling on his hands and knees out onto the trunk of a cottonwood tree that reached over the water. He hung one arm down and scooped up a drink.

"Sure is *sweet* water. What do they call this place?"

"Depends on who you talk to," I mumbled, glancing across at an Organ Pipe Cactus National Monument placard. "Well, the Park Service calls it Quitobaquito, after the Mexican *Quitovaquita*. And three hundred years ago, Padre Kino christened it *San Seguio*. But all I ever heard your people call it is *A'al Waipia*."

"A'AL WAIPIA? This is it?" He was stunned for a moment. "In all of my life I never thought I would get to see this place where we are standing! I just thought we were going to another place because those signs didn't say *A'al Waipia*. So this is where those little waters come up from the ground."

Those little springs, which the Papago call ṣonagkam, flow into a modest pond touted as one of the few authentic desert oases on the continent.

Out in a stretch of the Sonoran Desert where any sources of potable water are few and far between, *A'al Waipia* has been more than just a curious landmark. It is a critical *watermark* that has literally served and saved thousands of lives over the centuries.

Listen to the crusty explorer Karl Lumholtz soften under its touch around 1910:

> . . . The little stream of crystal clear spring water at Quitovaquita is smaller than a brook, but it seemed much alive as it hurried on in its effort to keep the dam full. As I had been long unaccustomed to seeing running water, and for twenty days had drunk it more or less brackish, the tiny brook seemed almost unreal and was enchanting in its effect. It was also a delight to indulge in my first real wash for nine days.

Where a spring bubbles up in the desert, water-loving plants cuddle around it. Thus *A'al Waipia* has been an ecological oasis, a spot of lush riparian growth. As such, it attracts vagrant and migrant bird species from the seas, seldom seen in the arid interior.

And that's what the Park Service plays on—they offer us a cool, shady sanctuary where we can sit and watch birds, and ponder over a little pond filled with endangered desert pupfish.

But, unfortunately, *A'al Waipia* is no more than a shadow of what it once was. To sense its historic significance, one must go to another place. Off the beaten track. A true Sonoran oasis, thirty miles to the south of Organ Pipe Cactus National Monument, in old Mexico. *Ki:towak*.

My pickup truck bounded along over the washboard road. Amadeo, Remedio, and I pointed out plants to each other as we went—the bristle-topped senita cactus, heavy-trunked ironwood trees, and odorous, yellowish-green croton shrubs.

We edged over a rise, and all of a sudden the desert was whisked away—palms and cottonwoods reached above the horizon, and teal splashed up into the air. Amadeo grabbed his field glasses—a white-faced ibis down on the mudflats of the pond, and a couple of pigs foraging in the saltgrass.

Remedio sighed, knowing that this place was the place he had heard of: *"Ki:towak."*

We parked the pickup beneath a towering California palm; then Remedio and I walked over toward an old Papago house sitting almost on the edge of the pond. Amadeo took off in the other direction to survey birds—he would sight twice as many species that afternoon as we had seen at *A'al Waipia* in the morning.

A lean old Papago quietly greeted us as we approached his house made of saguaro ribs and organ pipes. Luis Nolia had been born nearby and was now the oldest living resident of the Papago

281

settlement. A descendant of the semi-nomadic Sand Papago, he had as a child gathered the sweet underground stalks of sandfood in the dunes to the west, and eaten them like *carne machaca*. His family's women had crushed mesquite pods in bedrock mortars, and the men had transformed themselves into animals by wearing the fur masks and hoods of the *Wi'igita* ceremony.

Decades ago, Luis himself had been lured to the U.S. by wages to be earned picking cotton. After his wife died and his sons had grown up and settled in Arizona towns, he became lonely for the old ways and returned to Sonora. Now he grows summer field crops, keeps an orchard and a few animals, and is knowledgeable about the many medicinal plants with which the oasis springs are blessed.

Luis too blesses the oasis, for his work keeps it healthy. He is proud of the way the springs flow unencumbered by debris—he has dug out the fallen sediment so that the streamlets run clear from their source. Every summer, Luis plants squash, watermelon, beans, and other vegetables. His plowing and irrigation encourage at least six species of wild greens which he harvests at various times. Various medicinal plants and Olney's tules (for which *Ki:towak* is named) grow in the irrigation ditches.

Luis is appreciative of trees, too, and his plantings literally rim the oasis and field edges of *Ki:towak*. The willow cuttings that he stuck into the pond bank grew quickly into saplings and now stabilize the earth. He harvests willow branches to make leafy crosses that hang on the walls of every Papago household at *Ki:towak,* but keeps his own supply alive. Elderberry, salt cedar, date, and California palm are planted near his house to provide shade. Wolfberry, mesquite, and palo verde form a hedge on his field edge, and thorny brush is piled between the shrubs to discourage stray cattle from entering.

He has dug fig and pomegranate shoots from the base of ancient, abundantly-bearing trees, and transplanted them out to more open areas in the orchard where they can thrive. He offered Remedio and me a bag of figs and a few white-seeded, rusty-shelled pomegranates to savor.

Earlier in the day, walking on the west side of *A'al Waipia,* Remedio and I came upon a sight that made him sick inside. There were fruit trees—or the remnants of them—still putting out a few leaves (but no fruit) in an overgrown mesquite bosque. At least five pomegranate shrubs were dead, and another eighteen were dying. Just a few rangy, unpruned figs were left.

"Poor things, such old trees left with no one to help them!" Remedio lamented. He wondered if they could be dug up and given to a family who would care for them.

They were a reminder that until 1957, *A'al Waipia* had been a *peopled* oasis just as *Ki:towak* is today. Looking down at the foot of the dying fruit trees, we saw irrigation ditches running along in much the same pattern as those which still function in the field/orchard at *Ki:towak*.

For *A'al Waipia* was formerly a Papago settlement too, with six-and-a-half acres of crops, and more of orchards.

Geographer Ronald Ives put it simply: "It is reasonable to believe that this settlement, situated at a perennial spring, has been continuously occupied since man came to the area."

The archaeological record bears out the suggestion of long inhabitance—seven distinct prehistoric and historic sites have been found on the U.S. side of the *A'al Waipia* area, and at least two major ones south of the border fence.

From the 1830s onward, there are records of the names of Papago inhabitants who lived at the oasis. After 1860, Mexicans and Americans came to stay at *A'al Waipia* too, some even inter-marrying with the Sand Papago there. At least eight adobe houses were raised, a store and a mill were built, the pond deepened, and the ditches improved. But while visitors and buildings came and went, two Papago were patriarchs of the place for well over a century: Juan José and José Juan Orosco.

Juan José lived at the springs off and on from the 1860s until at least 1910, when Karl Lumholtz reported that he was well over a hundred but still in command of his faculties. José Juan Orosco, famous both as a medicine man and hunter, followed the older man as patriarch of *A'al Waipia* for several more decades. After Organ Pipe Cactus National Monument was established in 1937, Orosco's grazing rights were recognized, and he used them until his death in 1945.

The Park Service and the Papago disagree about what happened after that. The more-or-less official story is that José Juan Orosco's son Jim agreed to let the Park Service "condemn" his holdings in Organ Pipe Cactus National Monument, including "his" place at *A'al Waipia*. In return for the land and improvements he claimed by way of squatter's rights, Jim Orosco was given $13,000.

Some of the Papago tell versions of an altogether different story. One version claims that Jim Orosco never actually had exclusive rights to the place; he simply stayed there with the real caretaker, an old man called *S-Iawuis Wo:da*—Worn Out Boot—who was being paid by descendants of Juan José. The tellers of this version insist that *A'al Waipia* wasn't Jim Orosco's to sell: legally, all descendants of the two patriarchs should have been consulted. A great-granddaughter of Juan José lamented, "The old people knew it was wrong, but they didn't say too much when it happened."

Whether or not Jim Orosco could legally surrender Papago rights to *A'al Waipia,* it is clear that at the time the Park Service did not look upon the resident Papago as assets to the Monument. Orosco is reputed to have gone on drinking binges, and at one time, there was a Park Service sign on the road to the springs that said, "Watch Out for Deer, Cattle, and Indians."

The Papago farmland in the Monument was condemned without Congressional order, and without consultation with the Papago Tribe. By 1962, the National Park Service had destroyed all sixty-one structures remaining at *A'al Waipia* and the Growler Mine, wiping away most of the signs of human history in the Monument.

Bob Thomas of the *Arizona Republic* later commented on the Park Service's superficial commitment to its mandate of preserving ". . . various objects of historic and scientific interest." In a 1967 article entitled "Price of Progress Comes High," Thomas wrote:

> . . . Near Quitobaquito on the Organ Pipe National Monument a few years ago a government bulldozer knocked down the home of the late José Juan, a Papago Indian who lived there all his life. In doing so, workmen churned up the only known stratification of human habitation between Ajo and Yuma.

He added that the Papago:

> . . . distrust the government's promises to protect the park's treasures. In the past, the government has unknowingly or unfeelingly destroyed historic and prehistoric artifacts in the area.

By this destruction, the Park Service gained a bird sanctuary to provide tourists with a glimpse of wild plants and animals that gather around a desert water source.

Or so they thought. For an odd thing is happening at their "natural" bird sanctuary. They are losing the heterogeneity of the habitat, and with it, the birds. The old trees are dying. Few new ones are being regenerated. There are only three cottonwoods left, and four willows. These riparian trees are essential for the breeding habitat of certain birds. Summer annual seedplants are conspicuously absent from the pond's surroundings. Without the soil disturbance associated with plowing and flood irrigation, these natural foods for birds and rodents no longer germinate.

Visiting *A'al Waipia* and *Ki:towak* on back-to-back days three times during one year, ornithologists accompanying me encountered more birds at the Papago village than at the "wildlife sanctuary." Overall, we identified more than sixty-five species at the Papago's *Ki:towak,* and less than thirty-two at the Park Service's *A'al Waipia.* As Dr. Amadeo Rea put it, "It is as if someone fired a shotgun just before we arrived there. The conspicuous absences were more revealing than what we actually encountered."

When I explained to Remedio that we were finding far fewer birds and plants at the uninhabited oasis, he grew introspective. Finally, the Papago farmer had to speak:

"I've been thinking over what you say about not so many birds living over there anymore. That's because those birds, they come where the people are. When the people live and work in a place, and plant their seeds and water their trees, the birds go live with them. They like those places, there's plenty to eat and that's when we are friends to them."

I think that Remedio would even argue that it is natural for birds to cluster at human habitations, around fields and fence-rows. I'll go even further. It's in a sense natural for desert-dwelling humans over the centuries to have gathered around the *A'al Waipia* and *Ki:towak* oases. And although they didn't keep these places as pristine wilderness environments—an Anglo-American expectation of parks in the West—the Papago may have increased their biological diversity.

So if you're ever down in Organ Pipe Cactus National Monument and visit the Park Service wildlife sanctuary of Quitobaquito, remember that an old Papago place called *A'al Waipia* lies in ruin there. Its spirit is alive, less than forty miles away, in a true Sonoran Desert oasis. There, the irrigation ditches are filled with tules, and they radiate out from the pond into the fields like a green sunburst. *Ki:towak.*

BARRY
LOPEZ

Barry Lopez believes that "the most flattering letter you can get as a nonfiction writer is one that says, 'you wrote so well about A, which I really don't know much about or really care much about, that I understood B, which is what I care about.' Your book has set forth its ideas in such a way that people who are not interestd in the subject can use the ideas to throw more light into their own worlds."

Lopez is talking about his *Arctic Dreams: Imagination and Desire in a Northern Landscape* (1986), a book that is about a place where few people will go, but a book resonant with ideas. Lopez always has an abundance of ideas he is musing about, but he sees himself as a storyteller: "Landscape and animals are the oldest components of story."

Born in 1945 in New York, Lopez lived in southern California until he was eleven; his "childhood images" are from the Mojave and the Sonoran deserts. He and his wife Sandra live now in dark, wet woods along the McKenzie River in Oregon.

Lopez went to a Jesuit high school in New York City and then to Notre Dame. He tried out graduate school at the University of Oregon, but he knew by then he wanted to write, not go to school. "I want to just read and write. And in order to write, I travel. I am a writer who travels."

Like many writers in this anthology, he publishes widely in magazines, which is how he began his career in the early 1970s. His first published book was a collection of short, essaylike fiction, *Desert Notes: Reflections in the Eye of a Raven* (1976). *River Notes: The Dance of Herons* (1979) and *Winter Count* (1981), maintain this concise and resonant voice; though they are fiction, they create such vivid landscapes, with dreamlike images communicated with such straightforwardness—that they make readers define anew their ideas about reality. *Crossing Open Ground* (1988) collects ten years of landscape essays.

Lopez's 1978 book *Of Wolves and Men* won the John Burroughs Medal. *Arctic Dreams* won the National Book Award. "What has been given to me," Lopez has said, "is, by extension, being given to a larger group of people to deepen this kind of inquiry, which is to ask: what really is the relationship between the human mind and place? What can we learn from examining that relationship about how to lead better lives?"

The piece reprinted here comes from *Arctic Dreams*—from the chapter that joins the section introducing land and animals and the section that tells of human history. Lopez says, "'The Country of the Mind' is not a topic of the Arctic; it's the topic of a chapter for which the Arctic is the context."

The Country of the Mind

The daily cycle of tides rising and falling on the narrow beaches of Pingok Island during the open-water season is hard to read. In this part of the Arctic Ocean, where the Beaufort Sea washes against the north coast of Alaska, the vertical rise of the tide can be measured with a fingertip. On a windless day one can see reflected clouds undisturbed at the rim of the ocean's surface. It is possible to stand toe-to at the water's edge and, if one has the patience, see it gain only the heels of one's boots in six hours. Another peculiarity inherent in the land. In the eastern Arctic, at Ungava Bay and in embayments of the Canadian Archipelago, the tides are more substantial, rising up to 40 feet.

Pingok Island lies at 70° 35'N and 149° 35'W, a few miles off the north Alaska coast, some 30 miles east of the Colville River delta. It is the westernmost of the Jones Islands, a stretch of barrier islands that protects a shallow area along the coast called Simpson Lagoon, favored by migrating ducks.

This particular part of the arctic coast was little visited by Westerners until recently. Prudhoe Bay, where oil was discovered in February 1968, is 40 miles to the east; on clear days black clouds from flare-offs in those oil fields are faintly visible on the horizon. To the southwest a few miles, at Oliktok Point on the mainland, is an operational DEW line station. Pingok Island itself carries traces of modern inquiries into the region: the ubiquitous detritus of industrial reconnoitering and military exercises; and refuse from recent Eskimo and scientific encampments—strands of yellow polyethylene rope, empty wooden boxes and white gas tins, and outboard motor parts.

The most noticeable, man-made features of this island—it is about four and a half miles long and a half-mile wide at several points—are a shed and two pale yellow clapboard buildings that stand at its west end, and a cluster of coastal survey markers erected on the east end. For parts of two summers I lived with several marine biologists in one of the small, one-room buildings at the edge of the western beach. We were at sea most of the time; but on "weather days," when the

288

inevitable August snow squalls or heavy seas prohibited our working efficiently from a small boat, I walked the island's tundra plain.

This is an old business, walking slowly over the land with an appreciation of its immediacy to the senses and in anticipation of what lies hidden in it. The eye alights suddenly on something bright in the grass—the chitinous shell of an insect. The nose tugs at a minute blossom for some trace of arctic perfume. The hands turn over an odd bone, extrapolating, until the animal is discovered in the mind and seen to be moving in the land. One finds anomalous stones to puzzle over, and in footprints and broken spiderwebs the traces of irretrievable events.

During those summers I found, too, the molted feathers of ducks washed up in great wrack lines, in heaps, on the beach. Undisturbed in shallow waters on the lagoon side, I found hoofprints of caribou, as sharp as if the animals had stepped there in fresh clay only a moment before. They must have crossed over in late spring, on the last of the ice. I squatted down wherever the evidence of animals was particularly strong amid the tundra's polygon fractures. Where Canada geese had cropped grass at the edge of a freshwater pond; at the skull of a ringed seal carried hundreds of yards inland by ice, or scavengers; where grass had been flattened by a resting fox.

I saw in the sea face of a low bench of earth along the beach the glistening edge of an ice lens that underlay the tundra. The surface layer of plants and dirt overhung it like a brow-thatch of hair. I tried persistently but without success to sneak up on the flocks of feeding geese. I lost and regained images of ptarmigan against the ground, because of their near-perfect camouflage. I brought back to the cabin to set on a shelf by my bed castings of the landscape, to keep for a while and wonder over. The fractured intervertebral disk of a belukha whale. The prehistoric-looking exoskeleton of a marine isopod. And handfuls of feathers. Tangible things from my gentle interrogations, objects to which some part of a pervasive and original mystery still clung.

In the sometimes disconcerting summary which is a photograph, Pingok Island would seem bleak and forsaken. In winter it disappears beneath whiteness, a flat white plain extending seaward into the Beaufort Sea ice and landward without a border into the tundra of the coastal plain. The island emerges in June, resplendent with flowers and insects and birds, only to disappear again in a few months beneath the first snowstorms. To a Western imagination that finds a stand of full-crowned trees heartening, that finds the flight and voice of larks exhilarating, and the sight of wind rolling over fields of tall grass more agreeable, Pingok seems impoverished. When I arrived on the island, I, too, understood its bleak aspect as a category, the expression of something I had read about or been told. In the weeks during which I made some passing acquaintance with it, its bleakness was altered, however. The prejudice we exercise against such landscapes, imagining them to be primitive, stark, and pagan, became sharply apparent. It is in a place like this that we would unthinkingly store poisons or test weapons, land like the deserts to which we once banished our heretics and where we once loosed scapegoats with the burden of our transgressions.

The differing landscapes of the earth are hard to know individually. They are as difficult to engage in conversation as wild animals. The complex feelings of affinity and self-assurance one feels with one's native place rarely develop again in another landscape.

It is a convention of Western thought to believe all cultures are compelled to explore, that human beings seek new land because their economies drive them onward. Lost in this valid but nevertheless impersonal observation is the notion of a simpler longing, of a human desire for a less complicated life, for fresh intimacy and renewal. These, too, draw us into new landscapes. And desire causes imagination to misconstrue what it finds. The desire for wealth, for revivification, for triumph, as much or more than scientific measurement and description, or the imperatives of economic expansion, resolves the geography of a newfound landscape.

In 1893, Frederick Jackson Turner read a paper in Chicago before the American Historical Association that changed the course of American historiography—the way historians understand how elements of the past are causally related. Turner's idea, which came to be called the Frontier Hypothesis, has become so much a part of the way we think about the country's past that it now seems self-evident. At the time he presented it, it was unheralded and unique.

Prior to 1893, most historians believed America had been shaped by the desire to separate itself from European influences, or by the economic and social issues that came to a head in the Civil War. Turner offered a third view, that America was shaped by both the fact and the concept of its westering frontier. The national character, so distinguished by enterprise, initiative, and hard work, he said, derived from an understanding of its citizens' experiences on the frontier. Historians generally accepted Turner's hypothesis, and have refined upon and extrapolated from it for nearly a century.

Turner's observation showed at least two things: the narrative direction that a nation's history takes is amenable to revision; and the landscapes in which history unfolds are both real, that is, profound in their physical effects on mankind, and not real, but mere projections, artifacts of human perception. Nowhere in North American history is this more apparent than in the westward movement of the nineteenth century. Politicians and promoters, newspaper editors and businessmen argued hotly over the suitability of the tallgrass and shortgrass prairies for farming. In most of these arguments the political cant of boomers and nay-sayers and the abstractions of agrarian theorists counted for more than the factual testimony of the land in the form of rainfall records or the statements of people who lived there.

Perhaps this is obvious. In the modern age, one of the most irksome, and ironic, of political problems in North America is the promulgation of laws and regulations from Washington and Ottawa that seem grossly ignorant of the landscapes to which they apply. We all, however, apprehend the land imperfectly, even when we go to the trouble to wander in it. Our perceptions are colored by preconception and desire. The physical landscape is an unstructured abode of space and time and is not entirely fathomable; but this does not necessarily put us at a disadvantage in seeking to know it. Believing them to be fundamentally mysterious in their form and color, in the varieties of life inherent in them, in the tactile qualities of their soils, the sound of the violent fall of rain upon them, the smell of their buds—believing landscapes to be mysterious aggregations, it becomes easier to approach them. One simply accords them the standing that one grants the other mysteries, as distinguished from the puzzles, of life.

I recall in this context two thoughts. A man in Anaktuvuk Pass, in response to a question about what he did when he visited a new place, said to me, "I listen." That's all. I listen, he meant, to what the land is saying. I walk around in it and strain my senses in appreciation of it for a long time before I, myself, ever speak a word. Entered in such a respectful manner, he believed, the land would open to him. The other thought draws, again, on the experience of American painters. As they sought an identity apart from their European counterparts in the nineteenth century, they came to conceive of the land as intrinsically powerful: beguiling and frightening, endlessly arresting and incomprehensibly rich, unknowable and wild. "The face of God," they said.

As I step out of our small cabin on Pingok Island, the undistinguished plain of tundra spreads before me to the south and east. A few glaucous gulls rise from the ground and drop back, and I feel the cold, damp air, like air from a refrigerator, against my cheeks. A few yards from the door, stark and alone on the tundra, a female common eider lies dead. A few more yards to the west, a bearded-seal skin has been expertly stretched between short wooden stakes to dry. A few yards beyond, a northern phalarope spins wildly on the surface of a freshwater pond, feeding on zooplankton.

A southwest wind has been blowing for two days; it's the reason we are ashore today. The sky threatens squalls and snow. I head south across the tundra toward the lagoon, wondering if I will find ducks there. In my mind is a vague plan: to go there, then east along the coast to a place where the tundra is better drained, easier walking, then back across the island, and to come home along the seaward coast.

In such flat terrain as this, even with the lowering skies, I brood on the vastness of the region. The vastness is deceptive, however. The journals of arctic explorers are full of examples of messages stashed out there with a high expectation of their discovery, because the prominent places in such a featureless landscape are so obvious. They are the places a human eye notices right away. And there is something, too, about the way the landscape funnels human movement, such that encounters with strangers are half expected, as is the case in a desert crossing. Human beings are so few here and their errands such a part of the odd undercurrent of knowledge that flows in a remote region that you half expect, too, to know of the stranger. Once, camped on the upper Yukon, I saw a man in a distant canoe. When he raised his field glasses to look at a cliff where peregrine falcons were nesting, I surmised who he was (a biological consultant working on a peregrine census) from a remark I had overheard a week before in a small restaurant in Fairbanks. He probably knew of my business there, too. Some of the strangeness went out of the country in that moment.

If the mind releases its fiduciary grip on time, does not dole it out in a fretful way like a valued commodity but regards it as undifferentiated, like the flatness of the landscape, it is possible to transcend distance—to travel very far without anxiety, to not be defeated by the great reach of the land. If one is dressed well and carrying a little food, and has the means to secure more food and to construct shelter, the mind is that much more free to work with the senses in an appreciation of the country. The unappealing tundra plain, I recall, is to its denizens a storehouse of food and instant tools.

As I thread my way southwest, along the margins of frost polygons, I am aware of the movement of birds. A distant speck moves across the sky with a loon's trajectory. A Savannah sparrow flits away over the ground. The birds come and go—out to sea to feed or to the lagoon to rest—on a seemingly regular schedule. Scientists say the pattern of coming and going, of feeding and resting, repeats itself every twenty-four hours. But a description of it becomes more jagged and complex than the experience, like any parsing of a movement in time.

The sound of my footfall changes as I step from damp ground to wet, from wet to dry. Microhabitats. I turn the pages of a mental index to arctic plants and try to remember which are the ones to distinguish these borders: which plants separate at a glance mesic tundra from hydric, hydric from xeric? I do not remember. Such generalities, in any case, would only founder on the particulars at my feet. One is better off with a precise and local knowledge, and a wariness of borders.

These small habitats, like the larger landscapes, merge imperceptibly with each other. Another, remembered landscape makes this one seem familiar; and the habits of an animal in one region provoke speculation about behavior among its relatives in another region. But no country, finally, is just like another. The generalities are abstractions. And the lines on our topographic maps reveal not only the scale at which we are discerning, but our tolerance for discrepancies in nature.

A tundra botanist once described to me her patient disassembly of a cluster of plants on a tussock, a tundra mound about 18 inches high and a foot or so across. She separated live from dead plant tissue and noted the number and kind of the many species of plants. She examined the insects and husks of berries, down to bits of things nearly too slight to see or to hold without crushing. The process took hours, and her concentration and sense of passing time became fixed at that scale. She said she remembered looking up at one point, at the tundra that rolled away in a hundred thousand tussocks toward the horizon, and that she could not return her gaze because of that sight, not for long minutes.

My route across Pingok seems rich, but I am aware that I miss much of what I pass, for lack of acuity in my senses, lack of discrimination, and my general unfamiliarity. If I knew the indigenous human language, it would help greatly. A local language discriminates among the local phenomena, and it serves to pry the landscape loose from its anonymity.

I know how much I miss—I have only to remember the faces of the Eskimos I've traveled with, the constant flicker of their eyes over the countryside. Even inside their houses men prefer to talk while sitting by a window. They are always looking away at the land or looking up to the sky, the coming weather. As I near the lagoon, pondering the identity of something I saw a flock of ptarmigan eating, I smile wryly at a memory: it was once thought that scurvy was induced in the Arctic by the bleakness of its coasts.

There are no ducks nearby in the lagoon. With my field glasses I can just make out the dark line of their rafts on the far side of the water, a lee shore. I settle myself in a crease in the tundra, out of the wind, arrange my clothing so nothing binds, and begin to study the far shore with the binoculars. After ten or fifteen minutes I have found two caribou. Stefansson was once asked by an Eskimo to whom he was showing a pair of binoculars for the first time whether he could "see into tomorrow" with them. Stefansson took the question literally and was amused. What the *inuk* probably meant was, Are those things powerful enough to see something that will not reach you for another day, like migrating caribou? Or a part of the landscape suitable for a campsite, which you yourself will not reach for another day? Some Eskimo hunters have astounding natural vision; they can point out caribou grazing on a slope three or four miles away. But the meticulous inspection of the land that is the mark of a good hunter becomes most evident when he uses a pair of field glasses. Long after the most inquiring nonnative has grown weary of glassing the land for some clue to the movement of animals, a hunter is still scouring its edges and interstices. He may take an hour to glass 360° of the apparently silent tundra, one section at a time.

You can learn to do this; and such scrutiny always turns up a ground squirrel, an itinerant wolverine, a nesting bird—something that tells you where you are and what's going on. And when you fall into the habit, find some way like this to shed your impatience, you feel less conspicuous in the land.

I walk a long ways down the beach before arriving at the place where the tundra dries out, and turn inland. Halfway across I find the skull of a goose, as seemingly random in this landscape as the dead eider in the grass by the cabin. A more thoughtful inquirer, someone dependent upon these bits of information in a way that I am not, would find out why. To the southwest I can see a snow squall—I want to reach the seaward side of the island before it arrives, in case there is something worse behind it. The shoreline is my way home. I put the paperthin skull back on the ground. Far to the east I see a dilapidated spire of driftwood, a marker erected in 1910 by Ernest Leffingwell when he was mapping these coasts. Leaning slightly askew, it has the aspect of an abandoned building, derelict and wind-punished. It is a monument to the desire to control vastness. It is a referent for the metes and bounds that permit a proper division and registry of the countryside, an assignment of ownership.

The western history of Pingok Island comprises few events. John Franklin, a young British naval officer, led an overland party almost this far west from the mouth of the Mackenzie in 1826, trying for a rendezvous at Point Barrow, 250 miles farther on. But with bad weather and the physical strain on the men, he "reached the point beyond which perseverance would be rashness," and turned back, in the fall of that year. Robert M'Clure put a party ashore in August 1850, to entreat with a small group of resident Eskimo. The Eskimos regarded *Investigator* as "a swimming island," wrote a thirty-three-year-old Moravian missionary with the shore party. "At every movement of the ship, though it was half an hour's pull distant, they showed fresh alarm and an electric shock, as it were, went through them all."

Late in the nineteenth century the island was visited by American whalers, who took on fresh water from its ponds. Stefansson abandoned the ill-fated *Karluk* nearby, in September 1913. (Caught in the ice, the ship, a brigantine whaler refitted as a scientific expeditionary vessel, later drifted far to the west, where it was crushed and sank, with the subsequent loss of half the party.) Traders and explorers like Leffingwell were also in the area in these years. In 1952 an archaeologist named William Irving made the first excavations of the island's prehistoric sites. A few years later, the DEW line station was built at Oliktok. In the 1960s the U.S. Navy built two sheds and a 10- by 18-foot cabin for the use of scientific field parties working out of the Naval Arctic Research Laboratory at Barrow, and later for the federal Outer Continental Shelf Environmental Assessment Program and other projects. These buildings also occasionally shelter Eskimos traveling along the coast, whence the bearded-seal skin pegged on the tundra by the door of the cabin we were using.

The aboriginal history of the island is much deeper and also more obscure. It is likely, because of the way it is situated, that the island has been used by hunting peoples for centuries, though probably not continually. Today, a dozen 400-year-old dwellings, outlined on the tundra by protruding driftwood logs and bowhead whale ribs, are the only traces of early occupation left. This stretch of North American coastline was apparently never heavily populated. The Pingok houses, however, constituted the largest prehistoric site on the north Alaska coast in 1981.

An astute archaeologist would have anticipated the remains of such an encampment here. Pingok Island gets its name from an Inupiatun word for "a rising of earth over a dome of ice." The reference is to a long sand dune on the seaward edge of the island which gives protection from storm surges. Such protection is rare along this coast. It would be noticed and taken advantage of. Hunters camped at Pingok would benefit, too, from early access to bearded and ringed seals around the mouth of the Colville, where freshwater ice begins to disintegrate before the sea ice breaks up. On Simpson Lagoon they would find migrating geese and ducks. Along the beach are great supplies of driftwood (from the Mackenzie River), some of the trees 30 or 40 feet long. There are large freshwater ponds on the island, and from here there is good access to runs of char and to herds of belukha whale, and, in September, to bowhead whales headed west for Bering Strait. On the landward side of the lagoon they could expect to see caribou.

Excavations at the Pingok houses indicate that the people living here between 1550 and 1700 hunted all of these animals, and also walrus and polar bear. Among the more intriguing objects unearthed are a polar-bear-tooth(?) fish lure, a child's miniature hunting bow, and a piece of caribou-antler(?) plate armor.

Standing by the remains of these houses, one is struck by the fact that in the Arctic so much human history lies undisturbed on the surface of the land. And by the contrast with a more easily retrieved past, such as our own. Here is a prong from a bird spear; here is a walrus-tooth pendant; but what were the ideas attached to these objects?

The remains of other, more modern Eskimo dwellings are also found on Pingok—sod and whalebone houses from the 1920s. In recent years ethnohistorians have visited the islands, bringing with them the people who once lived in these houses. The people were interviewed here with these artifacts and the landscape itself close at hand in order to plumb "the memory culture," the culture tied to tools no longer used and still most accurately described in the Inupiatun language, whose vocabulary is rapidly fading.

One Sunday afternoon in the summer of 1981 several Eskimos from the village of Barrow paid us a visit on Pingok. They were conducting a land-use survey, a complex process of land assessment used by native peoples to substantiate their aboriginal claims to certain sections of land. We talked a little about the Eskimo "history" of Pingok Island (as if, without such corroborative details, acceptable to the men who owned the maps, the place would remain vague and unclaimable). Only some sort of acceptable, verifiable history could save it for them now. We all turned, with more ease and enthusiasm, to talk of hunting. Such chance conversations as this, far from the villages, where political and racial tensions can be strong, are often cordial. No one is apt to pursue a point that might lead to disagreement, or to ask a question that might be construed as prying. It is always acceptable, and good, to speak of hunting. A great deal of information about the local landscape passes back and forth in this context. One feels here, sharing the details of animals' lives in the memories of those present, the authority for a claim to the land just as legitimate and important as the things found at a 400-year-old house.

After they left—they were traveling in a small boat toward the east (and to most foreign observers they would have seemed underdressed and poorly provisioned for their journey, a common impression)—we talked among ourselves about the Eskimos' cultural history. The men who left carried with them a borrowed historiography, a matrix they put down like a net in the undifferentiated sphere of time that welled up in their own traditional and unwritten history. It is a system they are becoming familiar, and handy, with. And there was great dignity and authority in the Eskimo women who sat on driftwood logs on Pingok Island, recalling into tape recorders the details of their lives from so many years ago. One could so easily imagine, as memory bloomed before the genuine desire of another to know, filaments appearing in the wind, reattaching them to the land, even as they spoke.

Land-use-and-occupancy projects have been conducted by and with native peoples throughout the Arctic, in furtherance of their land claims and to protect their hunting rights. These studies have revealed a long and remarkably unbroken connection between various groups of indigenous people and the particular regions of the Arctic they inhabit. It is impossible to separate their culture from these landscapes. The land is like a kind of knowledge traveling in time through them. Land does for them what architecture sometimes does for us. It provides a sense of place, of scale, of history; and a conviction that what they most dread—annihilation, eclipse—will not occur.

"We are here [i.e., alive now in this particular place] because our ancestors are real," a man once told an interviewer. The ancestors are real by virtue of their knowledge and use of the land, their affection for it. A native woman, alone and melancholy in a hospital room, told another interviewer she would sometimes raise her hands before her eyes to stare at them: "Right in my hand, I could see the shore-lines, beaches, lakes, mountains, and hills I had been to. I could see the seal, birds, and game. . . ." Another Eskimo, sensing the breakup of his culture's relationships with the land, the replacement of his ancient hunting economy by another sort of economy, told an interviewer it would be best all around if the Inuit became "the minds over the land." Their minds, he thought, shaped as they were to the specific contours of the land, could imagine it well enough to know what to do. Like most Eskimos, as the land-use-and-occupancy surveys made clear, he could not grasp the meaning of a life divorced from the landscape—the animals, the weather, the sound of ice, the taste and nourishment that came from "the food that counts."

In a long passage in *The Central Eskimo* (1888), Franz Boas describes the birth of a child and the types of clothing the child wears during the first few days of its life—a cap made of arctic hare fur, underclothes of bird feathers, a hood made of a caribou fawn with the ears attached. One is struck by the great efforts of the mother, especially, to confirm the child immediately in a complex and intricate relationship with "the land," the future source of the child's spiritual, psychological, and physical well-being. Nearly a hundred years later, Richard Nelson, a northern ethnologist, described a similar, modern Koyukon understanding of natural history in *Make Prayers to the Raven* (1983). While many things have changed, the evidence of continued intimacy with a local landscape—a practical knowledge of it, a sensitivity toward it, a supplication of it—is still clear. The incorporation of the land into traditional stories—evidence of close association with the land and the existence of an uncanny and mesmerizing conformity of human behavior in response to subtleties in the landscape—is also still evident. The people, many of them, have not abandoned the land, and the land has not abandoned them. It is difficult, coming from cities far to the south, to perceive let alone fathom the richness of this association, or to assess its worth. But this archaic affinity for the land, I believe, is an antidote to the loneliness that in our own culture we associate with individual estrangement and despair.

I move the glasses off Leffingwell's tower. On the ridge of sand dune along the beach to the north I spot an arctic fox. A great traveler in winter, like the polar bear and the wolf. In summer, when water intervenes in the fox's coastal habitat, he may stay in one place—an island like this, for example. The fox always seems to be hurrying somewhere, then stopping suddenly to sit down and rest. He runs up on slight elevations and taps the air all over with his nose.

The arctic fox's fur runs to shades of brown in summer, which blend with ivory whites on its underparts. (In winter the coat is gleaming white or a grayish blue to pale beige, which is called "blue.") As with any animal, the facets of its life are complexly engaging. The extent and orderliness of its winter caches and its ability to withstand very cold weather are striking. Also its tag-along relationship with the polar bear. It is the friendliest and most trusting of the North American foxes, although it is characterized in many expedition journals as "impudent," derided for its "persistent cheekiness," and disparaged as a "parasite" and a scavenger. Arctic foxes are energetic and persistent in their search for food. They thoroughly scour the coastlines over which they travel and, like polar bears, will gather from miles around at a source of carrion. If it's a cook tent they choose instead, and thirty or forty of them are racing around, tearing furiously into everything, an expedition's initial sense of amusement can easily turn sour or violent. Arctic foxes so pestered Vitus Bering's shipwrecked second Kamchatka expedition that the men tortured and killed the ones they caught with the unrestrained savagery one would expect of men driven insane by hordes of insects.

In his encounter with modern man in the Arctic, then, the fox's efficient way of life has sometimes gone fatally against him. (His dealings with modern Eskimos have fit more perfectly, though also fatally, with human enterprise. Once largely ignored, he became the most relentlessly pursued fur-bearer in the Arctic with the coming of the fur trade and the advent of the village trading post.)

I watch the fox now, traveling the ridge of the sand dune, the kinetic blur of its short legs. I have seen its (or another's) tracks at several places along the beach. I think of it traveling continuously over the island, catching a lemming here, finding part of a seal there, looking for a bird less formidable than a glaucous gull to challenge for its eggs. I envision the network of its trails as though it were a skein of dark lines over the island, anchored at slight elevations apparent to the eye at a distance because of their dense, rich greens or clusters of wildflowers.

Because the fox is built so much closer to the ground and is overall so much smaller than a human being, the island must be "longer" in its mind than four and a half miles. And traveling as it does, trotting and then resting, trotting and then resting, and "seeing" so much with its black nose—what is Pingok like for it? I wonder how any animal's understanding of the island changes over the year; and the difference in its shape to a gyrfalcon, a wolf spider, or a bowhead echolocating along its seashore. What is the island to the loon, who lives on the water and in the air, stepping awkwardly ashore only at a concealed spot at the edge of a pond, where it nests? What of the bumblebee, which spends its evening deep in the corolla of a summer flower that makes its world 8°F warmer? What is the surface of the land like for a creature as small but as adroit as the short-tailed weasel? And how does the recollection of such space guide great travelers like the caribou and the polar bear on their journeys?

A friend working one summer near Polar Bear Pass on Bathurst Island once spotted a wolf running off with a duck in its mouth. He saw the wolf bury the duck, and when the wolf left he made for the cache. He couldn't find it. It was open, uncomplicated country. He retraced his steps, again took his bearings, and tried a second time. A third time. He never found it. The wolf, he thought, must have a keener or at least a different way of holding that space in its mind and remembering the approach. The land then appeared to him more complicated.

One day, out on the sea ice, I left the protection of a temporary building and followed a bundle of electric cables out into a blizzard. The winds were gusting to 40 knots; it was −20°F. I stood for a long time with my back to the storm, peering downwind into the weak January light, fearful of being bowled over, of losing touch with the umbilical under which I had hooked a boot. Both its ends faded away in that swirling whiteness. In the 40-foot circle of visibility around me I could see only ice hummocks. I wondered what notions of "direction" a fox would have standing here, how the imperatives for food and shelter would affect us differently.

One can only speculate about how animals organize land into meaningful expanses for themselves. The worlds they perceive, their *Umwelten,* are all different. The discovery of an animal's *Umwelt* and its elucidation require great patience and experimental ingenuity, a free exchange of information among different observers, hours of direct observation, and a reluctance to summarize the animal. This, in my experience, is the Eskimo hunter's methodology. Under ideal circumstances it can also be the methodology of Western science.

Many Western biologists appreciate the mystery inherent in the animals they observe. They comprehend that, objectively, what they are watching is deceptively complex and, subjectively, that the animals themselves have nonhuman ways of life. They know that while experiments can be designed to reveal aspects of the animal, the animal itself will always remain larger than the sum of any set of experiments.

They know they can be very precise about what they do, but that that does not guarantee they will be accurate. They know the behavior of an individual animal may differ strikingly from the generally recognized behavior of its species; and that the same species may behave quite differently from place to place, from year to year.

It is very hard to achieve a relatively complete and accurate view of an animal's life, especially in the Arctic, where field conditions present so many problems, limiting observation. Many biologists studying caribou, muskoxen, wolves, and polar bear in the North are more distressed by this situation than they otherwise might be. Industry, which pays much of the bill for this arctic research, is less interested in the entire animal than it is in those aspects of its life that might complicate or hinder development—or, to be fair, how in some instances industry might disrupt the animal's way of life. What bothers biologists is the narrowness of the approach, the haste with which the research must be conducted, and, increasingly, the turning of an animal's life into numbers. The impersonality of statistics masks both the complexity and the ethics inherent in any wildlife situation. Biologists are anxious about "the tyranny of statistics" and "the ascendency of the [computer] modeler," about industry's desire for a "standardized animal," one that always behaves in predictable ways.

A Canadian scientist told me, "I hate as a biologist having to reduce the behavior of animals to numbers. I hate it. But if we are going to stand our ground against [head-long development] we must produce numbers, because that's all they will listen to. I am spending my whole *life* to answer these questions—they want an answer in two months. And anything a native says about animals, well, that counts for nothing with them. Useless anecdotes."

A belief in the authority of statistics and the dismissal of Eskimo narratives as only "anecdotal" is a dichotomy one encounters frequently in arctic environmental assessment reports. Statistics, of course, can be manipulated—a whale biologist once said to me, "If you punish the data enough, it will tell you anything." And the *Umwelt* of a statistician, certainly, plays a role in developing the "statistical picture" of a landscape. The Eskimos' stories are politely dismissed not because Eskimos are not good observers or because they lie, but because the narratives cannot be reduced to a form that is easy to handle or lends itself to summary. Their words are too hard to turn into numbers.

What the uninitiated scientist in the Arctic lacks is not ideas about how the land works, or a broad theoretical knowledge of how the larger pieces fit together, but time in the field, prolonged contact with the specific sources of an understanding. Several Western scientists, including anthropologist Richard Nelson, marine mammal biologists

John Burns, Francis Fay, and Kerry Finley, and terrestrial mammal biologist Robert Stephenson, have sought out Eskimo hunters as field companions in order to get a better understanding of arctic ecology. Nelson, who arrived in Wainwright in the early 1960s quite skeptical about the kinds of animal behavior the hunters had described to him, wrote a line any one of the others might have written after a year of traveling through the country with these people: "[Their] statements which seem utterly incredible at first almost always turn out to be correct."

I walk on toward the dune where the fox has disappeared. The inconspicuous plants beneath my feet, I realize, efficiently harbor minerals and nutrients and water in these acidic, poorly drained soils. They are compact; they distribute the weight of snow, of passing caribou and myself, so it does not crush them. The stems of these willows are shorter than those of their southern counterparts, with many more leaves to take advantage of the light. It may take years for a single plant to produce a seed crop. What do these plants murmur in their dreams, what of warning and desire passes between them?

It is beginning to snow a little, on a slant from the southeast. I walk on, my eye to the ground, out to the horizon, back to the ground. And what did Columbus, sailing for Zaiton, the great port of Cathay, think of the reach of the western Atlantic? How did Coronado assess the Staked Plain of Texas, the rawest space he ever knew, on his way to Quivira? Or Mungo Park the landscape of Africa in search of the Niger? What one thinks of any region, while traveling through, is the result of at least three things: what one knows, what one imagines, and how one is disposed.

What one knows is either gathered firsthand or learned from books or indigenous observers. This information, however, is assembled differently by each individual, according to his cultural predispositions and his personality. A Western traveler in the Arctic, for example, is inclined to look (only) for cause-and-effect relationships, or predator-prey relationships; and to be (especially) alert for plants and animals that might fill "gaps" in Western taxonomies. Human beings, further, are inclined to favor visual information over the testimony of their other senses when learning an area, and to be more drawn to animals that approximate their own scale. Our view is from a certain height above the ground. In any new country we want panoramas.

What one imagines in a new landscape consists of conjecture, for example, about what might lie beyond that near horizon of small hills, or the far line of the horizon. Often it consists of what one "hopes to see" during the trip—perhaps a barren-ground grizzly standing up on the tundra, or the tusk of a mammoth in the alluvial silt of a creek. These expectations are based on a knowledge of what has happened in this land for others. At a deeper level, however, imagination represents the desire to find what is unknown, unique, or farfetched—a snowy owl sitting motionless on the hips of a muskox, a flower of a favorite color never before reported, tundra swans swimming in a winter polynya.

Imagination also poses the questions that give a new land dimension in time. Are these wolverine tracks from *this* summer or the summer before? How old is this orange lichen? Will the caribou feeding placidly in this swale be discovered by those wolves traveling in the distance? Why did the people camped here leave this piece of carved seal bone behind?

The way we are disposed toward the land is more nebulous, harder to define. The reluctant traveler, brooding about events at home, is oblivious to the landscape. And no one is quite as alert as an indigenous hunter who is hungry. If one feels longing or compassion at the sight of something beautiful, or great excitement over some unexpected event, these may effect an optimistic disposition toward the land. If one has lost a friend in the Arctic to exposure after an airplane crash, or gone broke speculating in a northern mine, one might regard the land as antagonistic and be ill-disposed to recognize any value in it.

The individual desire to understand, as much as any difference in acuity of the senses, brings each of us to find something in the land others did not notice.

Over time, small bits of knowledge about a region accumulate among local residents in the form of stories. These are remembered in the community; even what is unusual does not become lost and therefore irrelevant. These narratives comprise for a native an intricate, long-term view of a particular landscape. And the stories are corroborated daily, even as they are being refined upon by members of the community traveling between what is truly known and what is only imagined or unsuspected. Outside the region this complex but easily shared "reality" is hard to get across without reducing it to generalities, to misleading or imprecise abstraction.

The perceptions of any people wash over the land like a flood, leaving ideas hung up in the brush, like pieces of damp paper to be collected and deciphered. No one can tell the whole story.

I must set my face to the wind to head west, back toward the cabin. I drop down to the seaward beach where I will have the protection of the dune. Oldsquaw and eider ducks ride the ocean swell close to shore in the lee of the storm, their beaks into the wind. Between gaps in the dune I catch glimpses of the dark tundra, swept by wind and snow. My thoughts leap ahead to the cabin, to something warm to drink, and then return. I watch the ducks as I walk. Watching animals always slows you down. I think of the months explorers spent locked up

302

in the ice here, some of them trapped in their ships for three or four years. Their prospects for an early departure were never good, but, their journals reveal, they rarely remarked on the animals that came around, beyond their potential as food, as threats or nuisances. These were men far from home, who felt helpless; the landscape hardly registered as they waited, except as an obstacle. Our inattentiveness is of a different order. We insist on living today in much shorter spans of time. We become exasperated when the lives of animals unfold in ways inconvenient to our schedules— when they sit and do "nothing." I search both the featureless tundra to my left and the raft of brown sea ducks to my right for something untoward, something that stands out. Nothing. After hours of walking, the tundra and the ducks recede into the storm, and my mind pulls far back into its own light.

A Lakota woman named Elaine Jahner once wrote that what lies at the heart of the religion of hunting peoples is the notion that a spiritual landscape exists within the physical landscape. To put it another way, occasionally one sees something fleeting in the land, a moment when line, color, and movement intensify and something sacred is revealed, leading one to believe that there is another realm of reality corresponding to the physical one but different.

In the face of a rational, scientific approach to the land, which is more widely sanctioned, esoteric insights and speculations are frequently overshadowed, and what is lost is profound. The land is like poetry: it is inexplicably coherent, it is transcendent in its meaning, and it has the power to elevate a consideration of human life.

The cabin emerges silently up ahead in the flowing snow as the storm closes in. It seems to rest within a white cave or at the far end of a canyon. Sound only comes now from what is immediately around me. The distant voices of birds are gone. I hear the gritty step of my boots in the sand. Splash of wavelets on the beach. Wind rushing over the cones of my ears.

Through a window yellow with light I see a friend at a table, whipping the end of a boat line with waxed thread. I will have hot tea and lie in my bunk, and try to recall what I saw that did not, in those moments, come to mind.

RICHARD
NELSON

Born in 1941, Richard Nelson grew up in Wisconsin and first went to Alaska as an undergraduate, assisting with summer research projects. He says of his year with Inupiaq Eskimo people in the North Slope village of Wainwright in 1964–65: "This experience made a different person of me and shaped the entire course of my life."

Nelson went on to earn a Ph.D. in anthropology, living with and writing ethnographies of Inupiaq people (*Hunters of the Northern Ice*, 1969), Gwich'in Athabaskan people (*Hunters of the Northern Forest*, 1973), and Koyukon Athabaskan people (*Make Prayers to the Raven: A Koyukon View of the Northern Forest,* 1983). In the latter book and in *Shadow of the Hunter: Stories of Eskimo Life* (1980), Nelson moved away from straight scientific writing to a more personal and sensory style. After more than twenty years of working alongside native hunters, listening to village elders, and struggling to understand their world, Nelson wanted to express "my own feelings about Koyukon people and their traditions, while also staying within the bounds of scientific description."

In 1982, he left his academic career to become a full-time writer. Living on the Northwest Pacific coast, he began spending time on an offshore island, exploring his relation to this natural community ("a place that wholly engages the heart and mind"), using the island as source for his family's food as much as possible. This experience led to *The Island Within* (1989), a "progress report" on his journey, which won the Burroughs Award and from which this essay comes.

Nelson uses the word "respect" frequently—"respect toward others, toward my surrounding environment, and toward myself." He keeps the island's location vague, to preserve both his and the place's privacy.

Hunting is central to Nelson's search for intimacy with wild creatures, but he hunts with the attitude of the Koyukon: "I won't say I absolutely believe there's a spirit in everything in the natural world, but I believe that I should behave as if there were." And then, "writing about those experiences significantly heightened and intensified them for me, integrated them more deeply into my life, and allowed me to relive them as I never could have done otherwise."

Always, Nelson works from this credo: "I am writing to serve my subject rather than to serve myself." His current project is a book about deer. He says, "I am interested in exploring my personal ecology. I live from deer; this voice has been fed from deer. I appreciate the fact that I am made out of the animal I love."

The Gifts of Deer

Cold, clear, and calm in the pale blue morning. Snow on the high peaks brightening to amber. The bay a sheet of gray glass beneath a faint haze of steam. A November sun rises with the same fierce, chill stare of an owl's eye.

I stand at the window watching the slow dawn, and my mind fixes on the island. Nita comes softly down the stairs as I pack gear and complain that I've slept too late for these short winter days. A few minutes later, Ethan trudges out onto the cold kitchen floor, barefoot and half asleep. We make no direct mention of hunting, to avoid acting proud or giving offense to the animals. I say only that I'll go to the island and look around; Ethan says only that he would rather stay at home with Nita. I wish he would come along so I could teach him things, but know it will be quieter in the woods with just Shungnak.

They both wave from the window as I ease the skiff away from shore, crunching through cakes of fresh-water ice the tide has carried in from Salmon River. It's a quick run through Windy Channel and out onto the freedom of Haida Strait, where the slopes of Kluksa Mountain bite into a frozen sky. The air stings against my face, but the rest of me is warm inside thick layers of clothes. Shungnak whines, paces, and looks over the gunwale toward the still-distant island.

Broad swells lying in from the Pacific alternately lift the boat and drop it between smooth-walled canyons of water. Midway across the strait a dark line of chop descends swiftly from the north, and within minutes we're surrounded by whitecaps. There are two choices: either beat straight up into them or cut an easier angle across the waves and take the spray. I vacillate for a while, then choose the icy spray over the intense pounding. Koyukon elders often told me it's wrong to curse the wind or complain about the cold, but this morning I do it anyway.

A kittiwake sweeps over the water in great, vaulting arcs, its wings flexed against the surge and billow of the air. As it tilts its head passing over the boat, I think how clumsy we must look. The island's shore lifts slowly in dark walls of rock and timber that loom above the apron of snow-covered beach. Approaching the shelter of Sea Lion Point, the chop fades and the swell diminishes. I turn up along the lee, running between the kelp beds and the surf, straining to see if any deer are grazing on seaweed at the tide's edge.

Near the end of the point is a gut that opens into a tight, shallow anchorage. I ease the boat between the rocks, with lines of surf breaking close on either side. The waves rise and darken, their edges sparkle in the sun, then long manes of spray whirl back as they turn inside out and pitch onto the reef. The anchor slips through ten feet of crystal water to settle among the kelp fronds and urchin-covered rocks. On a strong ebb the boat would go dry here, but today's tide range is only six feet. Before launching the punt, I pull the binoculars from my pack and warm them inside my coat so the lenses won't fog. Then I spend a few minutes scrutinizing the broad, rocky shore and the sprawls of brown grass along the timber's edge. A bunch of rock sandpipers flashes up from the shingle and an otter loops along the windrows of drift logs, but there are no signs of deer.

I can't help feeling a little anxious, because the season is drawing short and our year's supply of meat is not yet in. During the past few weeks, deer have been unusually wary, haunting the underbrush and slipping away at the least disturbance. I've come near a few, but these were young ones I stalked only for the luxury of seeing them from close range. Now that the rutting season has begun, there's a good chance of finding larger deer, and they'll be distracted by the search for mates.

A bald eagle watches from a tall hemlock as we bob ashore in the punt. Finally the bird lurches out, scoops its wings full of dense, cold air, and soars away beyond the line of trees. While I trudge up with the punt, Shungnak prances back and forth hunting for smells. The upper reaches are layered and slabbed with ice; slick cobbles shine like steel; frozen grass crackles underfoot. I lean the punt on a snow-covered log, pick up my rifle and small pack, and slip through the leafless alders into the forest.

My eyes adjust to the darkness, the deep green of boughs, and the somber, shadowy trunks. I feel safe and hidden here. The forest floor is covered with deep moss that should sponge gently underfoot. But today the softness is gone; frozen moss crunches with each step and brittle twigs snap, ringing out in the crisp air like strangers' voices. It takes a while to get used to this harshness in a forest that's usually wet and velvety and silent. I listen to the clicking of gusts in the high branches and think that winter has come upon us like a fist.

At the base of a spruce tree is a familiar white patch—a scatter of deer bones: ribs, legs, vertebrae, two pelvis bones, and two skulls with half-bleached antlers. I put them here last winter, saying they were for the other animals, to make clear they were not being thoughtlessly wasted. The scavengers soon picked them clean, the deer mice have gnawed them, and eventually they'll be absorbed into the forest again. Koyukon elders say it shows respect, returning animal bones to a clean, wild place instead of throwing them away with trash or discarding them in a garbage dump.

The long, quiet, methodical process of the hunt begins. I move deeper into the forest, ever mindful of treading the edge between protracted, eventless watching and the startling intensity of coming upon an animal, the always unexpected meeting of eyes. A deer could show itself at this moment, in an hour, in several hours, or not at all. Most of hunting is like this—an exercise in patient, isometric endurance and keen, hypnotic concentration. I lift my foot, step ahead, ease it down, wait, step again. Shungnak follows closely, as we work our way through a maze of windfallen trees, across the clear disks of frozen ponds, and around patches of snow beneath openings in the forest canopy. I remind myself there is probably a doe or a buck somewhere in this stretch of woods, perhaps close enough to hear a branch snap or a bough scratch against my clothes. Deep snow has forced the deer off Kluksa Mountain and Crescent Peak, so they're sure to be haunting these lowlands.

We climb a high, steep scarp that levels to a wooded terrace. After pausing to catch my breath, I stand atop a log and peer into the semi-open understory of twiggy bushes, probing each space with my eyes. A downy woodpecker's call sparks from a nearby tree. Several minutes pass. Then a huckleberry branch moves, barely shivers, without the slightest noise, not far ahead.

Amid the scramble of brush where I saw nothing a few minutes ago, a dim shape materializes, as if its own motion had created it. A doe steps into an open space, deep brown in her winter coat, soft and striking and lovely, dwarfed among the great trees, lifting her nose, looking right toward me. For perhaps a minute we're motionless in each other's gaze; then her head jerks to the left, her ears shift back and forth, her tail flicks up, and she turns away in the stylized gait deer always use when alarmed.

Quick as a breath, quiet as a whisper, the doe glides off into the forest. Sometimes when I see a deer this way I know it's real at the moment, but afterward it seems like a daydream.

As we move farther into the woods, I hope for another look at her and think a buck might have been following nearby. Any deer is legal game and I could almost certainly have taken her, but I'd rather wait for a larger buck and let the doe bring on next year's young. Shungnak savors the ghost of her scent that hangs in the still air, but she has vanished.

Farther on, the snow deepens to a continuous cover beneath smaller trees, and we cross several sets of deer tracks, including some big prints with long toe drags. I poke my fingers into one track and feel its edges: still soft and fluffy, with no hint of the crustiness that develops in a few hours when snow is disturbed in cold weather. The powder helps to muffle our steps, but it's hard to see very far because the bushes are heavily loaded. The thicket becomes a lattice of white on black, every branch spangled in a thick fur of jeweled flakes. We move through it like eagles cleaving between tumbled columns of cloud. New siftings occasionally drift down when the treetops are touched by the breeze.

I stop for a while, not to watch for deer so much as to catch my balance in this feathery mosaic of snow, with its distracting beauty and dizzying absence of relief. A Koyukon word keeps running through my mind: *duhnooyh*, "clumps of powdery snow clinging on branches." In the old days, pregnant women drank water melted from this snow, so their children would grow up to be nimble and light-footed. For the same reason, I heard people advise the young boys to drink water melted from surface powder, not from the dense, granular snow, called *tliyh*, which forms underneath during the course of winter. Koyukon elders sometimes told riddles to help teach their children these words, to test their cleverness, and to sharpen their attention to details of the natural world:

> *Wait, I see something: We are sitting all puffed up across from each other, in coats of mountain sheep skin.*
> *Answer:* duhnooyh.

Slots between the trunks ahead shiver with blue where a muskeg opens. I angle toward it, feeling no need to hurry, picking every footstep carefully, stopping often to stare into the jumbled crannies, listening for any splinter of sound, keeping my senses tight and concentrated. A raven calls from high above the forest, and as I catch a glimpse of it the same old questions run through my mind. It lofts and plays on the wind, then folds up and rolls halfway over, a strong sign of hunting luck. Never mind the issue of knowing; I'll assume the power is here and let myself be moved by it.

I turn to look at Shungnak, taking advantage of her sharper hearing and magical sense of smell. She lifts her nose to the fresh but nebulous scent of deer who must have come through here this morning. I watch her little radar ears, waiting for her to focus in one direction and hold it, hoping to see her body tense as it does when something moves nearby. But she only hears the twitching of red squirrels on dry bark. Shungnak and I have a very different opinion of the squirrels. They excite her more than any other animal because she believes she'll catch one someday. But for a hunter, they make distracting spurts of movement and sound, and their sputtering alarm calls alert the deer.

We approach a low, abrupt rise, covered with obscuring brush and curtained with snow. A lift of wind hisses in the high trees, then drops away and leaves us in near-complete silence. I pause to choose a path through a scramble of blueberry bushes and little windfalls ahead, then glance back at Shungnak. She has her eyes and ears fixed toward our left, directly across the current of breeze. She stands very stiff, quivering slightly, leaning forward as if she has already started to run but cannot release her muscles. I shake my finger and look sternly into her eyes as a warning to stay.

I listen as closely as possible, but hear nothing. I work my eyes into every dark crevice and slot among the snowy branches, but see nothing. I stand perfectly still and wait, then look again at Shungnak. Her head turns so slowly I can barely detect the movement, until finally she's looking straight ahead. Perhaps it's just another squirrel. I consider taking a few steps for a better view.

Then I see it.

A long, dark body appears among the bushes, moving up into the wind, so close I can scarcely believe I didn't see it earlier. Without looking away, I gently slide the breech closed and raise the rifle to my shoulder, almost certain that a deer this size will be a buck. Shungnak, now forgotten behind me, must be contorted with the suppressed urge to give chase.

The deer walks silently, determinedly along the little rise, never looking our way. Then he turns straight toward us. Thick tines of his antlers curve over the place where I have the rifle aimed. I remember the Koyukon elders saying that animals come to those who have shown them respect, allowing themselves to be taken, in what is both a physical and spiritual passage. At a moment like this, it's easy to sense that despite my abiding doubt there is an invisible world beyond this one, a world filled with power and awareness, a world that demands recognition and exacts a price from those who ignore it.

It is a very large buck. He comes so quickly that I have no chance to shoot, and then he is so close I haven't the heart to do it. Fifty feet away, the deer lowers his head almost to the ground and lifts a slender branch that blocks his path. Snow shakes onto his neck and clings to the fur of his shoulders as he slips underneath. Then he half lifts his head and keeps coming. I ease the rifle down to watch, wondering how much closer he'll get. Just now he makes a long, soft rutting call, like the bleating of a sheep, except lower pitched and more hollow. His hooves tick against dry twigs hidden by the snow. I can almost feel the breeze blowing against his fur, the chill winnowing down through close-set hairs and touching his skin.

In the middle of a step he raises his head all the way up, and he sees me standing there—a stain against the pure white of the forest, a deadly interloper, the one utterly incongruous thing he has met here in all his life. He reaches his muzzle forward and draws in the affliction of our smell. A sudden spasm stuns him, so sharp and intense it's as if his fright spills out into the forest and tingles inside me like electricity. His front legs jerk apart and he freezes all askew, head high, nostrils flared, coiled and hard. I stare at him and wait, my mind snarled with irreconcilable emotions.

Here is a perfect buck deer. In the Koyukon way, he has come to me; but in my own he has come too close. I am as congealed and transfixed as he is, as devoid of conscious thought. It's as if my mind has ceased to function and only my senses remain.

But the buck has no choice. He instantly unwinds in a burst of ignited energy, springs straight up from the snow, turns in midflight, stabs the frozen earth again, and makes four great bounds off to the left. His thick body seems to float, relieved of its own weight, as if a deer has the power to unbind itself from gravity.

The same deeper impulse that governs the flight of a deer governs the predator's impulse to pursue it. I watch the first leaps without moving a muscle. Then, not pausing for an instant of deliberation, I raise the rifle back to my shoulder, follow the movement of the deer's fleeing form, and wait until he stops to stare back. Almost at that moment, still moving without conscious thought, freed of the ambiguities that held me before, now no less animal than the animal I watch, my hands warm and steady and certain, acting from a more elemental sense than the ones that brought me to this meeting, I carefully align the sights and let go the sudden power.

The gift of the deer falls like a feather in the snow. And the rifle's sound has rolled off through the timber before I hear it.

I walk to the deer, now shaking a bit as accumulated emotions pour through me. Shungnak is already next to it, whining and smelling, racing from one side to the other, stuffing her nose down in snow full of scent. She looks off into the brush, searching back and forth, as if the deer that ran is somewhere else, still running. She tries to lick at the blood that trickles down, but I stop her out of respect for the animal. Then, I suppose to consummate her own frustrated predatory energy, she takes a hard nip at its shoulder, shuns quickly away, and looks back as if she expects it to leap to its feet again.

I whisper thanks to the animal, hoping I might be worthy of it, worthy of carrying on the life it has given, worthy of sharing in the larger life of which the deer and I are a part. Incompatible emotions clash inside me—elation and remorse, excitement and sorrow, gratitude and shame. It's always this way: the sudden encounter with death, the shock that overrides the cushioning of the intellect. I force away the sadness and remember that death is the spark that keeps life itself aflame: these deer we eat from, and the fish, and the plants that die to feed us.

It takes a few minutes before I settle down enough to begin the other work. Then, I tie a length of rope onto the forelegs, run it over a low branch, back down through a loop in the rope, and up over the branch again like a double pulley, so I can raise the animal above the ground. This done, I cut the dark, pungent scent glands from its hind legs, to prevent their secretions from tainting the meat. Next, I make a small incision through the belly skin, insert my hand to shield the knife blade from

311

the distended stomach, and slice upward to make an opening about a foot long. Reaching inside, I loosen the stomach and intestines, then work them out through the incision, pulling carefully to avoid tearing the thin membranes and spilling stomach contents into the body cavity. The deer's inward parts feel very hot, slippery, and wet, as I suppose my own would if I could ever touch them. Finally the viscera slide out onto the ground: soft, bladderlike stomach and flaccid ribbons of intestine; a gray, shining mound, webbed with networks of veins and lacy fat, steaming into the cold, saturating the air with a rich odor of plant mulch and body fluids.

Next, I roll up my jacket sleeve and thrust my arm deep inside the deer, until I feel the diaphragm, a sheet of muscle that separates the abdomen from the chest. When I slice through it, a thick, hot rush of blood flows down my arm and sloshes into the vacant belly. There is a hollow, tearing sound as I pull the lungs free; and reaching up inside the chest, I can feel the firm, softball-sized muscle of the heart. The lungs are marbled creamy-pink and feel like soft, airy sponge. As I lay them beside the other organs, I whisper that these parts are left here as food for the animals. Shungnak wants to take some for herself but I make her stay away. Koyukon elders say the sensitivity and awareness leave an animal's remains slowly, and there are rules about what should be eaten by a dog. Shungnak will have her share of the scraps later on, when more of the life is gone.

The inside of the deer is now empty, except for the heart and the dark-purple liver, which I've left attached. I tie a short piece of cord around the end of the lower intestine to keep the remaining blood from flowing out when I carry the animal on my back. Then I poke a series of holes in the hide along either side of the belly incision and lace it shut with another cord. After lowering the deer onto the ground, I cut through the "knee" joints of the forelegs, leaving them attached by a stout tendon, then slice a hole in the hock—a space between the bone and tendon of the hind leg—and I toggle the forelegs through these openings. This way I can put my arms through the joined legs and carry the deer like a pack—not a trick to be used if there is the slightest chance another hunter might be around and mistake my burden for a live animal.

I barely have enough strength to lift the buck and trudge along, but there is plenty of time to work back toward the beach, stopping occasionally to rest and cool down. During one of these breaks, I hear two ravens in an agitated exchange of croaks and gurgles, and I wonder if those black eyes have already spotted the remnants. No pure philanthropist, if Raven gave this luck to me, it was only to create luck for himself. I remember how difficult it was, at first, to accept the idea of a sanctified creature having such a contradictory personality. The Raven described by elders like Grandpa William was both good and evil, sage and fool, benefactor and thief—embodiment of the human paradox. When Joe Stevens described an American president of dubious character, he said, "Just like Raven."

Half an hour later, sweating and exhausted, I push through the low boughs of the beachside trees, lay the animal down, and find a comfortable seat on the driftwood. Afternoon sun throbs off the water, but the north wind takes every hint of warmth out of it. Little gusts splay dark patterns across the anchorage; the boat paces on its mooring line; the strait races with whitecaps. I take a good rest, watching a fox sparrow flit among the alders and a bunch of crows hassle over some bit of food near the water's edge. At this low tide, Sea Lion Point has expanded to a flat sill of rock reaching out several hundred yards from the island's shore. The point has such scant relief that higher tides reduce it to a fraction of this size. The anchorage is nothing more than a gouge in the rocks, closely rimmed with breakers and jagged boulders, so it's only accessible to small skiffs whose pilots are either reckless or foolish. Despite its barren appearance, Sea Lion Point has extensive tide flats, ponds, and beds of estuarine grass that attract congregations of birds, especially during the spring and fall migrations.

Today, hundreds of gulls have gathered on the outer reaches of the point, all sitting with their beaks into the wind. They appear sluggish and languid, as if their sole purpose is to huddle together against the chill. But they're also keeping watch. When the breeze slacks to a momentary calm, a black foil sweeps out from the forest's edge. The eagle leans sharply down, half folds its wings, banks toward the gulls, and builds speed, falling and blurred and sinister. Gulls and crows swirl up like a handful of salt and pepper thrown into the wind. Clusters of ducks spray off the water in opposite directions. Shorebirds dazzle over the tangled skeins of kelp. A close formation of oystercatchers babbles across the anchorage in front of us.

The eagle shears through the scattering swarm, looking ponderous and clumsy, oddly outclassed by its darting prey. Carried into a steep climb by its momentum, the eagle swings around and drops again, legs dangling, unsheathed talons gaping in the frosted air. But the birds have whirled away, leaving an empty void like the eye of a storm. Its voice mingles with the cries and wails of the gulls, a shrill complaint amid easy laughter. Finally the eagle flaps off to a high perch. Swaying back and forth, jerking and flexing its wings for balance, it watches the crows dwindle away over the rocks, the gulls float down onto the flats again. All of the birds seem calm and unhurried, as if nothing of significance has happened, as if the whole thing has been only a game. The hoary quiet of winter returns, and the wait begins once more.

Though I feel satisfied, grateful, and contented sitting here, much remains to be done, and at this time of year the daylight ebbs quickly. Hunters are allowed more than one deer, so I'll stay on the island and take another look around tomorrow. As we idle from the anchorage, we pass within a few yards of lovely surf peeling across a smooth, triangular reef. If I had a surfboard and wetsuit it would be impossible to resist, no matter how frigid the air and water might be. I stop to watch a few waves pour over the shoals like liquid silver; then I follow the shore toward Bear Creek. By the time I've anchored and unloaded the boat, the wind has diminished and a growing winter chill sinks down in the pitched, hard shadow of Kluksa Mountain.

Bear Creek cabin is situated in a thicket well back from shore, hidden in summer but easily seen once the leaves have fallen. I split some half-dry wood, which hisses and sputters in the rusty stove, then reluctantly gives way to flames. After the fire starts crackling, I walk down to the creek. Dipping a bucket into a clear pool, I notice a few salmon bones scattered among the rocks and pebbles. I'm surprised to see them, but also surprised that so little would remain from the hordes of fish I watched here this fall. I had a similar feeling recently, when I went looking for the sperm whale carcass that beached last summer near Tsandaku Point. At first it seemed the storm swells had washed it away, but then I found a bare vertebra and a rib among the rocks. Eventually, I came across the skull—about ten feet long and weighing hundreds of pounds—half buried in the driftwood. Six months after the whale came ashore, scavengers and decay had taken every bit of flesh, gnawed or carried off the smaller bones, and left only a few fragments to wash in the surge.

After fetching water, I carry the deer inside the cabin and hang it from a low beam. Better to work on it now than wait, in case tomorrow brings more luck. The animal is dimly lit by a kerosene lantern and a blush of daylight through the windows. I feel strange in its presence, as if it still watches, still glows with something of its life, still demands that nothing be done carelessly and no offensive words be spoken in its presence. Grandpa William told me that a hunter should never let himself be deluded by pride or a false sense of dominance. It's not through his own power that a person takes life in nature, but through the power of nature that life is given to him.

After sharpening the knife, I slit the deer's skin along the whole length of its underside and down each leg to the hoof. Then I peel the soft hide away, using the blade to separate it from the muscles underneath, gradually revealing the inner perfection of the deer's body. When the skinning is finished, I follow an orderly sequence, cutting through crisp cartilage, severing the leg joints, brisket, ribs, vertebrae, and pelvis, following the body's own design to disarticulate bone from bone. Everything comes apart smoothly and easily, as deer becomes meat, animal becomes food, the most vital and fundamental transformation in all of living existence. There is no ugliness in it, only hands moving in concert with the beauty of an animal's shape. While I work with the deer, it's as if something has already begun to flow into me. I couldn't have understood this when I was younger and had yet to experience the process of one life being passed on to another.

Before I lived with the Eskimo people, I had never hunted and had never seen how game is prepared. But I was immediately fascinated by their skill at taking an animal into its component parts. The Eskimos always watched me closely and found my mistakes entertaining. If I did something uncharacteristically well, someone was likely to look bemused and declare: "Accident." They were passionate hunters and incredibly hard workers. When they hunted walrus, it took only a short while to stalk the animals but many hours to butcher them. As we pulled the skin-covered boat onto the ice, someone was sure to say, "Well, the excitement's over. Now it's time for the real work." But somehow, it never seemed like work to me, this deeply engaged process of learning about animals from the inside and out, of binding my own existence more closely to the lives that sustained me.

By the time I went to live with Koyukon people, I could skin and butcher animals; but I knew little about the delicate matter of keeping a right mind while working with them. Sarah and Joe Stevens were especially scrupulous about treating each animal as a sentient being and butchering it according to the traditional pattern, which was not only a technique but also a ritual of respect. They made certain that no usable part was wasted or tossed carelessly aside, that the meat was covered to keep dogs and scavengers away, and that it was well cached so nothing would spoil. Once, I met Sarah carrying a platter of meat to her neighbor's house, with a piece of cloth over it. She explained, "It wouldn't be right to leave this open to the air, like it doesn't mean anything." In this and other ways, she treated meat as a sacred substance, a medium of interchange between herself and the empowered world in which she lived. It seemed that everything she did in relationship to nature was both an activity and a prayer.

When I've finished with the deer, I put two slices from the hindquarter in a pan atop the stove. Scraps of meat and fat boil in a separate pot for Shungnak. She whines impatiently, perhaps remembering her sled dog days, when she lived mostly on meat and fish and bones. As soon as she's been fed, I sit on a sawed log and eat venison straight from the pan. No meal could be simpler, more satisfying, or more directly a part of the living process. I also savor a deep feeling of security in having this meat, bringing it home to freeze or can for the year ahead—pure food, taken from a clean, wild place, and prepared by our own efforts. There is a special intimacy in living directly from nature, nourishing my body from the same wildness that so elevates my spirit.

I wish Ethan were here to share this meal, so I could explain to him again that when we eat the deer, its flesh becomes our flesh. Each time we eat we should remember the deer and feel gratitude for what it has given us. And each time, we should carry a thought like a prayer inside: "Thanks to the animal and to all that made it—the island and the forest, the air, and the rain . . ." I would tell Ethan that in the course of things, he and Nita and I are all generations of deer and of the life that feeds us. Like the deer, we also come from the island, and from the earth that sustains us and gives us breath.

Later, perched atop rocks near the mouth of Bear Creek, Shungnak and I look out over Haida Strait to the sea beyond. A distant winter sun sprawls against the horizon, thins to a mound of shivering flame, and drowns itself in the cold Pacific. The sky fades to violet, darkens, and relaxes, like a face losing expression at the edge of sleep. Silence hovers in the brittle woods.

A great blue heron glides down into the anchorage cove and stands motionless in the shallows, like the shadow of a pterodactyl against the Mesozoic sky. Every few minutes I notice the bird's stance and position have changed, but invisibly, like a clock's hands, so that I never actually see its legs move. Then I notice its head slowly lowering, its body tilting, its neck stretching forward. Suddenly it flashes out and draws back, and a fish wriggles on the dripping spear of its beak. The recoiling heron stands erect, flips the fish lengthwise, gulps it, and resumes hunting. Over the next few minutes, the bulge of the fish gradually moves down its serpentine neck.

I've watched herons many times before, admiring them as elegantly plumed, primeval works of art. But I never thought about their impeccable skill and patience as hunters. This event gives me a better sense of the way they live—the measured and timeless stalks, the penetrating eyes fixed at the water's edge, the shadows of prey moving below, the saber beak striking down, the sudden consummation of predatory impulse. Given a choice of birds, I would be a heron, or an owl, a falcon, an eagle. I love these quick, canny animals, perhaps because they seem closest to my own kind. To feel otherwise about predators would be like shrinking from the face in the mirror.

Dusk settles on the waters of Haida Strait, swallows the far peaks and inlets, drifts down through the surrounding forest, takes the island inside itself, and joins it with the sky. Sitting in the darkness, I feel overcome with gratitude and wish for a way to express it. Words seem frail and empty; offerings seem foreign and artificial. Perhaps just being here is enough, becoming wholly engaged with this place, touching it, eating from it, winding my life as tightly as possible into it. The island and I, turning ourselves ever more inside out.

Warm in my sleeping bag, I let the fire ebb to coals. The lamp is out. The cabin roof creaks in the growing chill. I drift toward sleep, pleased that there is no moon, so the deer will wait until dawn to feed. On the floor beside me, Shungnak jerks and whimpers in her dog's dreams.

Dawn. The cold fire of winter sun climbs a pallid wall of sky. Mountains stand out as sharp and clear as the sound of shattering glass. Clouds of steam rise above the open riffles of Bear Creek. The silver calm of Haida Strait is splotched with dark blue where an uncertain breeze touches down against it. Three goldeneye ducks drift in the anchorage, like smudges on a sheet of polished iron.

The temperature is twenty degrees, perhaps much colder back away from shore. Although it rarely drops to zero along this coast, sea humidity and gusty winds often intensify the chill. But even so, our winter is a far cry from that of Koyukon country, where temperatures average below zero for five months of the year and may hover at forty to sixty below for weeks. Not surprisingly, Koyukon elders treat cold weather as a conscious thing, with a potent and irritable spirit. They warn the younger ones to speak carefully about cold, lest they incite its frigid wrath. In the old days, children were even told not to throw snowballs, because the frivolity or annoyance could bring on bitter weather.

When Shungnak was born it was so cold that one of her littermates froze stiff. Thawed out behind the wood stove, he survived, although his tail eventually fell off. Perhaps because she grew up in that climate, and because the frozen landscape meant freedom and adventure for a sled dog, Shungnak still loves winter weather. As we walk back to the cabin, she prances around me, full of excited energy, anxious to get started with the day.

An hour later, I anchor the boat at Sea Lion Point. After paddling the punt ashore, I follow Shungnak to where our tracks from yesterday enter the woods. Just beyond the place of the buck, a pair of does drift at the edge of sight and disappear. For an hour we angle north, then come slowly back deeper into the woods, moving crosswise to a growing easterly breeze. In two places, deer snort and pound away, invisible beyond a shroud of brush. Otherwise there is nothing.

We keep on in the same direction, probing first through snowy thickets, then through heavy forest with bare, frozen moss underneath. In a dense maze of young spruce, I come face to face with a red squirrel, clinging to the trunk of a dead tree. Luckily, Shungnak has lagged a few yards behind. And instead of scurrying to a high branch, the squirrel stays put, bold, curious and confident, apparently unconcerned about my intentions. I inch ahead, wait, then move again, until he's so close I could ruffle his fur if I blew hard enough.

The squirrel twitches this way and that on his skinny white tree, first head up, then head down, leaning out as if to get a closer look at me. He sticks effortlessly to the smooth wood, or actually hangs from it by the tips of his curved claws. Never satisfied to simply observe, I wonder how his claws can possibly be so sharp, and what keeps them from getting dull? The squirrel spends a long minute checking me out, constantly in motion, scratching up the tree and back down again, jerking from one angle to another, stitching his little feet, shaking his frizzy tail, shivering his long black whiskers. Then he jumps to a spruce just as close but with a slightly different angle. I can see the crenulations of his nose, the fine hairs on his snout, the quick pumping of his ribs, and my face reflecting on his bright indigo eye. When he's seen enough, he turns and jitters to a place above my head. I can tell he's ready to burst into a chatter, as squirrels often do after some deliberation, so I edge past and leave him alone.

A short while later we follow a familiar stretch of trail through a copse of shore pines and cedars. I kneel down to examine a bunch of deer bones in the snowless patch under a tree. Darkened by age and half covered with moss, they're hardly visible any more. I first came across them several years ago, when they still had a blanched white color, with bits of clinging skin, cartilage, and tufts of fur. It looked as if the deer had died only a few months before, and because the nearby tree trunk was heavily clawed, I guessed a bear had either killed the animal or scavenged its carcass. I always looked at the bones when I passed by, but never touched them because I wanted to see how long it took an animal's remains to vanish from the forest floor. Each year, a few more bones were missing or were cloaked over by the moss.

Last summer, I walked through here with a friend and showed him the bones. He touched several of them and pulled one out from the moss. Both of us were stunned by what he found: a hind-leg bone that had been fractured in several places while the deer was alive. It was so badly shattered that a piece the thickness of my index finger had stuck out almost two inches from the wound, as indicated by a line where healing flesh had closed around it. The deer must have lived a long time after its terrible injury. Long enough so the fragments knitted themselves together, as if liquid bone had seeped into the wound and solidified as a porous, bulging, convoluted mass. Though gnarled and misshapen, the fused bone seemed almost as strong as a healthy one. But the deer's leg was considerably shortened and had a hollow ivory splinter piercing out from it. As we turned the bone in our hands, I marveled at the determination of living things, and of life itself, to carry on, to mend, and to become whole again after being torn apart.

What could have caused such a wound? It might have been a bad fall, an unskilled hunter's bullet, or a bear. Hardest of all to imagine was the agony this deer went through, the days and weeks of unrelievable pain, endured in solitude, through nights of rain and storm, burdened by the omnipresent danger of being discovered by a bear. Of course, it might have been another bear that eventually killed the animal. After my friend and I left the bones, the forest seemed less beautiful for a while, less a place of shelter than of violence and tragedy. At that same moment, some other animal was probably suffering toward death not far away—perhaps severed by an eagle's beak or broken by a bear, perhaps old and weakened, perhaps riven with disease—biting the moss in torment and fear. I thought, there is little mercy in nature, little to relieve the pain or loneliness of death.

Many of the tragedies found in the human world are also found here. Then I realized that loving nature meant loving it all, accepting nature exactly as it is, not idealizing it or ignoring the hard truths, not reducing it to an imaginary world of peace and perfection. How could I crave the beauty of the flame without accepting the heat that made it?

Shortly after noon we come into a narrow muskeg with scattered shore pines and a ragged edge of brushy, low-growing cedar. I squint against the sharp glare of snow. It has that peculiar look of old powder, a bit settled and touched by wind, very lovely but without the airy magic of a fresh fall. I gaze up the muskeg's easy slope. Above the encroaching wall of timber, seamed against the deep blue sky, is the peak of Kluksa Mountain, with a great plume of snow streaming off in what must be a shuddering gale. It has a contradictory look of absoluteness and unreality about it, like a Himalayan summit suspended in midair over the saddle of a low ridge.

I move slowly up the muskeg's east side, away from the breeze and in the sun's full warmth. Deer tracks crisscross the opening, but none of the animals stopped here to feed. Next to the bordering trees, the tracks join as a single, hard-packed trail, showing the deer's preference for cover. Shungnak keeps her nose to the thickly scented snow. We come across a pine sapling that a buck has assaulted with his antlers, scattering twigs and flakes of bark all around. But his tracks are hardened, frosted, and lack sharpness, indicating they're at least a day old.

We slip through a point of trees, then follow the edge again, pausing long moments between footsteps. A mixed tinkle of crossbills and siskins moves through the high timber, and a squirrel rattles from deep in the woods, too far off to be scolding us. Shungnak picks up a strong ribbon of scent, but she hears nothing. I stop for a few minutes to study the muskeg's raveled fringe, the tangle of shade and thicket, the glaze of mantled boughs.

Then my eye barely catches a fleck of movement up ahead, near the ground and almost hidden behind the trunk of a leaning pine—perhaps a squirrel's tail or a bird. I slowly lift my hand to shade the sun, stand dead still, and wait to see if something is there. Finally it moves again.

At the very edge of the trees, almost out of sight in a little swale, small and furry and bright-tinged, turning one direction and then another, is the funnel of a single ear. Having seen this, I soon make out the other ear and the slope of a doe's forehead. Her neck is behind the leaning pine, but on the other side I can barely see the soft, dark curve of her back above the snow. She is comfortably bedded, gazing placidly into the distance, chewing her cud.

Shungnak has stopped twenty yards behind me in the point of trees and has no idea about the deer. I shake my finger at her until she lays her ears back and sits. Then I watch the doe again. She is fifty yards ahead, ten yards beyond the leaning tree, and still looking off at an angle. Her left eye is visible and she refuses to turn her head away, so it might be impossible to get closer. Perhaps I should just wait here, in case a buck is attending her nearby. But however improbable it might be under these circumstances, a thought is lodged in my mind: I can get near her.

My first step sinks down softly, but the second makes a loud budging sound, like stepping on a piece of toast. She snaps my way, stops chewing, and stares for several minutes. It seems hopeless, especially out here in an open field of crispy snow with only the narrow tree trunk for a screen. But she turns away and starts to chew again. I move just enough so the tree blocks her eye and the rest of her head, but I can still see her ears. Every time she chews they shake just a bit, so I watch them and step when her hearing is obscured by the sound of her own jaws.

Either this works or the deer has decided to ignore me, because after a while I've come close enough so the noise of my feet has to reach her easily. She should have jumped up and run long ago, but instead she lies there in serene repose. I deliberate on every step, try for the softest snow, wait long minutes before the next move, stalking like a cat toward ambush. I watch beyond her, into the surrounding shadows and across to the muskeg's farther edge, for the shape of a buck deer; but there is nothing. I feel ponderous, clumsy-footed, out of place, inimical. I should turn and run away, take fear on the deer's behalf, flee the mirrored image in my mind. But I clutch the cold rifle at my side and creep closer.

The wind refuses to blow and my footsteps seem like thunder in the still sunshine. But the doe only turns once to look my way, without even pointing her ears toward me, then stares off and begins to chew again.

I am ten feet from the leaning tree. My heart pounds so hard I think those enchanted ears should hear the blood rushing in my temples. Yet a strange assurance has come into me, a quite unmystical confidence. Perhaps she has decided I am another deer, a buck attracted by her musk or a doe feeding gradually toward her. My slow pace and lapses of stillness would not seem human. For myself, I have lost awareness of time; I have no feeling of patience or impatience. It's as if the deer has moved slowly toward me on a cloud of snow, and I am adrift in the pure motion of experience.

I take the last step to the trunk of the leaning pine. It's bare of branches, scarcely wider than my outstretched hand, but perfectly placed to break my odd profile. There is no hope of getting any closer, so I slowly poke my head out to watch. She has an ideal spot: screened from the wind, warmed by the sun, and with a clear view of the muskeg. I can see muscles working beneath the close fur of her jaw, the rise and fall of her side each time she breathes, the shining edge of her ebony eye.

I hold absolutely still, but her body begins to

stiffen, she lifts her head higher, and her ears twitch anxiously. Then instead of looking at me she turns her face to the woods, shifting her ears toward a sound I cannot hear. A few seconds later, the unmistakable voice of a buck drifts up, strangely disembodied, as if it comes from somewhere underneath the snow. I huddle as close to the tree as I can, press against the hard dry bark, and peek around its edge.

There is a gentle rise behind the doe, scattered with sapling pines and bushy juniper. A rhythmic crunching of snow comes invisibly from the slope, then a bough shakes . . . and a buck walks easily into the open sunshine.

Focusing completely on the doe, he comes straight to her and never sees my intrusive shape just beyond. He slips through a patch of small trees, stops a few feet from where she lies, lowers his head and stretches it toward her, then holds this odd pose for a long moment. She reaches her muzzle to one side, trying to find his scent. When he moves up behind her she stands quickly, bends her body into a strange sideways arc, and stares back at him. A moment later she walks off a bit, lifts her tail, and puts droppings in her tracks. The buck moves to the warm ground of her bed and lowers his nose to the place where her female scent is strongest.

Inching like a reptile on a cold rock, I have stepped out from the tree and let my whole menacing profile become visible. The deer are thirty feet away and stand well apart, so they can both see me easily. I am a hunter hovering near his prey and a watcher craving inhuman love, torn between the deepest impulses, hot and shallow-breathed and seething with unreconciled intent, hidden from opened eyes that look into the nimbus of sun and see nothing but the shadow they have chosen for themselves. In this shadow now, the hunter has vanished and only the watcher remains.

Drawn by the honey of the doe's scent, the buck steps quickly toward her. And now the most extraordinary thing happens. The doe turns away from him and walks straight for me. There is no hesitation, only a wild deer coming along the trail of hardened snow where the other deer have passed, the trail in which I stand at this moment. She raises her head, looks at me, and steps without pausing.

My existence is reduced to a pair of eyes; a rush of unbearable heat flushes through my cheeks; and a sense of absolute certainty fuses in my mind.

The snow blazes so brightly that my head aches. The deer is a dark form growing larger. I look up at the buck, half embarrassed, as if to apologize that she's chosen me over him. He stares at her for a moment, turns to follow, then stops and watches anxiously. I am struck by how gently her hooves touch the trail, how little sound they make as she steps, how thick the fur is on her flank and shoulder, how unfathomable her eyes look. I am consumed with a sense of her perfect elegance in the brilliant light. And then I am lost again in the whirling intensity of experience.

The doe is now ten feet from me. She never pauses or looks away. Her feet punch down mechanically into the snow, coming closer and closer, until they are less than a yard from my own. Then she stops, stretches her neck calmly toward me, and lifts her nose.

There is not the slightest question in my mind, as if this was sure to happen and I have known all along exactly what to do. I slowly raise my hand and reach out.

And my fingers touch the soft, dry, gently needling fur on top of the deer's head, and press down to the living warmth of flesh underneath.

She makes no move and shows no fear, but I can feel the flaming strength and tension that flow in her wild body as in no other animal I have touched. Time expands and I am suspended in the clear reality of the moment.

Then, by the flawed conditioning of a lifetime among fearless domesticated things, I instinctively drop my hand and let the deer smell it. Her black nose, wet and shining, touches gently against my skin at the exact instant I realize the absoluteness of my error. And a tremor runs through her entire body as she realizes hers. Her muscles seize and harden; she seems to wrench her eyes away from me but her body remains, rigid and paralyzed. Having been deceived by her other senses, she keeps her nose tight against my hand for one more moment.

Then all the energy inside her triggers in a series of exquisite bounds. She flings out over the hummocks of snow-covered moss, suspended in effortless flight like fog blown over the muskeg in a gale. Her body leaps with such power that the muscles should twang aloud like a bowstring; the earth should shudder and drum; but I hear no sound. In the center of the muskeg she stops to look back, as if to confirm what must seem impossible. The buck follows in more earthbound undulations; they dance away together; and I am left in the meeting place alone.

There is a blur of rushing feet behind me. No longer able to restrain herself, Shungnak dashes past, buries her nose in the soft tracks, and then looks back to ask if we can run after them. I had completely forgotten her, sitting near enough to watch the whole encounter, somehow resisting what must have been a prodigious urge to explode in chase. When I reached out to hug her, she smells the hand that touched the deer. And it seems as if it happened long ago.

I walk slowly from the spot, letting the whole event roll through my mind again and again, remembering the dream that began many months ago, that I might someday touch a deer. After trying and failing with the naive little fawn earlier this fall, I'd begun to think the idea was farfetched, perhaps even foolish. But now, totally unexpected and in a strange way, it has happened. Was the deer caught by some reckless twinge of curiosity? Had she never encountered a human on this wild island? Did she yield to some odd amorous confusion? Then I realize I truly do not care. I would rather accept this as pure experience and not give in to a notion that everything should be explained.

Koyukon elders simply accept what comes to them. They teach that everything in the natural world has its own spirit and awareness, and they give themselves to that other world, without expecting voices, without waiting for visions, without seeking admission to the hidden realms.

I am reminded of something that happened the last time I hunted with Grandpa William. While we sat talking at the edge of a meadow, an unusual bird started singing and chattering in a nearby treetop. At first it looked like a small hawk, but there was something different about its color and shape. When I asked what it was, he listened closely to its calls, then took my binoculars and watched it for a long while, intrigued and perplexed. "I don't know," he muttered, mostly to himself; then he suggested a difficult Koyukon name I'd never heard before. Shortly, his interest darkened to concern: was the arrival of this strange bird a sign, an omen?

Suddenly he began addressing the bird at length in the Koyukon language, speaking in a soft, gentle voice. "Who are you," he wondered, "and what are you saying to us?" He walked out into the meadow, still talking, still trying to establish that the loquacious bird was something ordinary, not an ominous stranger. "Wish us good luck, whoever you are," he said. "Wish us well, and surround us—your grandchildren— within a circle of protection." By this time I'd lost interest in identifying the bird, and my whole attention was focused on Grandpa William: a man imploring mercy and protection from a bird, addressing a feathered emissary in a treetop.

Those moments epitomized everything I had learned from Koyukon people, everything they had tried to tell me about living in a natural world filled with spirit and power. I've had few experiences that so moved me. For how many thousand generations, I wondered, have people spoken and prayed to the natural beings around them, as a customary part of daily life? At any other time in human history, this event would be as ordinary as talking to another person. To me, Grandpa William represented the universal man beseeching the powers that pervade his living world, powers so recently forgotten among my own people. More than anything else, I wished it had seemed quite unremarkable for me, wished my ancestors hadn't forsaken what Grandpa William still understood.

Neither Grandpa William nor I ever knew what that bird was, though I later concluded it must be a young northern shrike. And if the bird did carry an omen, who was it for?

I stop in the shadows along the muskeg's upper edge, and think back over the years with Koyukon people. What stands out for me at this moment is a special wisdom of their tradition—to expect nothing of nature, but to humbly receive its mystery, beauty, food, and life. In return, Koyukon people show the same respect toward nature that is shown toward humans, acknowledging that spirit and sacredness pervade all things. If I understand correctly, their behavior toward nature is ordered around a few simple principles: Move slowly, stay quiet, watch carefully, be ever humble, show no hint of arrogance or disrespect. And if they follow one overarching commandment, it is to approach all life, of which humans are a part, with humility and restraint. All things are among the chosen.

As I reflect on the experiences of yesterday and today, I find an important lesson in them, viewed in the light of wisdom taken from the earth and shaped by generations of elders. Two deer came and gave the choices to me. One deer I took and we will now share a single body. The other deer I touched and we will now share that moment. These events could be seen as opposites, but perhaps they are identical. Both are founded on the same principles, the same relationship, the same reciprocity. Both are the same kind of gift.

Koyukon elders would explain, in words quite different from my own, that I moved into two moments of grace, or what they would call luck. This is the source of success for a hunter or a watcher; not skill, not cleverness, not guile. Something is only given in nature, never taken.

Well soaked and shivering from a rough trip across the Sound, we pull into the dark waters of Anchor Bay. Sunset burns on the spindled peak of Antler Mountain. The little house is warm with lights that shimmer on the calm near shore. I see Nita looking from the window and Ethan dashes out to wait by the tide, pitching rocks at the mooring buoy. He strains to see inside the boat, knowing that a hunter who tells his news aloud might offend the animals by sounding boastful. But when he sees the deer his excited voice seems to roll up and down the mountainside.

He runs for the house with Shungnak, carrying a load of gear, and I know he'll burst inside with the news. Ethan, joyous and alive, boy made of deer.

V

TESTIMONY

"How can I invite the heart of mountains and streams into myself? How can I carry the forest into the city? I know from my own life this is a matter of survival, not aesthetics."

KIM STAFFORD, *ENTERING THE GROVE,* 1990

STEPHEN TRIMBLE

"The landscape interests me not as an object of research, but as a home," says Stephen Trimble. Born in Denver in 1950, Trimble roamed the West with his family during childhood, with his field geologist father as guide. In the last year of a liberal arts education at Colorado College, Trimble read Ed Abbey's *Desert Solitaire*, "transforming my notion of what could be done with what I loved most—poking around in the West." Within a year, he was working as a park ranger—just like Abbey—at Arches National Park. During several seasons as a ranger/naturalist in Utah and Colorado, Trimble served his apprenticeship as a writer and photographer, publishing small interpretive books about national parks. In 1979, he wrote his "first book long enough to bind with a spine," *The Bright Edge: A Guide to the National Parks of the Colorado Plateau*.

After completing a master's degree in ecology and a stint as director of the Museum of Northern Arizona Press ("my only full-time permanent job ever"), Trimble began

328

free-lancing full-time in 1981. He has continued to both write and photograph. Trimble says, "I grow restless if I do too much uninterrupted writing—I want to go out and have the simple and joyous emotional recharge of taking pictures. At the same time, I feel understimulated and anchorless if I am constantly in the field and cannot get back to my desk and tinker with words.

"I would be a less effective writer if I gave up photography; I would lose a special visual insight. And the words and research and verbal knowledge inform my photographs with a storytelling dimension they would otherwise lack."

Trimble sees the progression of his books as an effort to understand his home, layer by layer. He began with the land, with books about the Colorado Plateau, the Rocky Mountains, and the Ansel Adams Award-winning *The Sagebrush Ocean: A Natural History of the Great Basin* (1989). For five years in the 1980s, he lived in Santa Fe, New Mexico, exploring a new layer: the world of Southwest Indian people. Trimble says, "I am not a trained ethnographer; I see myself as a listener." His "listening" led to a book about Pueblo pottery, a children's book, and culminated with a major work, *The People: Indians of the American Southwest* (1993). With each project, he has extended the boundaries of his home.

Trimble now lives in Salt Lake City with his wife and two children. His exploration of what his coauthor, Gary Nabhan, calls "the natural history of fatherhood" led to their book (from which this essay comes), *The Geography of Childhood: Why Children Need Wild Places* (1994). "Now," he says, "I am hard at work on a novel. Maybe, just maybe, I am old enough and aware enough to be able to penetrate that next layer of understanding and *invent* a revealing story."

The Scripture of Maps, the Names of Trees

Never Summer Range, Snowy Range, Wind River Range, Horse Heaven Hills. Rabbit Ears Pass, Togwotee Pass, Lolo Pass, Chinook Pass. Longs Peak, Pikes Peak, Grand Teton, Mount Rainier. Some people remember from childhood the names of cherished baseball or football players. Others can recite still the multisyllabic names of molded plastic dinosaurs. My own remembered litany consists of place-names.

"Stevie, where do we turn?" my father would ask as I stared at the land charted in the folds of paper on my lap. My father, of course, knew which road to take at the next town. He also knew that I loved being trusted to help find our way.

During my childhood in the fifties and sixties, the best highway maps of the western states came from Chevron stations, in places like Rawlins, Wyoming, and Pendleton, Oregon. Stacked in wire racks coated with dust cemented in automotive grease, the maps waited for discovery, state by state. Down the road, I studied the air-brushed mountain ranges and passes and asked my father if I had them right: "Are those the Absarokas?" "The next bridge ought to be the Sweetwater River; tell me when we get there." He pointed out Big Southern Butte and Twin Buttes, landmarks of the Snake River Plain, and I tried to find them on the map.

I made lists, tallying the states I had passed through, the national parks I had visited. I searched the maps for the little red squares that marked "Points of Interest"; my questions about these cued my father's steering-wheel lectures about western history and geology. "Maryhill Museum: what's there, Daddy?" "What happened at Big Hole Battlefield?" "The map shows Crystal Ice Cave ten miles off the highway. Do we have time to go?" From my father, I learned about the Lake Bonneville flood, the differences between gneiss and schist, the route of Lewis and Clark down the Clearwater, and how Chief Tommy Thompson used to fish for salmon at Celilo Falls on the Columbia River before the dams.

My father, Don Trimble, worked as a field geologist for the U.S. Geological Survey for more than thirty years. Every summer, our family left home in Denver to spend the three months between school terms in a town near his field assignment, renting a house near whatever quadrangle he was mapping. In my infancy and early childhood, his mapping projects circled Portland, Oregon. In summers from second grade through high school, we lived in southeastern Idaho, in Aberdeen and in Pocatello.

Setting out across the western states, my mother drove our Dodge, my father the government Jeep. When I rode with my mother, we looked for music on the AM radio, hoping for jazz. When I rode with my father, he told me stories. When we all rode together during vacations, we alternated between these diversions. Boxes of gear for the summer's field season filled the rear of the Jeep and the trunk of the car: dishes, clothes, cameras, map cases, a Brunton compass, rock sample sacks sewn from white canvas and permanently scented with acrid basin and range alkali dust, my bicycle (and, once, in the rear window, my pet-store turtle in a Skippy peanut-butter jar, forgotten and inadvertently boiled when we stopped for lunch one day).

Connection to the natural world can begin with snakes, shells, or stars, birds, beetles, or blackberries. For me, connection started with the land itself, the bones and ligaments of the naked Earth exposed on the rocky surface of the arid West. Geography seeped into me, a bedrock awareness of landscape and place. I learned to pay attention to the flow of scenes framed by the car window. The relatively slow pace and the familiar earthbound perspective of driving made that progression of landscapes comprehensible in a way that today's commonplace airplane travel prohibits.

And I kept staring at the maps. The arbitrarily legislated shapes of national parks printed on them defined the places visually before I ever saw their vistas and wildlife. Today, I close my eyes and conjure these shapes, connected by the meandering red and blue lines of preinterstate highways, and I can feel the sweaty folds of paper in one hand, the hot rush of air beating and lifting against the other, stretched outward through the open window of the old Dodge.

My childhood in the West—where geology overwhelms biology, lightly vegetated landscape commands attention, and weather is intense—surely gave me an edge in mental map-making. Research suggests that we have map-making genes strung along our DNA, promoting our ability to integrate and organize our experiences of geographic space. Such mental map-making skills clearly gave our hunter-gatherer ancestors an evolutionary advantage.

In researching human development of such cognitive maps, a University of Pittsburgh team of psychologists emphasizes that children and adults begin their descriptions of environments with landmarks. Recognizing landmarks comes, on average, with the full development of the brain, after about four years of age. Route-finding, remembering a sequence of landmark relationships, comes later. Experience with my own children belies this model: at one and a half, my son chanted, "Mama, mama," as soon as we came within five blocks of home on our return from the baby-sitter at the end of the day. At two, his sister delightedly spotted her personal landmarks on our drives around town: "There's Kevin's office [our pediatrician]. That's Mom's old work. There's Bartie's house."

Many studies indicate that accurate mental map-making increases with active participation—with walking through an environment. Tying together a sequence of places works even better if those sites are connected by stories. One's own stories surely work best—the ones we create while walking to school or exploring a neighborhood gully. Even pre-schoolers can learn such sequences—following a baby elephant through a model jungle, for instance, if they learn the elephant's route in association with a story. Only older children can construct maps in which routes and smaller regions fit within some larger frame of reference. Recognizing printed maps as representations of places can happen as early as three years of age, though grasping their complexities proceeds slowly after that, in parallel with developing symbolic and spatial skills (just as Piaget, the dominant theorist of child development, would predict).

My own childhood mental maps of the West were fuzzy except when I could connect them to the stories I knew: the unfolding of our journeys from Denver to Pocatello, the paths of Lewis and Clark or the Oregon Trail traced on the maps, geological "creation legends" absorbed from my father. We learn our homeland from stories, just as we learn nearly everything from stories. Anthropologist Keith Basso has noted that Apache children in the Southwest constantly hear their elders link landscape features with the ethics of living correctly as an Apache. Listen to Benson Lewis, a Cibecue Apache elder:

> *I think of that mountain called "white rocks lie above in a compact cluster" as if it were my maternal grandmother. I recall stories of how it once was at that mountain. The stories told to me were like arrows. Elsewhere, hearing that mountain's name, I see it. Its name is like a picture. Stories go to work on you like arrows. Stories make you live right. Stories make you replace yourself.*

Each summer evening, my father came home, picked the wood ticks from his clothes, and showered off the dust and reek of sagebrush. After dinner, hunched over the kitchen table, he painstakingly inked his penciled field notes about contacts, dips, and faults onto more permanent mylar maps. In winter, in his Denver office, he worked with the maps still more, writing about the geologic history he had untangled from the land he had walked over. Like most children, what mattered to my father mattered to me. Children notice everything, and with my father as guide, I noticed the land. Once noticed, incoming sensory details needed organizing; they needed names. I grew up valuing the names of landforms, paying attention to where I was in the continent, believing in maps as Scripture.

The first six years of life work their subtle power on us throughout our lives. We remember few specifics. But our bedrock emotional security—our trust—comes from this time. We spend our first years striving to develop what psychologists call "a sense of competence." This drive for mastery—of grasping, crawling, walking, talking, and play—leads to astonishingly rapid and broad learning.

Recent research has surprised us with how emphatically our behavior and personalities are hard-wired by genetics. We start with our general emotional outlook on the world fixed by the magical code of our genes. The bent of personality that makes a girl or boy receptive to natural history may well be something we cannot instill, but rather something with which an individual starts. Nevertheless, genes work in context. No personality or process is independent of environmental and social dimensions.

Seeing with a naturalist's eye is neither eccentric nor artificial. Human brains evolved in the natural world, not in a clinic or lab. Infants prefer patterns a little different, but not *too* different, from those they have seen before. And so, once children learn "bird," they may naturally move on to observing the differences between a robin, a sparrow, a blackbird, and a jay. Infants push out toward the adventure of the unknown, but only so far: the security of the known tempers their reach. This tension between the old and the new, safety versus growth, dominates much of infancy and childhood.

Tiny humans begin their journeys in the haven of family—a safe place, we hope. They test their wills against the giants, the grown-ups, as they struggle to define unique relationships to the world. Each moves from there into the land, adventuring. The expanse of sky and ocean and prairie humble and overwhelm. Nowhere, it seems, do human concerns matter less. And yet, nowhere else is the simple fact of our existence so exhilaratingly clear. Nowhere do so few trivializing and demeaning assaults on egos exist. Nowhere do humans matter more.

By forging connections with plants, animals, and land, by finding ways to experience some relationship to the Earth, individuals can gain a sense of worth. Herein lies security. Edith Cobb, in analyzing the roots of creativity in great thinkers, found that many had experienced a pivotal childhood "discontinuity, an awareness of [one's] own unique separateness and identity, and also a continuity, a renewal of relationship with nature." Cobb marveled at what can grow from this paradox: ". . . a delighted awareness that knowing and being are in some way coincident and continuous . . . and that this kind of knowing is in itself an achievement of psychological balance."

The natural world does not judge. It exists. One route to self-esteem, particularly for shy or undervalued children, lies in the out-of-doors. If, as psychologist Jean Baker Miller asserts, the model of seeking identity by "developing all of one's self in increasingly complex ways, in increasingly complex relationships" is desirable, nature is a wonderful place to seek. The sun, the wind, the frogs, and the trees can reassure and strengthen and energize.

The diversity of creatures astonishes us. Cone-nosed kissing bugs and star-nosed moles. Narwhals, sharks, fireflies, and bats. Pythons, tortoises, sequoias. Venus flytraps and black rhinos. Paramecia, amanita; saguaro, tupelo. The endless forms generated by evolution subconsciously reassure us of our own validity. No matter that we differ a bit from our peers: difference is the norm. Understanding difference empowers us to grow and to care. The variety of organisms helps to teach tolerance.

The Earth enfolds people in storm or warm sun, in the glory of light filtering through the canopy of deep woods, or in the eddying flow of rivers—without regard for whether we say the right words, wear the right clothes, or believe the right dogma. We are simply human beings setting out into the sanctuary of fields, woods, desert. We have to pay attention, certainly, or we will find ourselves in danger; nonetheless, the land releases us from competition. Such acceptance restores us for the social fight.

"Mine, Mine!" resounds through the halls of preschool. In growing toward autonomy, children assert their control through possession. The thrill of *discovering* objects adds uniqueness and intimacy to the act. One friend's son collected "dinosaur eggs" all over the West—an increasingly heavy crate of stream cobbles and river gravel. Rocks, bugs, feathers, bones—many of us remember our treasures.

The summer I was eight, I caught frogs. We were in Aberdeen that field season, a tiny town on the western shore of American Falls Reservoir, surrounded by Idaho potato fields strewn with sprinkler pipe and, beyond, the black lava of the Snake River Plain. What mattered to me, however, was the grid of ditches that lined every street and allowed the mostly Mennonite and Mormon families of the town to flood-irrigate their lawns on hot summer mornings. With two buddies, Tony and Billy, the sons of my father's field partner, I searched for frogs.

The frogs were tiny—young leopard frogs. Adult frogs must have lived nearby, but I remember only the delicate animals an inch long. I lay on the banks and peered under the plank bridges where footpaths crossed the ditches. The silver surface of the water mirrored the hazy cloudless summer sky. With luck, a small amphibious head would break the surface, two bulbous eyes peering off to the sides. I lay in wait, then lunged. I harassed far more individuals than I caught, but the captures excited me as much as the first kill must for a boy in a hunting culture. I plunked the little frogs into empty coffee cans to take them home for a night, and then returned them in the mornings, sluggish but surviving.

Simply discovering that the frogs lived in those ditches in our front yards brought the wildness of other beings into my life. I acted with a child's need to handle and possess them. Each frog was distinct, but I had yet to develop any curiosity about the contrasts among groups of frogs, the species and genera and taxonomic identities of the forms of life that writer Henry Beston called "other nations."

Specificity floated into my consciousness one fall midway through elementary school in the form of golden leaves drifting out of the blue Colorado sky to land on the ground surrounding the church a few doors from our house. We lived in west Denver, where suburbs began their climb to the foothills of the Front Range. Our house lay on the plains, with an old farmhouse across the street, a horse pasture behind us, and undeveloped lots scattered through the neighborhood. The Baptist church buildings stood alone on an otherwise empty block. This was hardly pristine short-grass prairie, but native cottonwoods still dropped clouds of cottony seeds on the spring winds and the signature bird of the Great Plains, magpies, flapped from tree to tree.

I went out to collect leaves for a school project, ironing them between two sheets of wax paper and enlisting my mother to type labels for me. I remember the thrill of appropriating the object, the first step, and then, at the next level, of harvesting the power of its name. This was a new kind of knowledge: cottonwood, catalpa, silver maple, boxelder, locust, elm. These sources of power lay around unclaimed and unowned, there for the taking.

Native peoples who still depend on the land for sustenance acquire such power earlier. Their lives depend on attending to the behavior of their prey animals and on their knowledge of medicinal and food plants. This leads to power in an elemental way—an *appreciation* of the power of other lives.

In developing what the philosopher and ecologist Aldo Leopold called the "land ethic," regard for the wilderness often comes last. First comes a child's involvement with vacant lots, ditch creatures, and the leaves of "weed trees"—discovering what environmental psychologists Rachel and Stephen Kaplan call "nearby nature." Such comparatively mundane experiences lay the foundation for what can develop into Edith Cobb's ideal, "a living ecological relationship between . . . a person and a place"—topophilia, rootedness, placeness, knowing where home is.

Found objects from nature can define a home and nurture self-esteem. Think back to your feelings as a child: wandering, you find wonders, identify them (sometimes), take them home to your room, show them off to friends, and protect them. No one has a conch shell or chip of obsidian or fragile wisp of snakeskin or sack of chestnuts or nub of deer antler just like yours. Your possession is unique; thus, you are unique. Annie Dillard writes eloquently of such experiences in *An American Childhood.* Here is her description of a 1919 dime she once discovered in her alley:

> *Treasure was something you found in the alley. Treasure was something you dug up out of the dirt in a chaotic, half-forbidden, forsaken place far removed from the ordinary comings and goings of people who earned salaries in the light: under some rickety back stairs, near a falling-down pile of discarded lumber, with people yelling at you to get away from there.*
>
> *. . . In spring I pried flat rocks from the damp streambed and captured red and black salamanders. . . . In the fall I walked to collect buckeyes from lawns. Buckeyes were wealth.*

The key then is to plant the buckeye, feed the salamander, invent a story about the dime. With these acts of extension, children begin to cultivate relationship—and the concomitant risks and rewards of sharing, of giving, of love. By moving beyond simple ownership they avoid the trap of permanently linking their self-esteem with what is only the first step—acquisition. Eventually, the discovery suffices for power; observation serves as possession; and we leave these objects where we find them, transcending the old dead-end of human domination over nature.

Experiencing this adventure firsthand is crucial. Piaget himself said, "In order for a child to understand something he must construct it for himself, he must reinvent it . . . if in the future individuals are to be formed who are capable of creativity and not simply repetition." British planner Robin Moore, in his fine study of children's use of neighborhood spaces, *Childhood's Domain,* speaks of the powerful "qualities of openness, diversity, manipulation, explorability, anonymity, and wildness" offered by what he calls "rough ground" in the midst of urban and suburban communities:

> *The indeterminacy of rough ground allows it to become a play-partner, like other forms of creative partnership: actress-audience, potter-clay, photographer-subject, painter-canvas. The exploring/creating child is not making "art" so much as using the landscape as a medium for understanding the world by continually*

destructing/reconstructing it. Where is this vital activity to be carried on if every part of the child's environment is spoken for to meet the economic, social and cultural needs of the adult community?

Several surveys reveal where children make their contacts with the natural world; they often name the same key ingredients: small places, trees, and water (brooks and frog ponds). Children favor small places close by—with dirt, trees, bushes, and "loose parts"—to build and dabble in (at their best, within a hundred yards of their homes). Trees, as writer Colin Ward describes them in the British countryside, "can be climbed and hidden behind; they can become forts or bases; with their surrounding vegetation and roots, they become dens and little houses; they provide shelter, landmarks and privacy; fallen, they become part of an obstacle course or material for den-building; near them you find birds, little animals, conkers [chestnuts], fallen leaves, mud, fir cones and winged seeds; they provide a suitable backdrop for every conceivable game of the imagination."

Middle childhood comes after what psychologists call the five-to-seven shift. The brain is finally fully developed. Children become capable of far more sophisticated learning, what Piaget calls "concrete operations." In other primates, this shift leads right into puberty. Humans, however, have postponed the hormonal rush until the teenage years, opening up a six-year interval when childhood brains receive and learn in a uniquely fresh, receptive, and playful way. Edith Cobb emphasizes the potency of this time when children are "in love with the universe" and poised "halfway between inner and outer worlds." Here, she says, lie "latent power and purpose, the seeds of the writer's art, the painter's vision, the explorer's passion." As Melvin Konner notes in his fine book, *Childhood:* "These are the years when the child is seen by societies throughout the world as a vessel into which knowledge, skill, and tradition—in short, culture—can be steadily and reliably poured."

Writer and educator Paul Shepard speaks of "the ark of the mind," a lovely phrase. "A decade, from the beginnings of speech to the onset of puberty, is all we have to load the ark." With animals, with plants, with place, with sunrises and moonsets. With wildness.

This is the decade that I have remembered in my stories of maps and frogs, the years in which I was a vessel for my teachers, family, and peers to fill. I vividly remember the last travels of those years, park by national park, snapshot by snapshot. The year I turned thirteen, my ability to focus on the same experiences disappeared for a time beneath a haze of hormonal pyrotechnics.

Adolescents take whatever we have given them,

and run. On this mad dash, they seem to close their eyes and run in unpredictable directions. As parents, we try to keep them from running off cliffs. We hope they will climb mountains and not be trapped in one or another disastrous morass. But our control is disappearing fast. Our children are off on their own journeys, carrying with them whatever we have given them, knowingly or unknowingly.

Adolescence for me was too early to feel a part of the naturalist and philosopher Joseph Wood Krutch's "great chain of life," to understand the land in context, with an awareness of interdependencies. I neither knew enough nor had sufficient experience. Adulthood legally begins at eighteen or twenty-one for good reason.

I turned twenty-one and entered adulthood in several of the usual ways in 1971. And I returned once more to learning the names of trees.

For a college field project in an introductory botany class, I censused conifers along a thousand vertical feet of switchbacking trail on St. Charles Peak in Colorado's Wet Mountains. I learned to identify six evergreens in addition to the deciduous aspen I already knew, and analyzed their elevational distributions and limits. Never before had I noticed the specificity of trees in their environment. Now, I did. Douglas fir. Engelmann spruce. Common juniper. White fir. Subalpine fir. Limber pine. Seedling, sapling, adult.

In retrospect, knowing where an awareness of trees has led me, I see the discovery of their lives as just as pivotal as any other landmark of maturing. For the first time, I passed into what biologist E. O. Wilson calls "the naturalist's trance," when my connections to other creatures mattered as much as my humanity. I saw details with a little of the attentiveness of a writer. My ability to see and understand beyond my personal boundaries passed a crucial threshold.

Ever since, I have seen these trees as my friends. When they grow along my path, I reach out to them, draw their needles through my hands, and smile. I say their names, an acknowledgment of kinship—like a formal genealogy, another chapter of Scripture. *Pseudotsuga. Picea. Juniperus. Abies concolor. Pinus flexilis.*

"*Pseudotsuga.* Douglas fir. I am here, too."

Not every child has the predilection to become a naturalist. And it may take time to develop. For instance, I did not see beyond my adolescent self until twenty-one.

I started lucky—with a secure middle-class background, decent self-esteem, curiosity, and a privileged education. My childhood experiences with maps and geography and exposure to open country gave me an underlying understanding of environment ready to populate with animals and plants. Not until chance encounters in my last years of college, however—my best friend's passion for field biology, another friend's stories of working as a park ranger, encountering mentors—did I begin to *see* as a naturalist, watching the telephone poles for raptors, carrying binoculars to identify the warblers in riparian woods, learning the telltale characters of borage, mustard, and sedge. Only then did I come to believe that natural history was as important as civil rights, American literature, or having a romance in one's life. And not until I began to read the literature of natural history could I articulate my belief in the Earth as grounding and faith and guide.

None of us can predict or control the career or avocational choices of our children. All we can do is introduce, try to prevent prejudice, battle gender stereotypes, teach by the example of our own attention and wonder. All we can do is recite from the Scripture of maps and field guides. Give names to the mountains and rivers, give names to the trees. Give voice to the emotions that storms and tundra flowers, young bison and soaring ravens can pull from us.

As parents, we can take our children with us to the land. We can be there with them as they climb on rocks, play in streams and waves, dig in the rich soil of woods and gardens, putter and learn. Here, on the land, we learn from each other. Here, our children's journey begins.

LINDA
HASSELSTROM

In 1952, when Linda Hasselstrom was nine years old, she moved with her parents to a ranch near Hermosa, South Dakota, on the plains east of the Black Hills. She acquired her first horse, began writing in a journal, and planted permanent roots. After a brief foray away from Dakota and into academia (yielding an M.A. in American literature), she and her first husband returned to the ranch in 1971. The marriage ended, but the ranch, a place of "security and solitude," led to a career centered on words.

Publishing came first. At times working from basement and bunkhouse offices, Hasselstrom published twenty-three books under her Lame Johnny Press imprint (named after a horse thief hanged nearby) and for eleven years edited *Sunday Clothes,* a Great Plains journal of the arts. Showcasing 300 other writers and artists left little time for her own work, however. By the mid-eighties, she closed down the magazine and press, saw the first of her poetry books into print, and received a National Endowment for the Arts poetry fellowship.

340

In 1987, eight years into a solid marriage to George Snell, Hasselstrom's goal of bringing her life "into a circle: writing, laboring at riding, branding, gardening, taking care of the land, and doing it all with an awareness of how these things fit together" came to fruition. She published three books: *Windbreak, Going Over East,* and a book of new poems. Her stance as rancher/environmentalist unsettled both extremes in the land-use debate. *Windbreak: A Woman Rancher on the Northern Plains* created something of a cult among urban readers who romanticized her ranch life. She notes, though, that before the cult, the writing awards, and the profile in *LIFE,* came twenty-four rejections from publishers who did not see a market for *Windbreak.*

Beginning in 1988, Hasselstrom's life again changed dramatically. Her husband died from side effects of the treatment that had cured his Hodgkin's disease. Snell's death figures strongly in the essays and poems of *Land Circle: Writings Collected from the Land* (1991) (from which the following pieces come). In the next two years, her elderly father, unable to accept his independent daughter's need to write, insisted she choose between writing and the ranch; Hasselstrom sadly moved to Cheyenne, Wyoming. Blows kept coming: her best friend died, her father died, and she moved her mother to a nursing home. She leased her ranch and rented out her ranchhouses.

Hasselstrom "admits to living in Wyoming now, but I own the ranch, consider myself a South Dakota writer, and carry license plates from both states on my car." She is completing a book of essays that again grows from the land, especially animals: "Though I feel sometimes my writing is *too* personal, experience is the only research I really trust." The new book is about "loss—leaving everything that made up my world for the last twenty years." Her agent reminds her that the essays really are about survival.

Aurora Borealis and Bells

On March 13, 1989, seven months after George died, I woke from a dream in which he and I were driving somewhere. As usual, I took a few minutes to realize he wasn't lying beside me, and will never lie beside me again. Often when I can't sleep I read awhile, or walk on the deck looking at the stars and talking to George; under the deep star-filled sky I can believe he is listening. I got up to open the curtains and noticed that the sky was light in the east; I checked the time: one, too early for false dawn, and the moon had set in the west some time before. I couldn't explain the light, but its soft glow made the garage and the van beside it, the woodpile, the whole hillside, completely clear.

Barefoot, in my long flannel nightgown, I went to the frost-covered deck and realized I was seeing the Northern Lights more brilliantly than ever before. Waves of blue and white swept up from the northern horizon to a spot almost directly overhead, to meet in a whirlpool of mingled lights. I got a blanket and put on the moccasins George had made me, lined with sheepskin and faced with buffalo hair. My activity woke the dog and together we returned to the deck. The light began to flow faster, in long streaks of blue and white; each streamer of light seemed to flare more brightly as it moved toward the south, then diminish as it slid back north. All at once the pulses paused, and I began to gather the blanket around me, thinking the show was over.

Then the northern horizon glowed red, and a single shaft of crimson light shot up, to end at the whirlpool above me. Then came another, and another, until the whole sky pulsated like a heart with red veins. I felt like crouching and gibbering, as a primitive woman might have done. Behind the fires in the sky, I could see stars shining calmly in the deep blue.

Once before I had seen red Northern Lights, when George and I were newly married, and lived in a small apartment attached to my parents' house. As we all stood in the yard, my father spoke of watching red lights as a young man. His mother said they foretold war; World War I broke out not long after. I tried to dismiss the idea.

I rose, knowing that I had to call neighbors to share this. When I looked back, I saw two large and two small ovals melted in the frost where the dog and I had sat. Margaret sleepily answered the phone, and looked out of the window even before I hung up, thanking me while she struggled into her robe. Lawrence didn't sound nearly as grateful, but he did say that even in the trees he could see the lights.

Back outside, I walked to the waist-high stone cairn on the hill north of the house; George and Michael had stacked several blocks of sandstone there one hot summer day partly because they were in the way of our house-building crew. Later, as we added other stones, George confessed to wishing he could build a tall stone tower, where he could think and watch the countryside. I love the word "cairn" because of its Scottish origins; George wanted to play bagpipes strutting on this hillside, but his lungs were too weak from radiation. I have a photograph of him wearing the kilt, walking into the sunset. I leaned against the stones, and wrapped the warm wool around me. Frodo explored for a few minutes, then crept into a fold of the blanket and sat staring up, apparently as fascinated as I was.

The lights seemed to streak up the sky faster and faster, changing from red to blue to white to green without pause. Once or twice I craned my neck to look behind me as a single spear of green shot from the vortex of light to the southern horizon. I heard a coyote howl once to the east, a gentle creaking as the wind pushed softly at the snow fence, and the rustle of grass nearby. No cars passed on the highway. No other person was within a mile of me. The wind was gentler than breath.

I thought of the town twenty miles away, where people wasted the night in sleeping while this glory went on overhead. I wanted to wake them all—dial numbers at random, shout, "Northern Lights!" and dial again. But most of their homes are so surrounded by the artificial glare of security lights and streetlamps that they would see nothing; some, fearful of the gadget through which anyone can speak to us at any time of day or night, might call the police to report a crank caller. Later in the spring I would hear Larry Woiwode, a wonderful North Dakota novelist, say that the only difference between North and South Dakota is that southern Dakotans don't see the Aurora. I wanted to tell him this story, but he was surrounded by talkers braver than I am. I kept thinking of people I might call, but I couldn't bring myself to leave the hillside for the house, which suddenly seemed a clumsy way to disfigure a perfectly good grassy hillside.

Then the light wind died completely; all sound ceased. I leaned back against the pile of rocks, expecting thunder, expecting the sky to open and a goddess to descend, expecting a majestic voice. Nothing happened, only the cadences of light continued. I considered drowsing. I felt not only safe, but as if I were in a great cathedral, watching a performance so holy that no harm could even be thought about me. The dog's warmth against my thigh was comforting, but so was the chill stone my husband had set at my back, and the hard ground under me, the roughness of the deer antler on which my left arm rested. I could faintly smell the Scotch several friends had spilled here in an old ritual: pouring some on the stones for ghosts and gods. Other friends, noticing that the cairn was surrounded by antlers I'd found on prairie walks, have begun to leave gifts: beads, lovely bones, feathers, carvings. I've tucked sage, used by the Lakota for purifying a sacred space, into crevices in the rock, and hidden personal treasures from my life with George inside. One massive set of elk antlers given me by a friend arches over the stones, and the cairn feels like an altar or shrine. I think all of us who have contributed think of it as a spot that reminds us of George. I often watch the stars there, or pray.

Then I heard a distant tinkling, like bells. I sat straight, and thought carefully. I might fantasize about a miracle, but this was serious. Rationally, I considered the possibilities. No wind blew. The sound did not resemble any other I had heard that night: the snow fence, the coyote, the breeze. It came again, louder, just as a curtain of green light swept the entire width of the sky from north to south and died overhead at a line that seemed to divide the sky above me. The universe—isn't it odd that there are no synonyms for "sky" when it is the most obvious element of the prairie?—seemed to vibrate with green light. Again and again, the horizon glowed green and the curtain swept, sky-wide, to the center of the bowl over me. Each time green flushed the sky, the bells rang, the sound softening to a gentle tinkle as the light died. I shut my eyes—I could still see a green glow—and waited for an image of what might make that sound. All I could visualize were dozens of tiny glass bells, rung softly by delicate hands somewhere in the darkness.

I pictured the pasture that lay around me: each post, each strand of wire, each tree, the electric pole. I have sat and paced for hours on this hillside, in all weather. Nothing exists here that could make that sound. No one else was near; no cars passed. The sound could not exist, and yet it did.

Finally I remembered reading that arctic explorers insisted they could hear the Northern Lights; they had given no description of the sound, which they declared to be audible only in the far north.

I watched the lights and listened to the bells until three, drifting into waking dreams in which I thanked George for the lights. Do good souls get to orchestrate the light show? Do several spirits play the lights together, like an orchestra? George could be happy in such company; he had a good sense of rhythm, and a beautiful voice, but he had a hard time staying on the same notes other folks were singing. In a soundless orchestra, his abilities would be apparent.

No one to whom I've told the story can think of a rational explanation for the bells; all but my closest friends look at me with doubt clouding their eyes. I'm satisfied without a logical explanation; some things should remain mysterious.

On July 10 of the same year, reading *Smithsonian*, I learned that sunspots large enough to contain seventy earth-size planets, had come into view on the sun in mid-March. Giant solar flares erupted; the sun threw radiation and billions of tons of matter tens of thousands of miles into space. The earth's upper atmosphere was struck by solar particles carrying electrical currents that created magnetic fields; among the results were interruptions of power and communications, garage doors spontaneously rising up and coming down—and the phenomenon we know as the Northern Lights. One official said all those effects were nothing; a really big solar flare produces enough energy to light a big city for 200 million years.

Not a single scientist mentioned bells.

In Defense of the Common Sunflower

I grew up, like any well-trained child on the prairie, knowing sunflowers were my natural enemies, because they were not allowed in the garden. Hoeing was a job I attacked with vigor because I hated it: hated the heat, sweat pouring down my face, my glasses steaming up and sliding down my nose, my straw hat scratching my forehead, the calluses on my hands. (I changed my attitude when I discovered girls were supposed to wear dresses, but that calluses and a pocketknife were my ticket to the boys' mumbletypeg games on the playground.) Most of all, I hated weeds, including sunflowers.

Weeds were the prison bars between me and a long horseback ride on the prairie where I could explore every pile of rocks, wade in every trickle of cool water. The weeds fenced me in under the watchful eye of my parents; if I destroyed the weeds, I could escape into the prairie to become an Indian maiden or warrior, a Hun raider on the steppes, or anything else my imagination could devise. Unfortunately, the best I could do was to clear a row or two, and hope my parents would let me ride anyway.

Sunflowers were not quite as annoying as buffalobur and Canada thistle, both of which left stickers in my gloves, jeans, and fingers for weeks. But the tiny sunflowers I missed one day grew like Jack's beanstalk; the next day they would reach my knees. The thick, tough stems couldn't be chopped; they had to be pulled, one by one. Leave them a week, and their hold on the earth was greater than my strength.

We shed many of our childish prejudices as we grow older, particularly if we have children who can be forced to hoe the garden because work is as good for them as it was for us. By the time I planted my own garden, I had modified my views on sunflowers; because they attracted birds, which ate insects when they weren't nibbling sunflower seeds, I left them on the edges of the garden, and among the squash and corn.

When George and I built our house in 1981, construction left the earth bare around it. For a while after native grasses have

346

been destroyed or damaged, unpleasant weeds like fireweed and cheat grass proliferate. A rather predictable and orderly succession of plants, some of which we call weeds, will follow before native grasses return. Instead of planting fragile lawn grass, we walked around the yard using our pocket-knives to lift tiny plugs of buffalo grass, no more than a half-inch in diameter, tucking them into bare spots where they would spread. We left the sunflow-ers, because they were a vivid green in the bare brown earth. They grew tall, blossomed, and began to make seed, attracting goldfinches by the dozens. From our windows, we could watch the birds hanging upside down, digging seeds out of a brown head, sometimes almost invisible among the golden blooms. But in the hot summer of 1990, I came to appreciate sunflowers for a reason I could never have guessed.

Every day seemed hotter and drier than the day before, but rains in June made the grass and hay grow; by the end of the month, everyone was frantically mowing the biggest crop of hay we'd had in five dry years. By July, when no rain had fallen for three weeks, unmown hay and dry pasture grass became a fire danger. My father wanted a record of the tallest brome grass he'd seen in eighty years, so I photographed my Bronco in the hay field: the brome reached to the middle of the windows. I began tying my horse to posts around my house, where he'd alternately graze and trample the grass, creating a bare strip, a fire guard. In spring, I planted only enough new trees to replace the dead ones in my windbreaks. In my father's junk piles I found discarded hoses, used duct tape lavishly to patch the worst holes and longest splits, then fitted them into my existing system, inventing "leak irrigation." In several spots I planted two new trees below older ones, reasoning that water would soak down to them. But the higher trees and the dry heat absorbed every drop of moisture. I taped short pieces of broken hose over leaks to water the lower trees. Friends who walked through the windbreak tried to understand how the system worked, but remained bewildered. Jerry, an engineer for the Wyoming Highway De-partment, said, "If this mess works, you could get a job in the hydrology department any day."

My trees survived except where I failed to provide shade during their first weeks. When I couldn't write another line, I would walk the winding hoses, carrying duct tape and pieces of hose, patching leaks that weren't watering trees, or that were drawing too much water. Once or twice I wound the tape around myself, the dog, the hose, and the waist-high grass, and had to cut us all loose. Collapsing in helpless laughter didn't stop the leaks, but it did me good.

I had known from the day we buried George that growing flowers on his grave would be difficult. Most of the soil that covers the prairies is only an inch or two deep, light gray because it doesn't collect much humus; the soil raised to the surface when a grave is dug in the ceme-tery is much worse, a thick yellow clay. When thoroughly soaked, it is as impossible as molasses to scrape off shoes; when it dries, water can only seep through cracks. In the spring after George's death, many plants on his grave were crushed when we set his headstone in concrete. By then, temperatures were reaching nearly one hundred daily, so I planted a few bright flowers for temporary color, and began collecting more permanent plantings from the prairie. Walking through the pastures and the outer edges of the cemetery, I collected native plants, using a belt knife to slice narrow fragments containing roots out of the earth. In this way, I tucked into the grave earth both female and male sage, the latter called "winterfat" by the Lakota because it kept the bison fat against the cold; snakeroot (purple coneflower); buffalo grass, johnny jump ups (Nuttall violets); and other plants that might be tough enough to live there. In late summer, I carried old plastic bags in my pockets every day, collecting wildflower and grass seed to plant on the grave.

By September of 1989, a year after George's death, the only living things on the grave were irises my cousin had grown on a dry ranch in the southern Black Hills, a pincushion cactus from our home hillside, and the tough, prickly weeds and creeping jenny I pulled daily. But tucked into every crack in the yellow earth, and nestled between chunks of sandstone I'd placed at the foot of the grave to slow the rain runoff were dozens of seeds from penstemon, gayfeather, purple prairieclover, rush skeleton-plant, woolly verbena, scarlet gaura, and several varieties of mint. I'd planted a bed of catnip at home for George's cat Phred; he died the day after George did, and I didn't have the presence of mind to tuck his long gray body into the grave with George's. That fall, a sprig of catnip appeared on the grave. Oh, and the onion.

Not long after the soil was heaped over George's coffin, a tiny shoot of green appeared in the bare yellow earth on the north side. I laughed when I first saw it, suspecting what it was. Someone who visited the grave with me stared, and finally asked gingerly, "Did you plant that onion?"—wondering, I'm sure, what obscure symbolism it held. But I had nothing to do with it; I interpreted it as George's ironic comment on the futility of planting flowers on this hillside, and perhaps on why we plant flowers on graves at all. When, rarely, I picked wildflowers for a bouquet, he'd say, "Oh, you killed some flowers." The onion thrived, and I watered it daily, along with the wild garlic a friend had dug at a rendezvous in the Bighorns after I told her about the onion; "for George's grave," she said. She may also have heard George's tale of how he and his friends always got a day off from school when the first wild onions appeared. Children who had eaten a double handful of wild onions exuded fumes unbearable indoors until the odor had worn off.

Still, the weather soon became too hot for the young native plants to survive without help. No water is available at the cemetery, so I began filling two five-gallon jugs with water and hauling them to the grave each evening. I missed our long evenings together, sitting on the deck to watch the sunset colors drop over the land, coyotes hunting the valley below the house, bats swooping for insects around the eaves. Sometimes we'd sit quietly for more than an hour, perhaps holding hands; the only sound was the clink of ice cubes in gin and tonic. George often said it was his favorite time of day. I'd watch his face as the sun made it glow with golden light, then hid it in shadow; his gaze grew remote, his eyes narrowed as if he were looking a long way. To the mountains, I thought then. Now I think he was looking into a much greater distance.

At first, I dreaded the trips to the cemetery. I spent the day alternately sweating in the heat at ranch work, writing in my cool basement, adjusting the water to my trees, and watching for fire. But I knew the drought would kill all the plants if I didn't help.

The first hot afternoon I hauled water to the cemetery, Frodo ran barking ahead of me as I carried the hose. I remembered the game he and George had played, and turned the water on again. He grabbed the end of the nozzle, trying to bite the rushing water, yelping and ducking away when it shot up his nose, then dashing in again. In moments he was soaking wet and I was laughing. I thought of leaving him behind, but I knew he'd gallop down the road behind me, collecting dirt until he was a mud ball, barking frantically. I put old blankets over the seats, and we started to the cemetery. I always played cheerful music on the drive, and still arrived crying; song lyrics take on new meanings when you sing them on the way to your husband's grave, and it was painful to see that mound of earth settling an inch or two a week.

When I arrived, I'd tuck the car into the lilac bushes so that other cars could pass, and drag each jug up the hill. By mid-June, the irises at the foot of the grave were tall and green, a spot of freshness I could see from the highway. I was looking forward to their blooming, hoping they would be nearly as blue as his eyes. Next, I'd water the Stella D'Oro day lily I'd ordered from a specialist in plants for arid regions; George had once noticed that summer's prairie flowers are often yellow, and its bloom would be a rich gold. An evening primrose would bloom at night, and I'd mulched runners from wild strawberry plants, hoping they'd create a woven mat of green to hold water until grass grew. The red dragon's tongue and silver sedum I planted as ground covers were spreading and adapting well; by the end of the summer, I could break off sprigs to put on other family graves.

One day, after the water was gone, I sat down beside the stone, leaned my shoulders against it, and sobbed. I cried even harder when the dog stuck his nose under my arm, crept into my lap, and licked my cheek. Then I sat up, my shoulder against the rough granite, and looked out over the valley while the hot wind dried my tears. The silhouettes of tombstones, like my own, stretched out before me, turning deep black as the sun dropped behind the cemetery hill. Slowly, they darkened the store, the school, the playground where I'd played tag, the houses, the hayfields where I'd learned to operate a mower. Pickup engines revved as local men pulled up to the store and the bar; a few housewives came for groceries. I thought bitterly that the shadow of the cemetery hung over them all, whether they knew it or not.

Finally, I turned and watched the sun drop behind the Black Hills, and the sky grow gold, then red, then a dark rose. I didn't leave until it was so dark I accidentally put my hand down in the greasy yellow mud. I smeared most of it onto a patch of buffalo grass on my way to the car, but when I opened the door to let the dog in, his feet were wrapped in sticky yellow balls; I realized he'd walked across the grave. By the time I'd scraped most of the mud off him, my hands, shirt, face, and steering wheel were covered with it.

The next night I came to the cemetery a bit later, and when I'd finished watering, I deliberately sat down, my arm across the shoulders of the stone, and watched the sunset. It wasn't quite like sitting on the deck to watch it with George, but it was the best I could do. That may have been the day I first noticed the sunflowers. I'd ruthlessly yanked them out during the first summer, but their tiny leaves were a deep green; in the summer hell it seemed criminal to destroy anything alive. One grew in the precise center of the grave. Michael had named me Sunflower when he noticed that most of the people who went to rendezvous had "mountain names." No one else called me Sunflower, but I liked the name's reminder of mountain sunflowers in Sunlight Basin where we'd loved to camp.

By midsummer, the prairie was brown, dust rolling up in choking clouds behind mowers of ranchers determined to cut anything that might furnish cattle feed next winter. My visits to the cemetery had become the best part of my day, as our evenings on the deck had once been. I'd dig up plants with my Green River knife, watch the sunset, write in my journal, talk to George, or cry while I pulled weeds. Often I'd take a long time to dole out ten gallons of water, watching with satisfaction as it soaked into the yellow earth, waiting for the evening primrose to open.

Sometimes I walked among the other graves, studying different methods of decoration. Covering black plastic with crushed white quartz seems another death to me. As the plastic breaks down, weeds spring up between the white stones, which seem unnatural among the prairie colors. Other people dig up the burial plot every spring, to plant a fresh batch of annual flowers; the blooms are attractive, but by midsummer weeds have smothered the fragile flowers. What will happen when there is no one to replant them?

Many people bring artificial flowers to the cemetery on Memorial Day; they add an air of gaiety for a while. But many people never come back to remove them; as hot summer days pass, the flowers droop, come apart. Plastic roses and daisies blow out into the prairie grass, where they lurk for years, like fragments of some radioactive disaster. Other folks turn the soil each spring, resolutely pulling each plant, so that the grave always looks horribly fresh. In the corner of one family plot is a huge cedar tree; but the caretakers asked me not to plant trees or large bushes that would hamper their mowing or eventually tumble headstones. Some of the graves have not been tended for fifty years or more; slowly the buffalo grass has crowded out the weeds, and wildflowers have seeded themselves. These graves comforted me most, as if the people in them had truly gone back to the earth, and to the nurturing of its best plants.

One evening I found wild irises growing by the hundreds near the cemetery gate, where the caretakers had never mowed. For several evenings, I dug clumps from the center of the crowd and planted each bulb separately on the grave, knowing they would spread. We'd bought four cemetery plots, since my parents had never purchased any for themselves, so I planted irises around the edges of all of them. If I live a normal life span, I will be buried there after I have no family left to tend my grave. With the irises, I defined the outline of the rectangle, hoping holes for my parents and myself can be dug without destroying the flowers. I carried extra dirt to fill in the old, unused road that slants across that part of the hillside. Cemetery officials had planted steel bars to block the path, but ruts remained, and heavy rains washed them deeper.

One day, after an evening rain shower, I went to the grave and was shocked to see deep car tracks across it. No one could have failed to see the grave and the huge headstone in daylight; perhaps teenagers using the hill for nocturnal nature study had simply made a mistake. Still, Jerry brought three more narrow slabs of granite, and buried them nearly four feet in the earth one hot day. "If they drive over those," he remarked quietly, "they'll leave their transmission behind so you'll know who they are."

A few nights later, I arrived to find that the hired caretaker had driven his riding mower along the edges of every grave; at George's he had neatly chopped off the day lily, nearly ready to bloom. The irises outlining the grave plots had been slashed a half-inch above the earth, and were turning brown in the heat. The next day I drove to a pasture where blocks of sandstone had been piled when a railroad culvert was dismantled, and brought back the largest ones I could lift, tucking them along the sides of the grave.

When the sunflowers on the edges of the grave began to crowd plants I preferred, I cut or pulled some of them, but most remained. Some evenings I was surprised, after the day's heat, to find damp patches of earth; eventually I realized that the huge sunflower leaves provided shade. One evening, as I sat by the tombstone, I saw two tiny toads hopping among the strawberry runners. Their little hides were glistening green; when I was six years old and saw such toads in the garden, I believed they were fairies, who changed shape when a human child *almost* saw them. I wondered if I'd ever told George about that fantasy, and what he would say. He had been a dreamy child, too; perhaps he would have told me some story of his own childish ideas, some tale I will never hear. After that, I nearly always saw the little toads in the evenings.

When I passed the cemetery, as I do every time I go to town, reflex made me glance up at the grave; I could always pick it out, even from a half-mile away, because it was framed in deep sunflower green, stunning in the dusty brown countryside. In the precise center of the mound stood "my" plant, nearly eight feet tall, covered with yellow blooms that swayed in the slightest breeze. The headstone was partially hidden at first; eventually, the stem was bare nearly four feet up from the ground.

I smiled each time I passed, imagining that George would approve of the sunflower's incongruity, see it as a sign of hope, with the onion blooming beside it. Like us: the sunflower and the onion, the pair people thought had nothing in common. Each day we were together our bond grew stronger, and our trust in one another's different strengths increased.

One evening in early fall, as I turned toward the cemetery from the highway, I realized something was missing. Sometimes it takes a while, even in looking at a familiar scene, to find the anomaly, or perhaps my mind refused. I had parked and turned off the ignition when it hit me: the sunflower was gone. All the sunflowers were gone. I could not breathe.

Eventually, I got out of the car and walked closer, carrying the jugs. At the grave's foot the irises were blooming, deep blue heads against the dark granite; they were not the color of George's eyes. Much of the mound was covered with silver and red sedum. At the center was a hole at least eight inches deep. Framing it, one on each side, were two large footprints.

I dropped the jug, inhaled, looked around. The riding power mower had cut all the weeds short; they were burned crisp from the day's sun. Around the grave, sunflower stems stood short and gray. Footprints, knee prints, and jagged holes in the earth showed how difficult it had been to cut and pull the tough stems. Because my watering had created a thick clay cap over the grave, the sunflowers didn't come neatly up from a central hole; each plant's roots tore out a large chunk of earth and all the plants growing in it, plants I'd tended so carefully all summer. The center was the worst; the two big feet had crushed irises, smashed sedum, broken the globe of a nipple cactus that would have bloomed a vivid pink in another day. I'd set small, flat stones on edge to protect two Rocky Mountain columbines; the stones had been knocked away, the fragile stems were broken.

Clods of earth were scattered everywhere, with limp plants hanging from them.

I noted grimly that the destroyer hadn't moved the sandstone block; the day lily was blooming in the middle of the destruction. Then I sat down and began picking up the broken clods, tucking them back into the earth and sprinkling a little water on each plant, knowing the day's heat had probably already killed them. I didn't see the tiny toads; perhaps he'd stepped on them, or sprayed them with poison.

The longer I worked, the angrier I got. I could see the rigidity that shielded George's face when he was angry. I wanted to scream profanity over the town, to smash the face of whoever had done this; George never raised his voice, or threatened. But sometimes, to the surprise of those who didn't understand that stillness, he acted swiftly and decisively. As I muttered, the dog gave up his meandering around other tombstones and came to lie at the foot of the grave, staring toward the stone.

The irises around the grave had been cut short again; this time they might not survive. The shells and stones I'd collected on walks by the sea in California and Scotland had been stepped on, ground into powder. While grumbling, I planned a ten-foot-high fence around our grave sites—until I realized such a fence would be ugly, and make tending my plants difficult. I pictured myself roaming the prairie, digging prickly pear cacti by the dozens, but they spread in dry weather or by mowing; I would be condemning the entire cemetery to their intrusion. I thought of putting up a sign: "Do not mow this grave." But I didn't want to disrupt the stone's beauty. I tried to consider the caretaker's point of view; presumably, it was he who had done this. He probably thought I was an elderly widow, too weak to pull the sunflowers, and too dumb to pull them while they were small. Surely he didn't mean to be so destructive, to cause the pain in my heart. But how could I prevent him from doing it again?

A half-hour after dark, when I knew my parents would be worried about my absence, I was calm enough to stop at their house to explain my delay, and to practice what I would say to the caretaker. My father, who never let hired men swear in my presence, grew very still, as George might have, and said, "We'll put up a fence so the son of a bitch can't get in there." I explained why I didn't want a fence; he offered to help me do anything I thought was necessary, but I said I'd handle it, and drove home.

By the time I'd made enough calls to discover who took care of the cemetery, I was crying again, and the woman who had hired the caretaker assured me that she would explain my outrage to him, being certain he understood that I wanted him never, ever, to touch anything on my husband's grave again, no matter what it was. She seemed to understand; she is a widow.

Then I sat down to study my outrage. Sunflowers, after all, are common, even here, even in a drought. The cemetery is a public place; now that volunteers have arranged for regular care, logic dictates that people who use it submit our will to that of the majority. I cannot remove the plastic flowers and quartzite chips that offend me from someone else's graves, nor should I condemn what another thinks beautiful.

We are easily moved to defend the helpless: let a child fall down a well, and we send help, without condemnation, or questions about how the mishap occurred. Let whale or grizzly habitat be threatened, and we expend ink by the gallon to defend it.

But what have we done for weeds lately? If we save only the exotic species, our world will be a vast concrete desert, with a few rarities under hothouse glass, exhibited for a price most of us can't afford to pay. The unique and uncommon draw defenders by the millions; who will defend the insignificant, the plain, the sunflower-strewn prairie used as a passageway by people accelerating toward the splendors of Yellowstone? If we don't save the commons, first there will be no setting for the wondrous; later, there will be no wonder.

It is winter now, ten degrees below zero; George has been dead two years. When I returned recently from a trip to Scotland, I sat on a snowdrift by his grave with my pockets full of slate picked up near a standing stone circle on the north coast, and shells from sea-whipped beaches. As I drew each one out and tucked it into the snow beneath the headstone, I told George about the trip, even though I had felt he accompanied me. A large shell I'd placed earlier held an unlit cigarette, a sign left by a friend who does Lakota prayer rituals here. The snow was drifted more deeply over the native plants and among the broken stems of the sunflowers than anywhere else in the cemetery; they are catching their own moisture for summer. The smooth lines of a white shell exposed by the wind are echoed by the curve of a drift.

Last summer, I collected two dozen canisters of wildflower seeds, including sunflowers; they wait under the stairs. During each warm day this winter, I have lifted another block of stone from some spot where it has fallen out of a wall, or a well curbing, and hauled it to the cemetery. Slowly, week by week, I am laying down a low and subtle wall. Inside it, I will plant wild sunflowers; stone and sunflower shade will protect ordinary prairie plants; the toads can stay; and the weeds will cheerfully renew themselves.

ROBERT MICHAEL PYLE

Photograph © Stephen Trimble

Readers of Robert Michael Pyle's nature writing are often unaware of his scientific, scholarly work. After earning his Ph.D. in ecology from Yale, Pyle became an international authority on butterfly and insect conservation, founding the Xerces Society to further this work and writing classic texts (including *The Audubon Society Field Guide to North American Butterflies,* 1981). He spent a year in England as a Fulbright Fellow, consulted with the government of Papua New Guinea, and worked as a land steward for The Nature Conservancy.

Born in Denver in 1947, he first came to Washington (where his mother was reared) as a college student and always felt he "had a gene" for the state. By 1982, he began to bridle at the constraints of science and administration that kept him "too far from the plants and flowers. I'd rather be outdoors. I'd rather be writing." He retreated to Washington again—to a pioneer homestead overlooking the verdant bottomlands surrounding Gray's River—and became a full-time writer. In the years since, he has created the more personal half of his work while continuing to use his expertise as a naturalist.

Pyle's first years of roaming outward from his old frame house in southwestern Washington's Willapa Hills led to *Wintergreen* (1986), a book of essays that looks for "beauty and biological interest" in these damaged, cut-over lands—"ugly on the surface" but full of "elusive and subtle pleasures." *Wintergreen* won the Burroughs Medal; differing subtitles for the hardcover and paperback editions (*Rambles in a Ravaged Land* and *Listening to the Land's Heart*) testify to Pyle's belief that the latter can come from the former.

A generous teacher, he commits time to several Northwest colleges each year. Pyle is wrestling with a novel and editing Vladimir Nabokov's writings about butterflies. And, inevitably, he continues to work with insect conservation issues. When children ask for his signature, he inscribes his butterfly and insect coloring books: "From Butterfly Bob."

The following essay comes from *The Thunder Tree: Lessons from an Urban Wildland* (1993), Pyle's memoir of his childhood in Colorado, where he fell in love with nature and butterflies along a Great Plains irrigation ditch called the High Line Canal. A 1989 Guggenheim Fellowship supported his fieldwork for *Where Bigfoot Walks: Across the Dark Divide* (1995). His publisher proudly asserts that the book "may not make you believe in Bigfoot, but it will make you believe in Bob Pyle."

Pyle sums up his quest for balance: "The greatest challenge is to write, remain active, communicate (teach, correspond), and live, while remaining a 'real' naturalist who goes frequently afield." Pyle believes that "Nature writers are diplomats for nature—writing to show personal, precise views of nature that can help us cope, can change us, can redeem lives."

A Grand Surprise

To think / that every brain is on the brink of
nameless
bliss no brain can bear / unless there be no great
surprise.
—VLADIMIR NABOKOV, "RESTORATION"

When I handle a cocoon or a chrysalis, or when I roll the euphonious names of butterflies around in my mouth, I often think of my mother. I see her working the warm spring soil, weeding among the violets. Her yellow sunsuit shows the skin of her back already brown. I am three or four, impatient, but I like the worms and the smell of Mother and the earth she disturbs with her garden claw. She reaches down with her other hand and pulls a small golden sphere out of the soil. I take it, roll it gently between my fingers, hold it to the sunlight. It is thin and papery nearly round, and shiny. Mother says, "That's a cocoon, Bobby. Something pretty will come out of it."

I don't remember if anything ever did come out. It was very likely a spider's egg case, anyway. But an egg and a pupa are much the same thing: life gathered within a papery caselet, to emerge when it's ready. This is why the butterfly farmers of New Guinea call their chrysalides eggs. A name, too, is like an egg. Compact, shiny, it contains the seed of meaning that the namer had in mind. My mother didn't have a name for what that cocoon contained, but her love of life and language made her want to know about such things. From that minute packet was to come a shared passion for cocoons and their contents. Years after that earliest of memories, we learned their names together.

Like a butterfly, a name goes through changes as it finds its final form. A butterfly's name has metamorphosed doubly, once in its own right, once in the skin of the creature that carries it. A giant swallowtail was once an orange dog. A monarch has not only been a tiny green

358

egg, a striped caterpillar, and a jade chrysalis with gold spots; it has also been a milkweed butterfly, a storm fritillary, and a wanderer.

Mom and I found that we both liked the old-fashioned names used by bygone authors and common country folk before the modern monikers arose. For example, the red admiral (*Vanessa atalanta*) had once been called the alderman or the nettle butterfly, and before that, the red admirable, or simply the admirable. This old name, a simple aesthetic judgment, seemed better suited to a butterfly (in spite of its scarlet epaulettes) than the current martial title. The other three American Vanessas also had their aliases. The painted lady (*V. cardui*) has been the thistle butterfly, Cynthia of the thistle, and the cosmopolite, for its worldwide occurrence. The alias of *V. annabella*, the West Coast lady, is the malva butterfly. The American painted lady (*V. virginiensis*) has gone by Virginia lady, marbled Cynthia, and Hunter's butterfly. The mourning cloak, too, was once considered a member of the genus *Vanessa*. Now known formally as *Nymphalis antiopa*, it has been called Camberwell beauty, Antiopa, yellow edge, willow butterfly, and best of all, the grand surprise.

Dutch still-life paintings of the eighteenth century often contained a red admiral, a painted lady, a mourning cloak, or another insect among their extravagant flower arrangements. I've often wondered if they were included to surprise complacent viewers as much as for their decorative value. Certainly these animals have surprised me again and again, as they have decorated my life.

Red Admirable

When I was nine, my mother went away.

One gray afternoon a Yellow Cab came to the house. "Grab a few things," Mom told me gently, "and come on. We're in a hurry."

"Are we going to Grammy's?" I asked.

"No," she said. "Farther."

The taxi took us to Stapleton Airport. "We're going to Seattle," Mom said. "I want to show it to you."

We were leaving the ticket counter, on our way to the gate to board a Constellation for Boeing Field, when my father found us. "Where do you think you're going?" he yelled, grabbing my hand. Mother held the other one firmly.

"We're going to Seattle," Mom said, "and you can't stop us."

"Try me," said Dad. He pulled me in the other direction.

I started crying, and people stopped to watch. Mom and Dad were yelling, and she was crying too. They struggled over me,

359

and a pair of glasses came off. A small crowd gathered. The tug-of-war went on until an airport official came and said we couldn't have a scene right there in the airport. Security arrived and we were led off to a quiet room. Mother left for Seattle alone, and I didn't see her for years.

In fifth grade, I had a penchant for monsters of all sorts. My teacher, Mrs. Frost, growing weary of this preoccupation, insisted that I do something else for a butcher-paper mural of a country scene. When even a vampire bat was rejected, I cast about for a subject among a pile of magazines Mrs. Frost had brought in for our inspiration. Between Jersey cows, red barns, and oak trees, the rural mural was nearly filled, so I was obliged to choose a small subject. A butterfly caught my eye. Coal black with fiery orange-red bars and blue spots, it was labeled "Red Admiral." I painted a passable likeness in tempera.

"That's so much nicer than those bloody-fanged things you always draw, Robert," said Mrs. Frost. I was very taken with it myself and wondered whether I might ever see such a beast in real life.

Butterflies soon swept my interest from both sea-shells and monsters, leaving Dracula in the dust. When Jack Jeffers and I went out at night it was to collect moths at lights and flowers rather than to play werewolf games. Still, a spooky sense accompanied certain dark forays, never more than on one particular night's walk along the High Line Canal.

We were netting pale gold miller moths on the yellow blooms of rabbitbrush along the ditchbanks when we approached a lone farmhouse. This big, blocky red brick building was well known to us by day. Occasionally we had trespassed to snatch apples from its orchard or swallowtails from its big old lilacs. No one had ever challenged us, but once or twice I was aware of someone watching, a woman, her face blurred by the back porch screen. Local kids had told us that this lonely mansion was occupied by a witch. Our recent obsession with monsters and such was fresh enough that we were not inclined to ignore the tale altogether.

We intended to slip quickly past the red house, but in the lonely corona of the pole lamp by the driveway, dozens of moths were dancing with bats. We shrugged and walked over to check it out. The moths circled crazily in the repetitious rendezvous, bats diving at them like dark meteors across their orbits, sometimes making contact, often not. A few moths alighted on the pole where we could reach them with our nets or collecting jars.

Looking over my shoulder, I saw the screened veranda of the old house standing out like a lighted island in the dark sea of night. Myriad moths flapped about inside this illuminated cage. Jack saw them too; our nets twitched nervously. We longed to get at the captive abundance of moths, but how? We didn't want her calling the sheriff—he'd already mistaken our nocturnal hunts for prowling once, and warned us not to bother anyone—and besides, what if she *were* a witch?

Nudging each other, we got up the nerve to open the gate and approach the porch. "Maybe we can sneak in and get a few without her seeing us," I whispered to Jack.

"You do it," he said, "and hand them out before she gets you." I turned the squeaky handle on the porch entrance just as the heavy front door of the house opened wide. Before we could turn and run, a grandmotherly voice asked, "Can I help you, boys?"

The woman stood in the doorway, clad in a flowered dress and looking younger than we expected. Her face was luminous in the porch light, her hair drawn back in a silver bun, her lips drawn back too, in a thin smile. Stammering, I asked, "May we catch some moths here, ma'am?" I added, "We have a collection . . . for school." I thought "for school" made it sound more official. Actually, that might have helped. Far from a witch, Miss Margaret Robina Smith, a retired teacher and nurse, was a generous benefactor of education. Furthering the schooling of young people was her greatest love, even if it involved moths. But we didn't know that.

She scrutinized us for a moment, then looked up at the moths flitting and banging all around us, smacking the windows, screens, and lights, making a nuisance. "By all means. Take 'em all." She chuckled, shaking her head.

Jack and I swung and batted away at the horde, their scale-covered wings loosing a drift of sparkling powder into the light. Soon we realized that they were almost all armyworm moths, a kind of noctuid miller that erupts in vast numbers some years. About an inch across, their fore wings were brown, their hind wings smoky gray. One was not unattractive; hundreds were boring. Our jars were full, our hunters' hunger for the kill overwhelmed. We finished, having made little impact on the moth population.

The woman had stood in the doorway, watching us, with folded arms. We showed her some of the prettier species: a red, white, and black tiger moth, a snowy ermine moth, a striated orange geometer, a yellow underwing. She displayed interest, and not a bit of witchiness. Then she told us to sit and brought out lemonade and cookies. The scents of hay and running water and the High Plains at night filtered through the broad screens as we sat and sipped and talked. Rich, dark wooden trim in the parlor of the red house showed through the lace curtains of the windows. Miss Smith said she was the daughter of the pioneer who had built the house and planted the cottonwoods. I asked if she would mind if we spread a bait of stale beer, brown sugar, and bananas on the tree trunks one night to attract red underwing moths, and if we might chase butterflies there by day. She said, "Why yes, you may." And then she said, brushing a miller off her face, "But I wish you could have caught a few more of these."

One day after school that fall, Jack and I returned to the red house. Miss Smith was apparently out. We explored the big barn next to the house and netted West Coast ladies on rabbitbrush. Then in the late-day sun, basking in the warm, nettle-strewn barnyard, flat-out against a paving stone, appeared a vision from that butcher-paper mural: cinnabar bands on ebony wings, lapis lazuli set in obsidian—a red admiral.

It proved one of the most difficult butterflies to catch. One of the Vanessas, a group known for both fleetness and keen perception, it would bask invitingly and then vanish at the first approach. These brush-footed butterflies are considered highly evolved and quite intelligent as insects go. *V. atalanta* has a reputation for landing on people in their gardens, in fact singling out the same person, among several present, to alight on repeatedly. I suspect these traits have as much to do with its being called "admirable" as its gorgeous looks. In later years, after I'd had the experience of one landing on me a few times and watching it up close, I grew incapable of collecting a red admiral ever again.

Unable to overwinter in freezing climes, red admirals appear from the south most years in May. Their larvae consume stinging nettles. In midsummer they occur among the nettle leaves in webs filled with their own black frass. There were nettles by the barn, and red admirals seemed to come back there each spring until Miss Smith died. Then her orchard and fields were cut up into lots and sold off, and the barnyard became a backyard. When the next spring's journey brought them back to a destination no longer nettled, the red admirables flew on.

Hunter's Butterfly

When my mother came back from the coast, after numerous adventures, another marriage, hard trials, and a lot of illness, she lived in my grandmother's house in East Denver. We got to know each other again by walking the flowered parkways and Denver's old parks—Cheesman, Congress, City—searching for the butterflies I had newly discovered. Bright diurnal moths, black-and-white eight-spotted foresters and rust-and-yellow nessus sphinxes, let us catch them on the sweet rocket. I was shy with her, but we both loved nature, and that quickly broke down the silence of years.

Helen Lee had always been a closet naturalist. A splendid gardener in her younger years, she later became an enthusiastic mushroom hunter, though too wary to eat much of what she found. She was a meditative angler, too. Looking over their smiling photos in fishing togs during their honeymoon at Grand Lake, I wondered why fishing hadn't been a better bond between Mom and my fly-casting father. Now we two were finding a bond in butterflies.

362

When Jack drifted off, Mom came home just in time to take his place as my field partner. Then, when I was in junior high school, Mother regained the house and custody of her children. My father and stepfamily moved elsewhere in town. Susan had grown and gone off to college and marriage. Tom, Bud, and I lived with Mom again in Hoffman Heights. She worked in downtown Denver as a banker's secretary.

On the weekends we went to the mountains with Bud when time and cash permitted, or we would dash off to the Denver Botanical Gardens, to Kiowa Creek on the plains, or to the High Line Canal. Most often, the old ditch was our destination. While it might not have had the checkerspots and fritillaries of Rabbit Ears Pass or the hairstreaks and skippers of the Front Range canyons, it offered its own surprises. Mom took her time and netted fewer individuals, but she often found rarities and novelties.

On a summer day in 1964, we parked Mom's blue and white 1959 Chevy in Fairmount Cemetery and headed southwesterly along the canal. Butterflies speckled its weedy shores. Mother lagged behind to attend to a Weidemeyer's admiral as I chased ahead after a swallowtail. When I turned back I could see her broad smile from a distance. Her curled hair was silver, tinted light brown. She wore baggy jeans and a loose print shirt with pockets for specimen envelopes. As I approached, she eagerly held out her net with a butterfly trapped in a fold for me to examine. "What is it?" she asked excitedly. "It looks like a painted lady, but it's different."

So it was. Mother had netted an American painted lady. Rarest of the four species of *Vanessa* in Colorado, its salmon wings stood apart for their bright pink bars, soft purple patches, and big blue eyespots. She'd spotted it gracing a canalside thistle.

"It's a Hunter's butterfly," I said. "Our first!" She laughed with the pleasure of it, then coughed. Mother looked older than her forty-eight years, except for her outdoor smile. The catch, new for the High Line list, was something to celebrate. It would go into our collection with a label commemorating the day. Hunter's butterfly—the name suited her. Sublimating her own problems in the search, Mother was a keen hunter. And her enthusiasm revived and fired my own, keeping it alive through the ennui of adolescence.

Next, we crossed over into a weedy zone where lupines grew among wild barley. This was the only lupine patch in the neighborhood. We knew that where you see the blue of lupine flowers, blue butterflies might be found as well. We thrashed through the weeds, sweeping the gray and azure domes of the lupine plants, and emerged with bits of blue in our nets—butterflies that look as if they were distilled directly from lupine blooms, skipping the caterpillar stage.

Back at the car, we didn't want to go home. As long as we were out, our nets and the sun kept the other side of life at bay:

no homework for me, no bills or pills for Mom. We followed a two-tailed tiger swallowtail past the crematorium and deeper into the cemetery before it disappeared. A dappled green place, cool on a hot day, Fairmount might well have been a park with an unusually sedate public. Rows of green ash and hackberry trees shaded the roads and paths among the grave sites. Magpies swooped between markers, calling urgent queries, grieving for all the good carrion buried deep out of reach.

We spotted the tiger's larvae, thumb-size snakes with blue and yellow eyespots, grazing brazenly on green ash trees, and the pale green caterpillars and chrysalides of hackberry butterflies cryptic against rough and pungent leaves of their same color.

The swallowtail reappeared, gliding like a great yellow sailplane between two ashes. A brown blur darted out from a white tombstone toward the two-tail and circled the swallowtail like a pestering bat. When the larger species left, the other returned to its perch. A hackberry butterfly, its narrow wings were warm brown above, tan and lilac below, with bright eyespots. Mother began to go for it, then noticed another nearby, and another. We chased them between two obelisks into a memorial glade, where the still air was alive with *Asterocampa celtis*. We'd hit the peak of their flight period in a year of unusual abundance: hundreds of hackberries swooped through the glade in agitated encounters, basked on the boughs of blue spruces, and decorated the headstones. Catching and papering one after another, still surrounded by the quick brown insects, we looked at one another and laughed, a little embarrassed by our gluttony.

Taking a break in the shade of a Colorado blue spruce, we pondered the incredible numbers of the hackberries and the contrasting scarcity of the Hunter's lady. Neither species was numerous in eastern Colorado when Europeans arrived. Common in the East, the Hunter's thins out westerly in spite of the near omnipresence of its pearly everlasting and pussypaw food-plants. Hackberries, confined to the foliage of the tree by that name, originally occurred sparsely in a few Front Range canyons. But *Celtis* trees were planted in large numbers as shade trees in the cities, enabling the butterfly to spread and, in places like Fairmount, to become preternaturally abundant. Hunter's butterfly, meanwhile, continued to follow its own peculiar rhythm of rarity.

I'd been sending hackberry butterflies, both dried and alive, to a specialist in Tennessee who was working out their biology. This was our rationale for collecting so many—not that we really needed one. We were gathering in the closeness that the years apart had stolen. Butterflies were our medium, the canal our meeting place. The low sun cast long shadows behind the carved stones, and we had to go home. But we'd be back.

Early the following summer, we were catching pair after pair of fresh little painted crescentspots in a low weedy field beside the canal at Chambers Road when a heavy thunderstorm arose. The crescents meant money for college, since I was now collecting professionally for a New York dealer. I was paid in pittances—four cents for a sulphur, ten cents for a swallowtail, maybe a dollar for a rare alpine satyr—and having Mom along added measurably to the bag. We both disliked the bloodthirstiness of commercial collecting, and it took a toll on our pleasure ("All those little lives," as John Haines remarked of his trapping days), but it made a way to be out together in places like this green meadow while working for my tuition.

Mother had finally managed to take me to Seattle, if only for a visit. I liked it and determined to go to the University of Washington. Once accepted, I had to earn my share of the fees. I'd given up another summer job in favor of the dealer's offer so I could spend my time chasing butterflies with Mom whenever her work permitted.

We hurried for home as the rain came in torrents. Once there, the power went out. Jack dashed across the street with a black swallowtail he'd found trying to take shelter in a little elm in the park. We studied her by candlelight and fed her sugar water as we listened to a battery radio. A tornado had struck Plum Creek, near the source of the High Line in Platte Canyon, sending a wall of water down the Platte River toward the city. Listening from our front porch we could hear the advancing flood several miles away. When the water fell back, the field of the painted crescents had been scoured by the overflowing canal. Railroad tracks lay twisted near the Thunder Tree, and the wooden flumes over Toll Gate Creek and Sand Creek had been destroyed. When it stopped raining and the sun came out, Jack made me let the swallowtail go in our yard. He was right. Having survived the storm, she did not need to die for a dime.

After my high school graduation, the Aurora house was sold. Mother, Bud, and I moved back to Grammy's house. Almost daily I drove up the Front Range canyons of Deer Creek, Turkey Creek, Coal Creek, and Boulder Creek to collect in such places as Genessee Park, Mother Cabrini Shrine, and Parmalee Gulch, then dropped back down with my full collecting bag before the mountain rains began. In the late sunshine I worked over my catch and my accounts outside on my balcony, surrounded by the redolence of Gram's tall mock oranges, as the evening came on and Denver cooled. The mountains that were the source of my bounty flamed off to the west. When my mother came home from the bank, we walked Detroit Street to Seventh Avenue Parkway and back on the alleys in between, where we had discovered a species of peace butterflying together.

Around the alley hollyhocks, we often saw painted ladies of one sort or another. All the Vanessas are immigrants, showing up here and there, leading adventurous lives. Also called the cosmopolite, the painted lady follows the thistles north, where we found her and her sisters on dandelion lawns in April or May. West Coast ladies appeared on the canal in Aurora each September or October, having come from who-knows-where. Late spring was the time to watch for red admirables, whose second brood would still be flying shortly before the frosts of fall.

But of Hunter's butterfly, we never found another.

The Grand Surprise

Flashing open its bark-black cowl, the butterfly revealed its inner wings—maroon and chocolate, blue spotted, with a butter-yellow border. I had never seen anything so beautiful. When my knees worked again, I gave chase. Gliding in and out of the dappled shade of cottonwoods, this chimerical creature appeared, disappeared, and reappeared. I followed it back and forth, its wings closing into ashes, spreading into live cinders, and seeming to pause before my eyes momentarily as if it were mounted on the air. I swung my net, disturbing it, and the big insect vanished over the towering crown of the Thunder Tree.

I knew it was a mourning cloak from one of my favorite books. *Nature's Wonders in Full Color* depicted an array of seductive butterflies and moths, including a dogface sulphur on clover, tortoiseshells among autumn leaves, and Cynthia moths on forsythia. Another bright photo showed an open mourning cloak resplendent against red sumacs. "The mourning cloak spends the winter tucked in a hollow tree," read the caption, "and is one of spring's first butterflies." Now here it was, beside *my* hollow tree, a grand surprise if ever I'd seen one.

I would follow mourning cloaks through the coming years. In March, on the first warm days, they would appear along the ditch among patches of dirty snow with their smaller relatives, the Milbert's tortoiseshells. In June I might find masses of their red-and-black spiny caterpillars making confetti of our backyard weeping willow. I reared them, and by midsummer the chrysalides, gray thorny objects like diabolical bullets, would be hanging upside down inside my jars and from the eaves of houses.

June would see the fresh new cloaks on the wing, and by schooltime their offspring would be bedding down for their winter-long sleep. Occasionally, on a sunny walk in winter, I would spot a tatty Antiopa dodging between the naked trees.

When Jack and I went out on the first warm days, hungry for butterflies, we would see dozens of cloaks. But we both had difficulty catching them, as they never seemed to sit, instead patrolling the canal bed up and down, down and back. We had many chances to swing at them as they sailed past, but without fail they swerved over the net rim at the last second, like a tennis ball with radar. Finally I thought of letting one get past me and then swinging from behind, and it worked. I had one at last! But when I removed the big female from my net, the big surprise was to find her pallid and worn. Hers were hardly the "purple-brown wings edged with creamy yellow" described by Brown in *Colorado Butterflies*. We should have known better. As we'd already seen with the tortoiseshells, these winter-nymphs emerge from their hibernatoria both early and diminished. Even a mild winter takes its toll on those velvet vanes. We would have to wait until June to see the insect that moved the English to call it the Camberwell Beauty.

Something about the mourning cloak made me averse to killing it. Like the red admiral, it held a kind of totemic value for me, having been among the ones that thrilled me from the start. Too, being big and exceedingly strong bodied, cloaks didn't die easily, and I hated to see them struggle. The spring individuals were never perfect enough for specimens, and the bright ones of high summer had a mission to perform. I wondered whether, growing a little jaded from the professional collecting, I was also getting soft. In any case, the cloak was so common that it brought a tiny price, entirely disproportionate either to the effort of catching it or its own great beauty. Letting one go was no loss at all.

I cashed in my butterflies and went off to college. Mother and Bud moved into a small apartment in East Denver near Gram's. Mom seldom drove her old Chevy but took a bus to work when she could, or a taxi. On foot, she carried her net only as far as City Park, and her mushrooming was confined to the lawns and old maple stumps of the parking strip.

By coincidence, the apartment stood across the street from the home of a young doctor who was also an avid lepidop-terist. When Ray Stanford saw the nets my mother had placed in the hands of neighborhood kids, he knew there must be another collector around. They met, and Ray helped maintain Mom's zest for the hunt in my absence.

When I came home for visits, we drove to the old haunts. Sometimes, with Ray and other friends, we organized expeditions in search of rarities. They always knew that Helen was likely to find them, though she couldn't walk far or fast. In March we chased green hairstreaks among the red rocks of the canal's headwaters. In June we found the once flooded field repopulated with painted crescents. I spent the summer in Denver working as a postman. Weekends afield with Mom, Bud, and my fiancée JoAnne were almost as before. We found the scarce Nokomis fritillary in a desert seep in western Colorado, and a gully alive with tens of thousands of blues thronging the milkweed bloom.

JoAnne and I married in 1966 and returned to Se-attle in the fall. I hoped the butterflies and mushrooms would invigorate Mom, though I knew her health was getting worse and her debts larger. The next summer I stayed in Seattle to work for the Sierra Club. But I was worried when Mother didn't write, and she sounded bad on the phone. I hitchhiked back through a hot, dusty August.

The sagelands of southern Idaho made Denver seem cool. I found my mother ill with an infection in her legs that GrandPop's cheeseweed poultice couldn't cure. Bud, only in junior high school, was at a loss to help her. I felt I should stay with Mom, having come so far to see her, but the close, smoky apartment, the disease, and her tears were oppressive.

I had been writing a piece on the ditch and hoped on this trip to find its terminus at last. I'd imagined Mother and me doing that together, and it being as it had been, but she couldn't walk. So I took Bud, who was depressed by her condition; after doing what we could to make Mom comfortable, we drove east with our friend Chuck Dudley.

Out on the High Plains northeast of Aurora, the canal vanishes into a many-headed hydra of prairie watercourses. Ditches and creeks and gullies all intersect, and all look about the same. The maps were unclear as to the ultimate end of the High Line. Bud, Chuck, and I planned to walk it right through.

As we drove north on Tower Road, a western kingbird flew into the Chevy's windshield. I recovered it, and we stroked its soft, hot, pale yellow breast. Blood ran from one corner of the bill. Stilling this voluble and lively prairie singer seemed a bad start.

We plowed through mud behind emerald frogs, crisscrossed the mile roads north of Colfax Avenue, and got all hot and sweaty following gullies stuffed with tumbleweeds. By midafternoon we still hadn't found the end and felt we'd better get back. But we stopped off at the A & W for a Frosty Mug before heading into the city, and when we finally got back to the apartment, Mom was lying on the floor, sobbing in pain and frustration. She'd left her chair, fallen, and couldn't get back up.

"Where have you been?" she cried as strong Chuck lifted her onto a couch. I told her about our day, but the stories that once would have cheered her seemed to make it worse. She didn't even say that she wished she could have been with us. I didn't understand what she knew: that she would not be walking on the canal with me again.

I ambled south along the High Line toward the scruffy lupine patch near the mausoleum and Masonic plot. Hackberry butterflies followed me along the row of stickery blue spruces. Golden skippers flew out from pink inflorescences of milkweed. A small mourning cloak haunted the hems of peachleaf willows overhanging the water, laying eggs in masses on the branches. I came to the spot where my mother had caught the Hunter's butterfly.

A bit beyond, I paused and looked through a gap in the blue spruce wall, into the cemetery. I saw the rounded granite stone that read PHELPS and walked over to it. My father and his father were buried a mile across the lawns, in a newer part of the cemetery where upright stones were not allowed. But here, in the old section, obelisks, stones rugged and smooth, and small mausoleums stood among mature trees.

Sitting on a curved concrete bench shaded by an arched juniper, I read the names of my grandmother's parents on the marker: "Hannah and Edward Phelps." Between them, Grammy and her sister, Aunt Helen, lay interred in their urns. Their names were inscribed on brass plaques attached to the memorial stone. "Grace Phelps Miller, 1894–1977," said one. The other, shiny and new, read "Helen Bailey Phelps, 1896–1989."

I went around to the other side, where the stone was blank but for a single brass plaque facing the canal above the grass and clover. "Helen Lee Pyle," it read, nothing else. Her legal name was Lemmon, but Grammy never recognized my mother's second husband. Her dates would have read 1916–1967. Mom's ashes were not there either: some had gone into Puget Sound, some to her favorite ravine in Seattle, and some had merged with the muddy flow of the High Line Canal.

Eventually I did find where that flow ends up, in a tiny tributary of Second Creek. It was in the autumn. I had Mom in mind as I wandered a lateral ditchlet to an empty, weed-filled little reservoir near the canal's mouth. Once it held a share of High Line water for a ghost farmstead signified by a few dead locusts. Dried-up salsola and goatsheads filled the parched hollow, and young cottonwoods lined the lee side, dying for want of water. A magpie nest six feet up in one of the saplings had been abandoned.

The reservoir had become an ad hoc firing range. Brass shells lay everywhere, and fragments of shattered black clay pigeons littered the dusty ground among fall leaves. A row of tall old cottonwoods along the inlet ditch had been shoved over; the only one left standing, a broken snag, was peppered with lead. No wonder the magpies departed.

The prone cottonwood trunks lay about like femurs in a boneyard, hardly rotting in the desiccating air. I walked out the longest one. Its bark had fallen away and little thornlike prickles covered the pale bare wood. I came to the root mass of the log, where old mud once surrounding burned-out roots had lithified into hollow casts. They were brown as the fields that stretched into the distance on all sides. The scene seemed lifeless except for horned larks lurching behind clods of earth and the contrails of jets scratching the blue sky.

The decaying base of the tree had the same charred smell as the Thunder Tree in storm. It took me back. Just then something dark blew out of the hollowed root wad. It was a grand surprise! Had it planned on spending the winter there? Mahogany vanes spread in an instant's hesitation, flashing sapphire spots and yellow bands before my eyes, then sailed away.

The mourning cloak flew northeasterly across the open plain. I watched it with my binoculars as far as I have ever followed a butterfly in flight, maybe a mile. Then it disappeared. Not below the horizon—I could see the curvature of the earth—but up above it. Into a point.

TERRY TEMPEST
WILLIAMS

Photograph © Stephen Trimble

Terry Tempest Williams
lives in Utah's Wasatch Mountains, immediately
adjacent to the Mormon Trail traveled by John
Henry Tempest in 1847, her ancestor six generations removed. She says, "I grew up five miles
from this house. I am a woman whose ideas have
been shaped by the Colorado Plateau and the
Great Basin. These ideas are then sorted out
through the prism of my culture—and my culture
is Mormon." Born in 1955, she says, "As children,
we listened to the stories of four generations
around the table. There is no separation between
spiritual and secular things in a Mormon household."

Williams finds guidance
in diverse places. The work of Helene Cixous,
French deconstructionist, guides her as a woman.
She treasures the poetry of Mary Oliver. Native
American writers like Simon Ortiz and Linda Ho-

gan "bring me back to center, to seeing the world whole, even holy. They remind us of *story*—the umbilical cord between past, present, and future." Other words that turn up frequently in Williams's sentences: politics, intimacy, reciprocity, courage, family, community, sacred.

For fifteen years, Williams has been associated with the Utah Museum of Natural History, first as an educator, now as naturalist-in-residence. With her dramatic public presence, she has become a widely read force in what she calls the Coyote Clan—those "who are quietly subversive on behalf of the land." Writers involved with the natural world are her kinfolk. She brings them together, encourages younger writers (especially women), fosters *community*.

Williams traces her work from the 1984 *Pieces of White Shell,* an account of her time teaching on the Navajo Indian Reservation that looks at the "stories we tell that evoke a sense of place" in Mormon and Navajo culture; to *Coyote's Canyon* (1989), a collection of stories set in southern Utah canyons; to *Refuge* (1991), "how we find refuge in change"; to *Desert Quartet* (1995), a prose poem that asks, "How do we make love to the land?" Of the latter, she says, "I wanted to explore the erotic landscape, to take it as far as I could." Such risk-taking characterizes her work, a fact recognized by her 1993 Lannan Foundation Fellowship in creative nonfiction.

An Unspoken Hunger: Stories from the Field (1994) collects her essays; two children's books pass on her enthusiasms to younger readers. *Refuge: An Unnatural History of Family and Place* established her voice. *Refuge,* from which these essays come, mourns the parallel losses of Williams's mother to ovarian cancer, the deaths of both her grandmothers, and the drowning of Williams's spiritual homeland, Bear River Bird Refuge, by the rising Great Salt Lake. She likens the alternating natural history/family structure of the book to breathing: "One informs the other; there is no separation." She says the same about her life.

Snowy Plovers

lake level: 4209.10'

The day the pumps were turned on, the lake did an about-face on its own. Great Salt Lake is receding, having dropped more than two feet from last year's lake level high of 4211.85'.

Where the water has pulled back, the land looks as though it is recovering from a long illness. Barbed-wire fences act as strainers. Sheets of algae and rotting vegetation hang like handmade paper and bobs of tangled hair.

A "bomb catcher" is being built in the West Desert. It is the newest component of the West Desert Pumping Project.

The United States Air Force has disclosed information from their own environmental assessment report: although most bombs exploded on impact during training missions conducted since World War II, some did not. There is a fear that unexploded bombs, including some in watertight containers embedded in the salt flats, might be dislodged by the pond water and float toward Great Salt Lake.

"Imagine a giant comb about eleven hundred feet long," says Brent S. Bingham, president of Bingham Engineering, Inc., the Salt Lake City company that has designed the bomb catcher. "It consists of twenty-two hundred fiberglass bars, five feet tall and six inches apart, that will span the spillway, preventing bombs from being carried into the lake by the stream of water pouring out of the new holding pond west of the Newfoundland Mountains."

Mr. Bingham told newspaper reporters today that no bombs have been seen floating in the pond, which is two and a half to three feet deep, but state officials don't want to take any chances.

Dee Hansen, Director of the Utah Department of Natural Resources, says, "The bomb catcher is not for major bombs. It's for phosphorous bombs and different types of bombs in canvas bags. . . . The Air Force experimented with a bunch of stuff out there. Most of it has probably deteriorated if it didn't explode. But the Air Force is pretty cautious, and we want them to be." He adds, "An explosive ordnance disposal unit from

374

Hill Air Force Base inspected the corridor for the twelve-mile-long Newfoundland Dike before construction began. They found some unexploded ordnance in the area, which were retrieved."

All I can see are thousands upon thousands of tumbleweeds cartwheeling over the surface of the water, beating the floating bombs to the strainer.

The West Desert Pumping Project is one of thirteen engineering efforts nominated for the 1988 Outstanding Civil Engineering Achievement Award presented by the American Society of Civil Engineers.

"The award recognizes engineering projects that demonstrate the greatest engineering skills and represent the greatest contribution to civil engineering progress and mankind," said Sheila Brand, spokesperson for the society.

We had several calls at the museum today from people who wanted to know if there had been an earthquake. According to the seismology station on campus, there had been no tremors.

It turns out the rattling vibrations were in the air, not the ground.

Atmospheric shock waves were generated when the air force exploded twenty-five thousand pounds of munitions near Great Salt Lake at 2:30 P.M.

Airman First Class Jay Joerz, with Hill Air Force Base public affairs, said, "Munitions are disposed of on a regular basis at the test and training range just west of the lake. Weather conditions must have been just right for the shock wave to carry so far. Yesterday we had another twenty-five–thousand-pound explosion and nobody noticed."

Snowy plovers have shown a 50 percent decline in abundance on the California, Oregon, and Washington coasts since the 1960s, due to the loss of coastal habitats. The National Audubon Society petitioned the U.S. Fish and Wildlife Service in March 1988 to list the coastal population of the western snowy plover as a threatened species. The present population estimate for the western United States, excluding Utah, is ten thousand adult snowy plovers, rising to thirteen thousand individuals after breeding season. A knowledge of inland population numbers and distribution is essential to our understanding of the status of the species as a whole. That's why we are counting them in Utah.

I have been combing the salt flats north of Crocodile Mountain for them since early morning. So far, my count is zero.

Margy Halpin, a nongame biologist leading the survey for the Utah Division of Wildlife Resources, and I are walking parallel to each other, maybe a half-mile apart. The distance between us feels greater than it is because of the intense heat and glare of the alkaline terrain.

I walk slowly, following the western shoreline of Great Salt Lake. Clay bluffs along the water's edge resemble Normandy: they have eroded into fantastic shapes, alcoves, and tunnels from past wave action. There are no footprints here.

Windrows of brine flies and ladybug carcasses twist along the beach. Otherwise, it is littered with limestone chips, which clamor like coins when walked upon. The heat is brutal. I pause to dip my scarf in the lake and tie it back around my forehead.

I turn west away from the lake and walk back across the salt flats. Another hour passes. I see movement. Two snowy plovers skitter ahead. Margy also has them in view—we motion each other simultaneously waving with our right hands. If they were not dashing across the white-brocaded landscape, they would be impossible to see. They are perfectly camouflaged.

Margy and I join each other and sit on the salt to watch them. I have to squint through my binoculars to shut out the light reflecting off the flats. Heat waves blur the plovers. They appear to be foraging on half-inch golden beetles. We pick up one of the insects close to us for a better examination of what the plovers are eating. The golden carapace is translucent, gemlike. We set the creature back on its course, and it skeeters away.

Snowy plovers are the scribes of the salt flats. Their tracks are cursive writing, cabalistic messages for the bird-watcher who cares enough to follow their eccentric wanderings.

We spot two more adult plovers with chicks. Two chicks. Margy and I check with each other to make sure.

"*Ku-wheet! Ku-wheet! Ku-wheet!*"

On this day, their calls are the only dialogue in the desert.

The snowy plover is considered to be an uncommon summer resident around the shores of Great Salt Lake, so our total count of six on June 11, 1988, is no surprise. They are listed as common residents of Pyramid Lake in Nevada and Mono Lake in California. Long-term distribution records show that snowy plover populations rise as Great Salt Lake retreats. More habitat supports more birds.

What intrigues me about these tiny white birds with brown bands across their breasts is how they manage their lives in such a forbidding landscape. The only shade on the salt flats is the shadow they cast. There is little fresh water, if any. And their diet consists of insects indigenous to alkaline habitats—brine flies and beetles.

Fred Ryser explains, in *Birds of the Great Basin*, how this "wet food, even during the driest and hottest time of year, contains much water of succulence . . . with each mouthful of food, the plover drinks."

To cool off, the snowy plover stands in the salt water and lets the brackish water evaporate from its body.

Another question rises with the heat of the salt desert. Why don't their eggs bake?

Snowy plovers nest in shallow scrapes, open and exposed. Some plovers will use brine fly pupal carcasses for a nesting bed, and then line them with small pebbles and shells. Both male and female snowys incubate the eggs; on hot days, such as today, they trade places frequently, alternating from sitting to standing (not so unlike us). Parenting plovers have been seen to soak in salt water and, upon returning to their clutch of eggs, will ruffle their wet feathers, sprinkling the eggs with water. An average clutch size is three eggs. Research suggests half the broods in Utah might fledge two young.

Margy and I share drinks from her canteen. I have a throbbing headache, which tells me I have been ignoring my own need for water. I fear I may be suffering from heatstroke and begin to worry about getting home. Too much exposure.

Before walking back through shoulder-high greasewood, I take a quick swim in the lake. The silky waters of Great Salt Lake cool my parched skin, even though the salt burns. This offers a momentary reprieve from my nausea. I lick my swollen lips and am careful not to rub my eyes.

I catch up with Margy and follow her through the maze of greasewood. We hear rattles and stop. It is the driest sound on earth. We take another path and walk briskly toward Crocodile Mountain.

Driving home alone on the solitary dirt road that winds around the lake, I am struck with delirium. I stop the car. Nothing looks familiar. I get out and heave violently behind the sagebrush.

The next thing I remember is waking up in a dark motel room in Tremonton, Utah. I call Brooke to see if he can tell me what happened. He is not home. Snowy plovers come to mind. They can teach me how to survive.

November 15, 1988. Lettie Romney Dixon passed away at noon from lingering illness. My grandfather, Sanky, has not left her side for months. Last night, I sat with them all night long. He held her hand and I held his. Mother felt near. Death has become a familiar landscape. I can smell it.

We prepare my grandmother's body. Her tiny arms stiff around her chest are like chicken wings because of Parkinson's disease. They have not been able to hold those she loved for years. This was the pain I could not embrace. Her blue eyes did. And now they are closed.

My uncle Don, from out of town, walks into the room. We hug. I see my mother's face in his and do not hear a word he says.

Once home, I split open a ripe pomegranate. Red juice trickles over my hand and spills onto my lap as I eat the tart, succulent seeds.

Mothers. Daughters. Granddaughters. The myth of Demeter and Persephone lives through us.

"This cannot be a coincidence, can it?" I ask my cousin Lynne, over the telephone. "Three women in one family unrelated by blood, all contract cancer within months of each other?"

"I have no idea, Terry. All I know is that my mother has breast cancer and her surgery is tomorrow."

"Is there a pattern here, Lynne, that we are not seeing?"

Lynne's voice breaks. "What I do know," she says, "is that I resent so much being asked of the women and so little being asked of the men." There is a long pause. "I'm scared, Terry. I'm scared for you and me."

"So am I. So am I."

Something is wrong and I can't figure it out—the egg collection at the Museum of Natural History. On first appearance, these clutches of eggs arranged in a nest of cotton move me. The size range and color differentiation is stunning, from the pink and brown splotching of a peregrine falcon's eggs to the perfectly white, perfectly round eggs of a great horned owl. And the smaller birds' eggs are individual works of art, canvases on calcium spheres—some spotted, some striped.

But when I hold one of these eggs, there is no gravity in my hand. A weightless shell. Life has literally been blown out through a pinhole.

It dawns on me, eggs are not meant to be seen. This collection is a sacrilege, the exposed medicine bundles of a tribe. These eggs are the hidden wealth of a species, tenderly guarded beneath the warm, bare brood patch of a female bird.

Secrets were housed inside these shells, enough avian lives to repopulate a marsh, even Bear River. But we have sacrificed them in the name of biology to substantiate the obvious, that we know where each bird comes from. These hollow eggs are our stockpile of evidence.

On my way home, I drop by to visit Mimi. She is painting on her easel in the dining room. She rinses her brushes and we sit in her turquoise study.

"What's on your mind?" she asks.

"Tell me what eggs symbolize?"

She runs her hand through her short gray hair. "For me, it is where life originates. In mythic times, the Cosmic Egg was believed to be held within the pelvis of the ancient Bird Goddess. Why do you ask?"

I describe my encounter with the egg collection at the museum, how disturbing it was.

"The hollow eggs translated into hollow wombs. The Earth is not well and neither are we. I saw the health of the planet as our own."

Mimi listened intently. She stood and turned sideways to switch on the lamp. It was dusk. I could not help but notice her distended belly, pregnant with tumor.

"It's all related," she said. "I feel certain."

"The total number of snowy plovers counted around Great Salt Lake was 487, with 26 young in 11 broods," I tell Mimi as we drive out to Stansbury Island. "Biologists figure we may have two thousand breeding pairs in Utah." She wanted to get out of the house for a change of view. Her strength is holding in spite of the cancer.

We had just seen four snowys scurrying between clumps of pickleweed.

Just outside Grantsville, thousands of Wilson's phalaropes and eared grebes were feeding in the median ponds adjacent to the freeway. No doubt a migratory stop.

In recognition of Great Salt Lake's critical role as a migrational mirror reflecting ducks, geese, swans, and shorebirds down for food and rest, the Western Hemisphere Shorebird Reserve Network has identified the lake as a crucial link in the chain of primary migratory, breeding, and wintering sites along the great shorebird flyways that extend from the arctic to the southern tip of South America.

By becoming part of the network, Great Salt Lake could gain international support for local conservation efforts and wetlands management. It has been nominated by the Utah Division of Wildlife Resources, the U.S. Fish and Wildlife Service, and Bureau of Land Management. And just recently, the Utah Division of Parks and Recreation, along with the Division of State Lands and Forestry, endorsed the nomination.

To qualify, a site must entertain in excess of 250,000 birds a year, or more than 30 percent of a species' flyway population.

Great Salt Lake qualifies. It hosts millions of birds in a season. Don Paul points out, however, that the lake qualifies on the basis of Wilson's phalaropes alone—flocks of 500,000 to 1,000,000 are not uncommon during July and August, when they are en route to South America.

The Western Hemisphere Reserve Shorebird Network has paired Great Salt Lake with Laguna Del Mar Chiquita, the salt lake in the Cordoba region of Argentina where the phalaropes winter. They are sister reserves.

"Think about one phalarope flying those distances," Mimi said, looking through her binoculars. "And then think about flocks of phalaropes, millions of individuals being driven on their collective journey. We go about our lives giving little thought, if any, to such miracles."

There is a chorus of wings navigating the planet. Twenty million shorebirds migrate through the United States each year to arctic breeding grounds in the spring and back to their wintering sites in South America. One bird may cover as many as fifteen thousand miles in a year.

Great Salt Lake is a refuge for these migrants. And there are certainly other strategic sites along the migratory path, essential to the health and well-being of those birds dependent upon wetlands. The Copper River Delta in Alaska, Canada's Bay of Fundy, Grays Harbor in Washington, the Cheyenne Bottoms of Kansas, and Delaware Bay in New Jersey are just a few of the oases that nurture hundreds of thousands of shorebirds.

Without these places of refuge, successful migrations would cease for millions of birds. None of these sites are secure. Conservation laws are only as strong as the people who support them. We look away and they are in danger of being overturned, compromised, and weakened.

Wetlands have a long history of being dredged, drained, and filled, or regarded as wastelands on the periphery of our towns. Already in Utah, there are those who envision a salt-free Great Salt Lake. A proposal has been drafted for the Utah State Legislature to introduce the concept of "Lake Wasatch." The Lake Wasatch Coalition would impound freshwater flowing into Great Salt Lake from the Bear, Weber, Ogden, and Jordan Rivers and other tributaries, by means of more than eighteen miles of inter-island dikes stretching through four counties between Interstate 80, Antelope Island, Fremont Island, and Promontory Point.

They see Lake Wasatch as fifty-two miles long and twelve and a half miles wide—three times the size of Lake Powell in southern Utah and northwest Arizona.

With 192 miles of shoreline, which unlike Lake Powell, is mostly under private ownership, there would be opportunities for unlimited lakeside development. Promoters already have plans for Antelope Island. They see it as an ideal site for a theme park with high-rise hotels and condominiums.

Lake Wasatch is a chamber of commerce dream. Finally, the Great Salt Lake would be worth something.

What about the birds?

Mimi turns to me, her legs outstretched on the sands of Half-Moon Bay.

"How do you place a value on inspiration? How do you quantify the wildness of birds, when for the most part, they lead secret and anonymous lives?"

Screech Owls

lake level: 4206.00'

Mimi passed away this morning at 5:10 A.M., June 27, 1989.

One week ago, she said to me, "You know, Terry, it's the strangest thing, I keep expecting to see an owl one morning."

"Have you ever seen an owl here?" I asked, looking out her bedroom window through the trees.

"No," she said.

"Have you ever heard one?"

"No, but I just keep thinking that one morning I will wake up to see an owl."

Four days later, I was lying next to her. We were talking. I took hold of her broad, square hand.

"Mimi, when you die, if there really is something beyond death will you send me a sign, so I will know you are fine?"

She looked at me with her eyes that always squinted when she smiled and laughed.

"It doesn't work. I asked my father the same thing and he never came back."

Jack sits quietly beside Mimi's body. A single candle burns on her dressing table. Reflected in the mirror, it appears as two.

Dad and Richard leave to call the family. I walk outside.

The sky is electric blue, the sycamore and horse chestnut trees are silhouetted in black. I walk down the porch, past the bedroom windows, to the privacy of the backyard. I think I hear the cooing of mourning doves above the lilacs. I look up to see them, but they are not doves at all.

They are owls. Two owls are circling each other on top of the telephone pole.

"*Dance. Dance. Dance,*" I hear Mimi say.

I stand below them. One screech owl turns, faces me, then flies. The other owl turns. We stare. It lifts its wings over its head, flutters them, then disappears in the direction of the other.

> *Ah, not to be cut off,*
> *not through the slightest partition*
> *shut out from the law of the stars.*
> *The inner—what is it?*
> *if not intensified sky,*
> *hurled through with birds and deep*
> *with the winds of homecoming.*
> *—Rainer Maria Rilke*

Lying in my hammock at home, the wind rocks me back and forth. It is all that is left to comfort me.

Mimi and I shared a clandestine vision of things. I could afford to dream because she could interpret the story. We spoke through the shorthand of symbols: an egg, an owl. And most of what we shared was secret, much like the migrations of birds.

If I am to survive, I must let my secrets out like white doves held captive too long. I am a woman with wings.

With Mother I buried my innocence. With Mimi I will bury my haven.

Auden echoes from the open grave, "Our dreams of safety must disappear."

The Division of Water Resources has officially turned off the pumps. Great Salt Lake is on its own. The flood is over.

The Bear River Migratory Bird Refuge is able to breathe once again at lake level: 4206.00'.

Avocets and Stilts

lake level: 4204.70'

The way to the Bird Refuge is clear for the first time in seven years. Great Salt Lake has retreated from sight, except for the faint line of silver on the horizon.

Refuge headquarters is unrecognizable. The buildings have been leveled. An old exhibit panel with a silhouette of a redhead flying over cattails stands akimbo among the wreckage. A partial title remains, reading simply, HISTORY—

Climbing over the rubble, spiders are everywhere. They are reinhabiting the Refuge. Their gossamer threads are binding it all together. Within minutes, I am draped with them. Even the avocets with their long, thin legs, sky blue, have silken strands trailing behind them.

The smell from the newly exposed land is ripe. Stilts walk on the cracked mud with skirts of brine flies around their red legs. Only a thin vein of water flows through the old canal, but volunteers have secured the banks. The U.S. Fish and Wildlife Service has promised $23 million toward the restoration of the Bear River Migratory Bird Refuge.

I turn. All at once, a thousand avocets take flight. More. Tens of thousands. A white and black flurry of birds circles me. The soft whistling of wings fills both time and space. I can no longer see the sky—above me, before me and behind me, avocets and stilts flock.

Oh, blessed wings.

In this moment, I realize how little I have hung on to for so long.

Brooke and I slip our red canoe into Half-Moon Bay. Great Salt Lake accepts us like a lover. We dip our wooden paddles into the icy waters and make strong, rapid strokes, north. The canoe powers gracefully ahead.

For two hours we paddle forward, toward the heart of the lake.

384

At the bow of the boat, I face the wind. Small waves take us up and down, up and down. The water, now bottle green, becomes a seesaw. We keep paddling.

The past seven years are with me. Mother and Mimi are present. The relationships continue—something I did not anticipate.

Flocks of pintails, mallards, and teals fly over us. There are other flocks behind them, undulating strands of birds like hieroglyphics that constantly rewrite themselves. Spring migration has begun.

We keep paddling. I have a turquoise and black shawl wrapped around me, protecting my face from the cold. This shawl is from Mexico, a gift to myself from the Day of the Dead.

I recall the impulse in me that said, "Go." I needed a ritual, a celebration to move me from death to life. I wore red for eight days—a simple cotton dress, drop-waisted and loose. I wanted no restrictions.

And when I entered the village of Tepotzlán, I bought flowers: gardenias, calla lilies, and lavender. But it was the marigolds that moved me. They were flames in the marketplace. Villagers plant seeds in May to be harvested for this occasion. They call them *cempaxuchil*, the flower with one thousand blossoms.

In the *mercado,* there was a man purchasing masks of jaguar, frog, and deer. I watched him. He knew something. I bought a mask for myself, an owl made of papier-mâché.

The man left. I followed him through the market. He bought loaves of bread, chicken, molé, tomatoes, clumps of cilantro, basil, and thyme. In the plaza, he stopped and abruptly turned around. Our eyes met. I pretended to be buying incense.

"*¿Habla usted Inglés?*" I asked.

"*Sí,*" he said shifting his rucksack of food and masks to his other shoulder. (It turned out he was a North American who left the States in 1969. He had not been back since.)

"What should I know about *el Día de los Muertos?*"

He looked at me long and hard.

"What do you want from them?" he asked.

"From whom?" I responded.

"From your Dead."

I looked away.

"There is a small adobe up the hill. Look for a turquoise door. The Dead will be there—five o'clock tomorrow afternoon, the eve of *el Día de los Muertos*. If you are to find it . . . you will."

I found the turquoise door. A white gauze curtain was billowing from the doorway. As soon as I walked in, an old woman with a long gray braid running down her back led me out, behind the adobe, and baptized me in lime water.

Once inside, I sat down on one of the four white pews with a dozen or more villagers. It was a small white room. A woman knelt in front of a white altar, reciting prayers. Candles were burning. Thirteen candles. The shrine was smothered with white gladiolas. From one corner to the other, white crepe-paper flowers were strung between straws. The villagers prayed out loud with the kneeling woman.

I folded my hands across my lap and bowed my head. I was filled with gratitude for the graciousness of these people, that I could sit with them. A wave of emotion crested in me and broke. I wept silently for all I had lost. I reentered my own landscape of grief with perfect recall.

Songs were sung. More prayers were offered. And slowly my individual sorrow was absorbed into a sea of collective tears. We all wept.

Two women and a child, all dressed in white, sat in straight-backed chairs adjacent to the altar. In time, each one rose, trembling uncontrollably, sucking air through clenched teeth. They were in a quivering, hissing trance. I watched the Dead enter their bodies. They became taller, more robust, and confident. One by one, I listened to their stories. I watched their hands gesture the past as a mother spoke through her daughter, a sister spoke through her sister, and a mother spoke through her son.

After each account, the trembling and hissing returned, until the spirits slipped out through the storytellers' teeth and the peasants were returned to themselves. They collapsed, exhausted, back into the large white chairs.

Their stories were not so unlike my own. It was the reverberation of tone I recognized, like a piece of music you return to again and again that awakens the soul. The voices of my Dead came back to me.

Wearing my owl mask, I danced in the cobblestone streets. Bonfires lit every corner. Townsfolk circled them warming their hands. Tequila poured through the gutters. In one glance, I saw both lovers and murderers kissing and knifing each other against doors. Puppet shows were performed in the plaza as firecrackers exploded at our feet. Costumed children paraded through the village, carrying illuminated gourds as lanterns. All night long there is the relentless clamouring of bells, and the baying of dogs.

Carrying a lit candle, I entered the procession of masked individuals walking toward the cemetery. We followed the pathway of petals—marigold petals sprinkled so the Dead could follow.

The iron gates were open. Hundreds of candles were flickering as families left offerings on the graves of their kin: photographs, flowers, and food; calaveras—sugared skulls—among them. Men and women washed the blue-tiled tombs that rose from the ground like altars, while other relatives cut back the vines that obscured the names of their loved ones. There were no tears here.

A crescent moon rose above the mountains, a blood-red sickle.

"*¿Por qué está aquí?*" asked an old woman whose arms were wide with marigolds.

I looked up and stood. "*Mi madre está muerta.*"
She points down. "*¿Aquí?*"

"*No, no aquí*"—not here. I try to explain in poor Spanish. "She is buried back home, *Los Estados Unidos,* but this is a good place to remember her."

We both pause.

The woman motions me to another place in the cemetery. I follow her until she turns around. She slowly sweeps her hand across five or six graves.

"*Mi familia,*" she says smiling. "*Mi esposo, mi madre y padre, mis niños.*" Then her hand moves up as she recklessly waves to the sky. "*Muy bonito . . . este cielo arriba . . . con las nubes como las rosas . . . los Muertos están conmigos.*" I translate her words. "Very beautiful—this sky above us . . . with clouds like roses . . . the Dead are among us."

She hands me a marigold.

"*Gracias,*" I say to her. "This is the flower my mother planted each spring."

My mind returns to the lake. Our paddling has become a meditation. We are miles from shore. In sight are four blue islands: Stansbury Island on our right, Carrigan Island to our left, and straight ahead we can see Antelope Island and Fremont.

My hands are numb. We bring in our paddles and allow ourselves to float. Brooke pulls out a thermos from his pack and pours two cups of hot chocolate. I spread cream cheese over poppyseed bagels. We eat.

There is no place on earth I would rather be. Our red canoe becomes a piece of driftwood in the current. Swirls of brine shrimp eggs cloud the water. I dip my empty cup into the lake. It fills with them, tiny pink spherical eggs. They are a mystery to me. I return them. I lean into the bow of the canoe. Brooke leans into the stern. We are balanced in the lake. For what seems like hours, we float, simply staring at the sky, watching clouds, watching birds, and breathing.

A ring-billed gull flies over us, then another. I sit up and carefully take out a pouch from my pocket, untying the leather thong that has kept the delicate contents safe. Brooke sits up and leans forward. I shake petals into his hands and then into my own. Together we sprinkle marigold petals into Great Salt Lake.

My basin of tears.

My refuge.

EPILOGUE:
REMEMBERING
EDWARD ABBEY

Ed Abbey died March 14, 1989. His friends took his body into the Sonoran Desert and buried him under an anonymous pile of rocks. On a warm Utah morning two months later, some five hundred people gathered to pay homage to Abbey. Under a clean blue sky, they convened in a slickrock bowl overlooking his beloved Arches National Park and its landscape of mountains, buttes, and hoodoos.

Abbey's family and closest friends were in the congregation. So were bedraggled boatmen from Moab, hikers and naturalists from all over the West, writers, physicists, lawyers, teachers, tourists, babies, and dogs.

Ken Sleight, environmentalist, river-runner, and guide (and model for Seldom Seen Smith in Abbey's *The Monkey Wrench Gang*), organized the memorial event and spoke first, telling river-running stories on Ed. C. L. Rawlins and Wendell Berry read poems. A string trio played Mozart, and friends of Abbey's sang folk songs. Dave Foreman, co-founder of Earth First! and a rousing wilderness preacher, likened Abbey to a mudhead kachina, the sacred clown of the Southwest. Terry Tempest Williams called the gathering "a rededication of purpose and place," a time "to acknowledge family, tribe, and clan." Everyone, both the writers who spoke and the readers who listened, saw it as a chance to honor a man who had taught them about the land, to make a gesture of respect toward a man who had gone, and to gather together and celebrate the things that Ed Abbey cared about.

May 20, 1989, near Moab, Utah

Dear Ed: . . . You were always able to put words together so all of us could understand them. It was exactly the way we wanted to say it, but we ourselves couldn't. You spoke for us.

Ken Sleight

I once wrote a book in the East and quoted Thoreau with whom Abbey has been compared. No comparison, Ed. Thoreau went to the wilderness and came home to mother's for dinner every night. Ed Abbey went to the wilderness, and he never came back

Ann Zwinger

He was a red hot moment in the conscience of the country. And I suspect that the half-life of his intransigence will turn out to be comparable to that of uranium.

We will miss him. The comfort is that when we need him, he will still be there.

Wallace Stegner
(in a letter read by Wendell Berry)

Ed left us two legacies, at least. His words of love and anger, and his person, in all its seeming contradictions. When I remember this man, I try to think carefully on just a *few* things. To read *Desert Solitaire* was to hear a clarion call, like *Silent Spring*. To know this man was to know a sort of integrity was not just admirable but humanly possible. . . .

Ed will always remind us to speak clearly, to assault the venality and duplicity, the cowardice, of those charged with administering our fate. And he will remind each of us in our humble and uncertain and complicated ways to do what we can to bring light, to view a world larger than our own, and to throttle the Beast.

We heard what you said, Ed. Thank you. Travel well.

Barry Lopez

The reason I'm here is that it's not the end, this isn't something you just sign off and close the book on and walk away from. . . .

Though most of "Hayduke" is buried under a pile of black rock someplace, his spirit lives on. We always had Abbey to walk point. I think that's a great hole, a great deficit. Without Ed walking point anymore, there is going to be more work for all of us.

Doug Peacock
(aka George Washington Hayduke)